Foundations of Advanced Air Mobility

Foundations of Advanced Air Mobility

Kevin A. Adkins, PhD

College of Aviation
Embry-Riddle Aeronautical University
Daytona Beach, FL, USA

WILEY

Published by John Wiley & Sons, Inc., Hoboken, New Jersey.

For general information on our other products and services or for technical support, please contact our Customer Care Department within the United States at (800) 762-2974, outside the United States at (317) 572-3993 or fax (317) 572-4002.

Wiley also publishes its books in a variety of electronic formats. Some content that appears in print may not be available in electronic formats. For more information about Wiley products, visit our web site at www.wiley.com.

Library of Congress Cataloging-in-Publication Data has been applied for:

Print ISBN: 9781394359295
ePDF ISBN: 9781394359318
ePub ISBN: 9781394359301
oBook ISBN: 9781394359325

Cover Design: Wiley
Cover Image: Courtesy of David A. Zink, zf L/Getty Images

To my wonderful parents, you have always been my unwavering cheerleaders and rock. Jennifer Ann, you have so often sacrificed so that I may walk ahead, but it's always been a larger-than-life joint adventure. Maya and Julia, I know that you will put good into the world that far outweighs the impact of any single book and this is what I truly look forward to waking up and seeing every day. And, finally, to all of my wonderful personal and professional mentors and colleagues throughout the years who shepherded my growth. I thank you each for your love and support and can only begin to repay it by paying it forward to the next generation.

To my students, my wish for you is that you get as dumb lucky in life as I have gotten.

As I look at things seen, all I see is the one that is unseen.

Contents

About the Author *xi*
Preface *xiii*
Acknowledgments *xv*
About the Companion Website *xvii*

1 What Is Advanced Air Mobility? *1*
1.1 Definition and Scope of AAM *3*
1.2 Drivers of AAM Development *6*
1.3 Key Stakeholders in AAM *9*
1.4 Summary *11*
Reference *11*

2 AAM Vehicles *13*
2.1 Types of AAM Vehicles *14*
2.1.1 The VTOL Aircraft *14*
2.1.2 The eCTOL/eSTOL Aircraft *19*
2.2 The eVTOL Design Considerations *21*
2.2.1 Design Drivers *21*
2.2.2 Enabling Technologies *22*
2.2.3 Fundamentals of eVTOL Design *22*
2.3 Propulsion Technologies *36*
2.3.1 Power, Torque, and RPM *36*
2.3.2 Distributed Electric Propulsion *38*
2.3.3 Hybrid Propulsion Systems *41*
2.4 Energy Storage and Management Systems *45*
2.4.1 Battery Basics *46*
2.4.2 eVTOL Battery Modules and Packs *50*
2.4.3 Battery Safety *53*
2.4.4 Battery Management Systems *55*
2.4.5 Battery State Indication and Display *55*
2.4.6 Fuel Cells *57*
2.5 Summary *59*
References *60*

3 Airspace Management and Integration *61*
3.1 How Air Traffic Is Managed Today *61*
3.1.1 The Structure of Controlled and Uncontrolled Airspace *62*
3.1.2 Separation Minima, Surveillance, and Communication *63*
3.1.3 Performance-based Navigation and Trajectory Conformance *63*
3.1.4 Limitations of Legacy ATC for Low-altitude High-density Operations *65*
3.2 Integrating AAM into the Skies: Automation, Predictability, and Flow Management *65*
3.2.1 AAM Operational Density and the Need for Predictable Behavior *65*
3.2.2 Automation as a Prerequisite for Airspace Integration *66*
3.2.3 Remote Supervision and the Ground-based Human Role *68*
3.2.4 Strategic Flow Management for AAM Operations *73*
3.3 Urban Airspace: Challenges and Engineering-based Solutions *75*
3.3.1 The Physical and Operational Characteristics of Urban Low-altitude Airspace *75*
3.3.2 Structured Routing and Corridors *77*
3.3.3 Vertiport-linked Airspace Geometry *81*
3.4 Communication, Navigation, Surveillance, and Separation Assurance *84*
3.4.1 Communication and Information Exchange *85*
3.4.2 Navigation Performance and Trajectory Conformance *85*
3.4.3 Surveillance and Cooperative Sensing *86*
3.4.4 Distributed Separation Assurance *87*
3.5 Summary *88*
References *88*

4 Infrastructure Considerations *91*
4.1 Vertiports *92*
4.1.1 Core Functions of a Vertiport *93*
4.1.2 Functional and Spatial Architecture of a Vertiport *95*
4.2 Urban Integration and Land-use Planning *101*
4.2.1 Compatibility *102*
4.2.2 Siting *103*
4.2.3 Community and Environmental Considerations *105*
4.2.4 Zoning and Regulation *105*
4.2.5 Connectivity and Accessibility *107*
4.2.6 Planning Process *108*
4.3 Charging and Refueling Infrastructure *109*
4.3.1 Electrical Charging Fundamentals *110*
4.3.2 Vertiport Electrical Architecture *113*
4.3.3 Charger Design and Interoperability *115*
4.3.4 Fast Charging, Managed Charging, and Energy Scheduling *116*
4.3.5 Battery Swapping and Alternative Energy-delivery Architectures *117*
4.3.6 Site-level Integration and Grid Interaction *118*
4.3.7 International Considerations *119*
4.4 Multimodal Transport Hubs: Connecting Ground and Air *120*
4.4.1 Physical and Spatial Integration *121*
4.4.2 Operational Coordination *124*

4.4.3 Urban and Regional Integration *125*
4.5 Summary *128*
References *128*

5 Regulatory and Policy Considerations *129*
5.1 The US Regulatory Framework for AAM *130*
5.1.1 Federal AAM Legislation *130*
5.1.2 Federal AAM Regulations *132*
5.1.3 FAA Regulations: Certification of AAM Vehicles *132*
5.1.4 FAA Regulations: Training and Certification of Pilots *140*
5.1.5 FAA Regulations: Certification of Operators *140*
5.1.6 FAA Regulations: Maintenance *142*
5.1.7 FAA Regulations: AAM Infrastructure *143*
5.1.8 The Role of State, Local, and Tribal Governments *144*
5.1.9 Accident and Incident Investigation *145*
5.2 The International Regulatory Framework *146*
5.2.1 China: The First Country to Type Certificate a Civil eVTOL *147*
5.3 Ethical Considerations *148*
5.4 Summary *148*

6 Operational Considerations *151*
6.1 Urban Passenger Mobility Concept of Operations *152*
6.2 Regional Cargo Concept of Operations *155*
6.3 Medical Response Concept of Operations *157*
6.4 Security and Cybersecurity Considerations *159*
6.4.1 Passenger and Facility Security *159*
6.4.2 Cybersecurity *161*
6.5 Community and Public Acceptance *163*
6.6 Summary *164*
Reference *165*

7 The Environmental Operating Context of AAM *167*
7.1 AAM and the Weather Environment *167*
7.1.1 The ABL: A New Environment for Sustained Aviation Operations *168*
7.1.2 The UBL and the Built Environment *173*
7.1.3 Spatiotemporal Weather Observations *177*
7.1.4 NWP for AAM *180*
7.2 Sustainability *183*
7.2.1 Operational Sustainability in AAM *183*
7.2.2 AAM and System-level Sustainability *187*
7.3 Aeroacoustic Considerations for AAM *188*
7.3.1 Origin of Sound *189*
7.3.2 Atmospheric Attenuation *191*
7.3.3 The Qualitative Nature of Sound and Noise *192*
7.3.4 Measuring Noise *193*

7.3.5 Noise Contour Maps *197*
7.3.6 Sources of Noise *198*
7.3.7 Designing to Reduce Noise *204*
7.4 Summary *207*
References *208*

8 Business Logic of Advanced Air Mobility *209*
8.1 Freight and Logistics Applications *211*
8.1.1 Surface-based Freight as the Primary Economic Baseline *212*
8.1.2 Conventional Aerial Freight Economics *213*
8.1.3 Network Structure and Routing Economics *214*
8.1.4 Hybrid Logistics Networks and Selective Aerial Integration *215*
8.1.5 AAM as a Timing and Reliability Enabler *215*
8.2 Passenger Transport: Regional and Urban Applications *217*
8.2.1 Regional Air Mobility *218*
8.2.2 Urban Air Mobility *220*
8.3 AAM for Public Good *221*
8.4 Emerging and Hybrid Business Models *223*
8.5 Summary *224*

9 The Future of Advanced Air Mobility *227*
9.1 Conditional Futures and System-level Constraints *227*
9.1.1 Scaling Without Centralization *230*
9.1.2 Trust, Transparency, and Human Oversight *231*
9.1.3 Energy, Infrastructure, and System Coupling *232*
9.1.4 Urban Integration and the Disappearance of the Airport *233*
9.1.5 Governing a System That Must Keep Changing *234*
9.2 Summary *236*

Index *237*

About the Author

Dr. Kevin A. Adkins is an aerospace engineer, atmospheric scientist, aviator, and professor at Embry-Riddle Aeronautical University in Daytona Beach, Florida. Kevin was born in Ann Arbor, Michigan, to Larry and Kitty Adkins, and his fascination with flight and love of the outdoors resulted in him building various air, land, and sea vehicles during his teenage years and completing a pilot certificate at the age of 17. These experiences propelled him to undertake studies at the University of Michigan in aerospace engineering where he earned a BA and an MEng in aerospace engineering. Following his service as an officer and pilot in the United States Air Force, Dr. Adkins spent 12 years in industry with distinguished service and leadership on a variety of transformational fixed-wing and rotorcraft development projects, while also undertaking advanced coursework in meteorology and atmospheric science at various institutions and earning a PhD in aerospace engineering from Mississippi State University. Dr. Adkins is a professor in the College of Aviation, and during the spring of 2023, he delivered the first collegiate course on Advanced Air Mobility (AAM) in the United States. Dr. Adkins also serves as the director of the UNcrewed Vehicle and Atmospheric Investigation Lab (UNVAIL) and the director of the Advanced Air Mobility Research and Innovation Lab (AAMRIL) at Embry-Riddle. Dr. Adkins holds numerous professional memberships and is an elected Fellow of the Royal Aeronautical Society.

Dr. Adkins is married to Jennifer Adkins and, of all of his accomplishments, he is most proud of his two wonderful daughters, Maya Hope and Julia Faith.

Preface

Advanced air mobility (AAM) is one of the most consequential developments in contemporary aviation, and the excitement surrounding the field creates a rare opportunity to cultivate new generations of thinkers, builders, and operators who are prepared to engage with aviation as a tightly coupled socio-technical system. The purpose of this text is to clearly and accessibly present the foundational principles of AAM, while remaining grounded in the physical and operational realities that shape the field. The intent throughout is to guide readers through the essential ideas that define AAM without assuming prior specialization, allowing for the development of a solid conceptual understanding that can support deeper study and engagement, regardless of how the field continues to evolve.

This book is written for a multidisciplinary, upper-level undergraduate audience, as well as for industry participants who seek a structured introduction to the multifaceted AAM ecosystem. Over the last several years, I have spoken with engineers, planners, policy professionals, pilots, and others whose responsibilities intersect with AAM or who expect to work closely with it. Those conversations reinforced that a text of this scope can serve readers well beyond a traditional classroom. As a result, explanations and frameworks are written to be accessible without sacrificing rigor, and technical discussions are introduced with deliberate pacing. A background in introductory physics may be helpful for certain topics, but foundational ideas are revisited as needed to establish common ground for all readers. The goal is to provide an inviting but disciplined entry point into a field that spans multiple complex domains.

This balance between accessibility and rigor is reflected in the overall structure of the book. Rather than serving as a reference manual or catalog of emerging technologies, the text emphasizes conceptual models intended to support durable, long-term understanding as AAM continues to mature. Core ideas related to vehicle architectures, energy considerations, airspace integration, infrastructure systems, and operational planning are developed with attention to their underlying relationships that extend beyond their current state. Mathematics is used selectively, primarily when first-principles reasoning provides the clearest insight. In this way, mathematical tools serve conceptual clarity rather than functioning as an end in themselves.

I taught the first class on AAM at Embry-Riddle Aeronautical University during the spring of 2023. Since the first offering, enrollment in the class has more than tripled. Much of this book is the result of this early evolution, and this expansion of work now supports a recently established minor in AAM, an exciting addition to the curriculum at Embry-Riddle. This sequence of courses has been designed to appeal to students across all of our colleges. Given the nature of our institution, many students enter the sequence with prior experience in small uncrewed aircraft systems. Numerous textbooks already thoroughly address this topic. For that reason, this text focuses on the vehicle classes positioned to anchor AAM operations at scale. These include larger eVTOL, eSTOL, and eCTOL aircraft intended for both urban and regional mission profiles. Throughout the text, attention remains centered on these platforms and on the systems, infrastructure, and operational frameworks required to integrate them effectively.

The perspective reflected in this book has also been shaped by my own academic and professional trajectory. My path into AAM reflects a long-standing curiosity about physical principles and social systems, including how these domains interact within complex sociotechnical systems, as well as rewarding collaborative work with colleagues whose expertise spans disciplines far beyond my own. Although I was not present for the earliest chapters of aviation history and did not witness the formative eras of commercial or high-speed flight, much of my lifetime has unfolded during remarkable achievements in space exploration. To now witness and, in a small way, participate in this transformational moment in aeronautical engineering inspires a deep sense of excitement and gratitude.

The intent of this book is not to predict the future of AAM, nor to assert that any particular operational vision will necessarily prevail. Significant uncertainty remains in how technically viable AAM concepts perform at scale, particularly with respect to economic viability, governance structures, system integration, and public acceptance. Instead, the text is written from the perspective that AAM is best understood through enduring conceptual frameworks that remain valid across a wide range of possible futures. Throughout the book, AAM is discussed in the present tense and treated as an active domain, reflecting the reality that sustained technical, operational, and institutional work is already underway. At the same time, the analysis is deliberately decoupled from assumptions that current approaches represent a final form.

Within this framing, AAM is examined as a system defined by relationships among vehicles, energy systems, infrastructure, airspace, and operations. The conceptual models developed here support structured thinking about AAM across a wide range of possible futures. In this context, a conceptual model refers to a simplified but structurally meaningful representation that captures essential characteristics and relationships, even as specific implementations evolve or fail to materialize. Such models are common in physics and engineering, where abstraction is used to make complex systems intelligible without claiming literal equivalence. Take, as a familiar example, the solar system analogy for atomic structure; this example is valuable, of course, not for physical accuracy but for its ability to convey organization, interaction, and scale.

Because AAM is an international endeavor and many countries are actively engaging in its development, this text is written with that perspective in mind. Of course, where appropriate, you'll see references to specific actions taken by manufacturers based in certain countries or concepts put forth by different civil aviation authorities. But this book is not intended to inform readers about AAM strictly from the perspective of one individual country.

Readers are encouraged to approach the material with this perspective in mind. The usefulness of the book does not depend on whether particular AAM use cases achieve widespread deployment, remain limited in scope, or are ultimately supplanted by alternative aerospace paradigms. The analytical frameworks presented here are intended to remain applicable to future systems that share similar characteristics, including distributed propulsion, close coupling between vehicles and ground infrastructure, high operational tempo, and operation within dense human environments. If the text succeeds in grounding readers in these fundamentals while encouraging curiosity and critical engagement, it will have fulfilled its purpose.

Acknowledgments

Writing a book is a substantial commitment of time, attention, and energy. Even with this level of individual commitment, this work would not have been possible without the support, guidance, and patience of many people along the way.

I would like to begin by thanking Embry-Riddle Aeronautical University, and especially the College of Aviation, for supporting a sabbatical during the fall of 2025 that allowed me the space to focus more fully on this work. I am particularly grateful to my Dean, Dr. Alan Stolzer, whose early support made the granting of that sabbatical possible. I am also thankful to Dr. Jim Gregory, Dean of the College of Engineering, who, alongside Alan, helped guide this first-time book author through the early stages of the process. The encouragement and perspective of these experienced authors and leaders in aviation were invaluable.

I am especially indebted to Trevor Simoneau, who provided thoughtful feedback on the earliest drafts of this manuscript. More importantly, Trevor has helped shape how I think about many aspects of advanced air mobility through years of conversation, first as a student and later as a colleague. Trevor's generosity with his time and thoughts, along with his genuine love for aviation, has influenced this work in ways that extend far beyond specific edits, and I continue to value every conversation with such a bright and reflective mind.

I am deeply grateful for the painstaking efforts of my student, Kayla Taylor, who served as an editor on early drafts of the text. Kayla brings exceptional talent as a scientist, engineer, and communicator. Even more impressive, she approached advanced air mobility without prior exposure to the field. That perspective proved priceless. Her careful reading, clarity of thought, and willingness to challenge assumptions strengthened this work in meaningful ways.

More broadly, I owe thanks to the many students who have contributed to the evolution of this material since I first taught the inaugural collegiate course on advanced air mobility in the spring of 2023. Conversations with curious, motivated students continue to shape how I explain, frame, and revisit these ideas. At the same time, I am continually grateful for the intellectually rich discussions I share with colleagues at Embry-Riddle. The opportunity to remain a learner while teaching is one of the reasons I cannot imagine a more rewarding profession.

It was also a privilege to collaborate with Lauren Lacey Haertlein, who authored Chapter 5 on regulations. Lauren's depth of experience and thoughtful approach brought an essential perspective to this book. She currently serves as Deputy General Counsel, Policy and Regulatory, at Joby Aviation, and previously held roles including Aviation Regulatory Counsel at Zipline and General Counsel and Director of Safety and Regulatory Affairs at the General Aviation Manufacturers Association. Lauren has served on the Board of Directors of the National Aeronautic Association and on the inaugural Industry Advisory Board for the Boeing Center for Aviation and Aerospace Safety at Embry-Riddle. From 2020 to 2022, she served on the Women in Aviation Advisory Board established under the FAA Reauthorization Act of 2018, and in 2018 she received the FAA's Friend of Aviation Safety Award. She earned her BA in biology and political science from Tufts University and her JD and MA in philosophy from Duke University, and she clerked for the Honorable James A. Wynn, Jr. on the US Court of Appeals for the Fourth Circuit. Lauren is also an active, instrument-rated commercial pilot. Her contribution significantly strengthened an important aspect of this text.

Beyond these aforementioned names, I received generous feedback from many colleagues and industry professionals who offered useful insights, critiques, and encouragement throughout the process. Their willingness to engage honestly with early ideas improved this work in so many important ways. I am grateful for their time and perspective. These individuals are acknowledged in alphabetical order by last name:

Dr. M. Ilhan Akbas, PhD, Associate Professor, Embry-Riddle Aeronautical University
Don Berchoff, Founder and CEO, TruWeather Solutions
Mr. Travis Billette, MSA, A&P, Associate Professor, Embry-Riddle Aeronautical University
Dr. Kyle Collins, PhD, Assistant Professor, Embry-Riddle Aeronautical University
Dr. Gustavo B. H. de Azevedo, PhD, Research Assistant Professor, Oklahoma State University
Dr. Victor Fraticelli Rivera, PhD, Assistant Professor, Embry-Riddle Aeronautical University
Mr. Daniel Friedenzohn, JD, MA, Professor, Embry-Riddle Aeronautical University
W. Trent Holder, Aviation Planner, Woolpert
Dr. Laxima Niure Kandel, PhD, Assistant Professor, Embry-Riddle Aeronautical University
Dr. Michael J. McCormick, PhD, Associate Professor, Embry-Riddle Aeronautical University
Dr. Maria Muia, PhD, Senior Aviation Research Scientist and Aviation Planner, Woolpert
Mr. Avinash Muthu Krishnan, MS, Assistant Professor, Embry-Riddle Aeronautical University
Tyler Seeholzer and the BETA Technologies team
Dr. Adrian Sescu, PhD, Professor, Mississippi State University
Dr. Ryan Wallace, EdD, Professor, Embry-Riddle Aeronautical University
Dr. Scott Winter, PhD, Associate Professor, Embry-Riddle Aeronautical University

About the Companion Website

This book is accompanied by a companion website.

www.wiley.com/go/adkins/aam

To access the website materials, click on the link under Related Resources on the book's page on wiley.com. You can also find this by searching www.wiley.com for the ISBN 9781394359295.

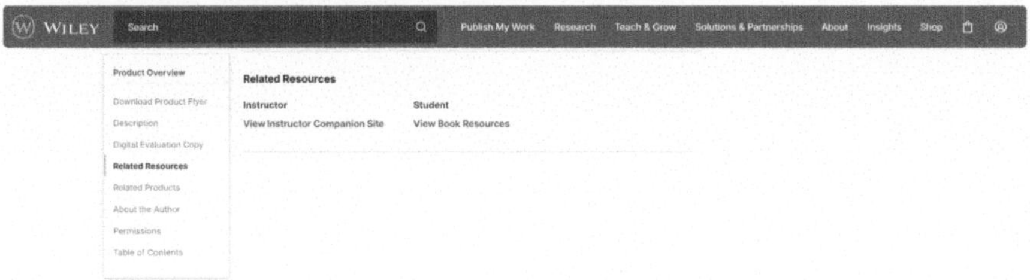

This website includes:

- Multiple Choice Questions

1

What Is Advanced Air Mobility?

Human progress in flight has always advanced at the intersection of imagination and the desire to enhance human life and transcend limitations. Long before the realization of powered flight became feasible, visionaries and humanists were already conceiving how humanity might one day rise above and travel freely through the air. Within that continuum, the earliest conceptual and technical roots of what is now termed *advanced air mobility* (AAM) can be traced to the imaginative literature of an earlier age.

In 1886, Jules Verne published *Robur le Conquérant* (Robur the Conqueror), a science fiction story introducing the world to the *Albatross* (Figure 1.1). A large, electrically powered multirotor airship capable of vertical flight, the *Albatross* was carried aloft by an array of electrically powered rotating propellers that distributed thrust evenly around the airframe. In modern terminology, Verne had imagined distributed electric propulsion and vertical takeoff capability, defining engineering principles that now guide AAM design. Verne's conception was not purely fantasy detached from reason; it was rooted in an early understanding of electromagnetism and lightweight structural composites. He envisioned the machine as both a technical and social revolution, capable of shrinking distance, transforming mobility, and altering humanity's relationship to the planet's surface and to one another.

Eighteen years later, in *Maître du Monde* (Master of the World), Verne advanced this vision further with the *Terror*, a new vehicle that could operate as an automobile, a speedboat, a submarine, and an aircraft. Its seamless transition among modes of travel prefigured the modern concept of "multimodal integration," which is a defining characteristic of today's AAM systems. The narrative also explored the societal implications of such capability, including public fascination, apprehension toward technological power, and questions of control over technology that remain just as relevant today as they were in 1904, as AAM evolves toward autonomous operations and broader adoption. Through Robur's machines, Verne's imagination extended beyond the mechanics of flight to consider its social and ethical dimensions, anticipating both the promises and perils that have accompanied every major transformation in human mobility.

Nearly eight decades later, *The Jetsons* carried the idea of personal aerial mobility into mainstream culture. The animated series portrayed a world in which families commuted in compact vertical takeoff vehicles between elevated homes and skyborne workplaces. Though playful, *The Jetsons* helped the public imagine and visualize several principles now central to AAM planning: distributed aerial access points, on-demand travel, automation, and a vertically organized urban form designed for aerial connectivity, to name a few examples. While *The Jetsons* analogy divides opinion, with some dismissing it as trivializing the complexity of real-world aviation, the series's

Figure 1.1 The *Albatross* from *Robur le Conquérant* by Jules Verne, published in 1886. *Source*: Léon Benett / Wikimedia Commons / Public domain.

portrayal of flight as an everyday utility continues to inspire designers and engineers who view AAM as the next stage of integrated mobility. This imagination remains essential to engineering progress. Without the capacity to envision what might be, technical advancement risks stagnation.

Yet, on the flip side, history teaches humility in predicting what the future of flight will bring. On October 9, 1903, *The New York Times* published an editorial titled "Flying Machines Which Do Not Fly," asserting that the practical flying machine might emerge "in from one million to ten million years," and concluding that human effort would be better spent elsewhere. Seventy-two days later, the Wright brothers demonstrated controlled, sustained powered flight at Kitty Hawk, North Carolina. However, even these pioneers underestimated how close they were to success. In correspondence from 1901, Wilbur Wright wrote to Orville that he believed powered flight might not be achieved for 50 years. A mere two years later, he and Orville achieved it. Such episodes illustrate the limits of foresight even among experts in nascent domains. They remind us that technological revolutions rarely proceed linearly; more often, they appear improbable, then emerge suddenly once the materials, methods, and collective will align.

Today, AAM stands at a similar threshold. It represents the latest expression of the same creative impulse that has always driven humanity to overcome the limits of distance and accessibility. The once-fictional vision has become a guiding framework for how this new era of aviation will connect people, places, and possibilities. Whether AAM achieves a fully realized vision of widespread on-demand aerial transport or evolves into a more specialized network of urban and regional services, its influence on aerospace, energy, transportation, and industry is already profound. Even if the complete vision of AAM were never to materialize, many of the innovations that it has catalyzed, such as distributed electric propulsion, advanced autonomy, and integrated energy systems, have permanently reshaped the trajectory of aviation. The imagination that once belonged to fiction has now become a technical discipline, and its realization, in every form, will continue to redefine how humanity moves through the air.

1.1 Definition and Scope of AAM

AAM represents the contemporary evolution of aviation that seeks to weave flight into the daily movement of people and goods across cities, towns, and rural areas alike. It reflects the growing convergence between aviation and the everyday mobility of people and goods, where the accessibility once limited to airports now extends directly into the places where people live and work. Rather than defining a single aircraft class, AAM characterizes a network of technologies and operational practices that extend aviation's reach from airports to the very fabric of urban and regional environments. The central objective of AAM is to enhance connectivity and sustainability by reducing travel times, lowering emissions, and improving access for communities of every scale.

Although AAM encompasses a wide range of technologies and missions, the concept first gained popularity under the term *urban air mobility* (UAM). UAM was initially developed as an aviation-focused response to increasing metropolitan congestion. Within this framing, electric aircraft with vertical-lift capabilities were envisioned to depart from compact sites and move passengers across crowded cities. Policymakers and engineers began exploring how advancements in electric propulsion, automation, and lightweight materials could make such flights on shorter routes viable between densely populated urban areas. The vision was to supplement surface transportation by using the third dimension of urban space, the airspace immediately above the city. The emphasis on vertical takeoff and landing (VTOL), electrified propulsion, and operations near urban centers was an ambitious but narrowly scoped framework. As research, demonstration projects, and policy initiatives advanced, it became increasingly clear that the underlying technologies that would enable UAM held far wider potential. The same systems capable of linking city districts could also connect regional towns and rural areas, deliver cargo, or support a wide variety of public services.

Recognizing that the urban label no longer captured the breadth of potential, the field adopted a broader term: AAM. Beyond a mere semantic refinement, this change in terminology signaled a transition from a narrowly defined urban service to a comprehensive framework for integrated aerial mobility. Simultaneously, the emphasis expanded from the aircraft themselves to the infrastructure, energy systems, and regulatory structures that would enable these vehicles to operate safely and at scale. AAM came to represent a national-scale mobility system rather than a metropolitan experiment.

AAM now encompasses urban, suburban, and regional operations, with three operational subsets illustrating this continuity. The first, still commonly referred to as UAM, focuses on passenger and cargo transport within metropolitan regions or between a major metropolitan region and a nearby suburban area using electric vertical takeoff and landing (eVTOL) aircraft designed for short, high-frequency flights. For example, a short-haul commuter flight between downtown Manhattan in New York City and Brooklyn, New York, a distance of 26 km driving and 15 km via air, would fall into the UAM category. UAM operations are not intended to replace existing modes of metropolitan transit. Rather, they will supplement them, offering faster and more flexible movement within and between adjacent cities.

The second operational subset, known as *regional air mobility* (RAM), extends aerial connectivity outward, linking communities across distances of roughly 80–800 km via electric or hybrid-electric aircraft. Consider, for example, a flight between London, England and Manchester, England, a driving distance of 346 km and 262 km via air. RAM operations can often use existing small airports, many of which have lost scheduled air service over time. By employing electric, hybrid-electric, or hydrogen-electric propulsion systems, regional operations can reduce operating costs, emissions, and noise while restoring access to air travel for communities that conventional commercial aviation has left behind. In this way, RAM aligns technological progress with social equity, renewing the promise of flight as a connector of people and economies.

The third subset involves low-altitude *small uncrewed aircraft systems* (sUAS). These aircraft, colloquially referred to as *drones*, provide a variety of services, such as small package logistics, inspection, and support for emergency missions. These sUAS services are typically conducted in very low-altitude airspace on the order of 100 m above ground level. Although the scale and payloads differ, the underlying logic for sUAS operations remains consistent with their larger counterparts, striving to use the airspace above communities more efficiently, safely, and responsibly. These low-altitude operations are integral to the overall AAM ecosystem because they provide a testbed for automation, autonomy, and community integration.

Collectively, as shown in Figure 1.2, UAM, RAM, and sUAS form the operational architecture of AAM, merging short-range logistics, intracity transport, and interregional travel concepts into a coherent transportation system. The system grows through mutual reinforcement, with each domain contributing to the maturity of the whole. That said, as we explore AAM throughout this text, our focus will be on larger vehicles within the UAM and RAM operational architectures. The reader is encouraged to reference the greater volume of literature on sUAS vehicles and operations [1].

The evolution from strictly UAM to the all-encompassing AAM moniker also reflects a maturation in how the field understands the role of aviation within broader society. Initially, UAM was envisioned primarily as an aviation-driven initiative seeking to solve a challenge within urban environments. AAM reframed that relationship by placing mobility at the center and treating aviation as one component within a multimodal transportation network (multimodal referencing existing forms of transportation: cars, buses, rail, etc.). Under this new framing, flight is a part of a continuous flow of movement that begins and ends on the ground. This transition has also ushered in a new measure for success. The litmus test for AAM has broadened from demonstrating vehicle feasibility to achieving systemic viability. This means aligning aircraft design, infrastructure planning, power generation, and community engagement around shared goals

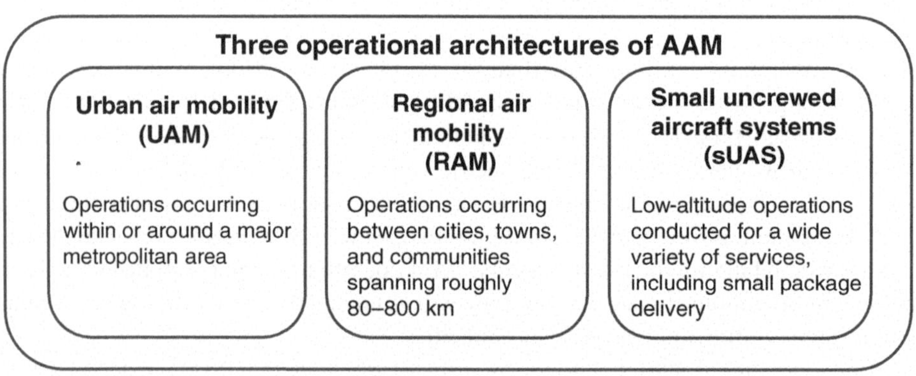

Figure 1.2 Three operational architectures of AAM.

of safety, equity, and sustainability. This results in electric propulsion, automation, and digital airspace management advancements, among others, no longer being viewed as discrete innovations but as interdependent elements of a larger ecosystem. All this is described as a *system of systems*.

A system of systems, with particular reference to the interdependence among these systems, defines AAM. AAM transportation hubs (called *vertiports*) are being designed as multimodal nodes that link air operations with ground transit and electrical-grid systems. Meanwhile, public agencies, private developers, and regulatory bodies are collaborating to ensure the growth of aerial mobility aligns with broader environmental goals and community expectations. AAM, in this sense, represents the intersection of disciplines advancing integrated mobility and bridges the traditional separation between aviation, surface transportation, energy systems, and digital communication networks, with key concepts and relationships illustrated in Figure 1.3.

AAM's current form thus reflects a convergence of technologies and evolving philosophies. With its evolution, certain characteristics have come to define its core tenets:

- *Launch and recovery versatility* encompassing vertical, short, and conventional takeoff and landing operations that allow adaptation to diverse environments.
- *Electrified and hybrid-electric systems* that reduce emissions, improve flight control authority, and enable distributed thrust configurations to enhance safety and reliability.

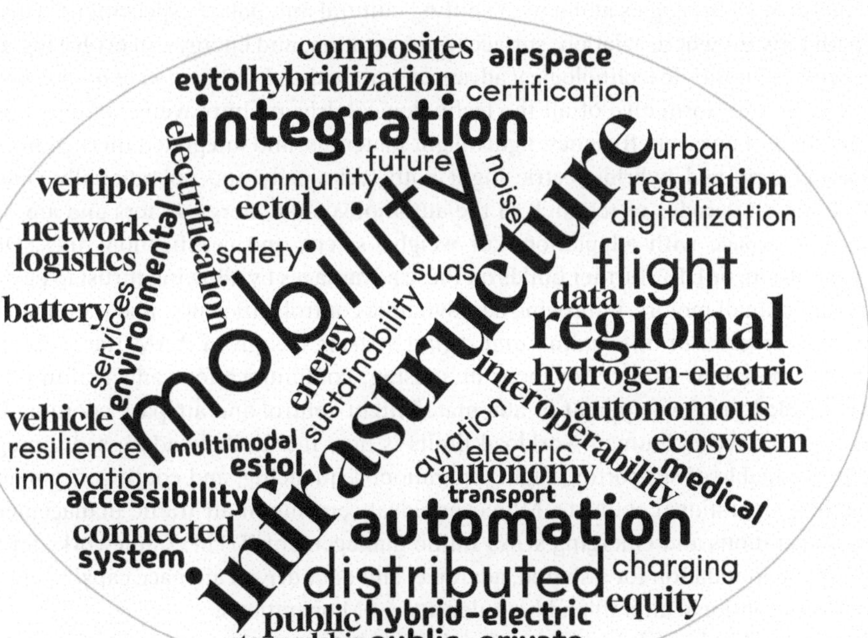

Figure 1.3 Keywords drawn from the conceptual and operational frameworks of AAM.

- A *distributed network structure* that disperses flight activity across many smaller nodes to shorten trip distances, increases operational resilience by reducing reliance on a few centralized hubs, and brings the benefits of flight closer to users.
- *Equitable and sustainable integration* within communities that ensures broad access to mobility while maintaining the environmental and social capacity of future generations to meet their own needs.
- *Digital and automated airspace* integration that enables the coexistence of high-density AAM operations with traditional air traffic and establishes the foundation for a more adaptive, digitally managed airspace system.
- *Interoperable infrastructure and energy systems* that allow for diverse vehicles, both aerial and surface, to share facilities efficiently.
- *Community integration and governance* that ensures the benefits of aerial mobility align with community priorities and public trust.

These seven characteristics have come to define AAM.

1.2 Drivers of AAM Development

The emergence of AAM results from the convergence of technological, environmental, economic, and societal forces that collectively enable a new form of aviation. AAM arises at a moment when innovation across propulsion, energy, digital systems, and infrastructure aligns with global priorities for sustainability, accessibility, and resilience. The industry represents the intersection of these matured technologies along with shifting cultural and policy expectations. This creates a unique path forward where aviation, surface transportation, and energy can evolve together.

At the center of multiple technological advancements is the *electrification of propulsion*. Perhaps this is the most transformative of all the technological drivers. Improvements in battery energy density, motor and inverter efficiency, lightweight materials, and integrated thermal management have made electric and hybrid-electric flight both technically and operationally viable. These advances build on decades of research in the automotive and energy sectors and are now transferred into aerospace with adaptations for weight, safety, and certification rules. Distributed electric propulsion, or DEP, further builds on these advantages by allowing thrust to be distributed across multiple small motors, improving aerodynamic control, efficiency, and redundancy.

Meanwhile, *progress in automation and digital systems* has given AAM the tools to manage high-density operations safely. Advances in sensing, communication, and real-time computation have enabled the development of automated flight control and airspace coordination. What was once constrained by human workload limits is now increasingly addressed through digital architectures capable of supporting flight separation, sequencing, and contingency management within defined operational contexts. Moreover, advancements in air traffic management (ATM) for crewed operations and emerging sUAS traffic management (UTM) frameworks for sUAS are establishing the foundation for a shared, adaptive, and data-driven airspace capable of accommodating both conventional aircraft and high-density AAM operations.

Another key technological driver is *materials and manufacturing*. Integrating composite structures, additive manufacturing, and modularity into aircraft design has reduced airframe weight while improving the strength-to-weight efficiency of load-bearing structures. These advances allow for complex geometries, embedded electrical and thermal pathways, and repairable modular components that make AAM aircraft easier to build and maintain. Together, electrification, automation, and advanced materials have incremented the technological foundation of modern aviation.

Alongside these aerospace technological advances, *energy infrastructure* is undergoing its own transformation as utilities and planners adapt grids, energy storage, and charging systems to meet the growing demands of homes, vehicles, and industry. Although not a lead motivator, AAM stands as an adjacent beneficiary of these broader upgrades because electric aviation depends on reliable, high-capacity energy systems. The same infrastructure that supports electric vehicles on the ground is gradually expanding in scope to include aviation, linking the energy networks that power both surface and aerial mobility.

Outside of technological innovations, the second major driver of AAM development comes from *environmental* and *sustainability* imperatives that are reshaping the global transportation sector. Aviation, though responsible for a smaller percentage of total emissions than ground transportation, remains a highly visible environmental concern because it relies significantly on fossil fuels and consumes large amounts of energy relative to the distance traveled and payload carried. The electrification of flight is a direct response to this challenge. AAM supports international goals for decarbonization by replacing combustion-based propulsion with electric or hybrid systems that can be charged using renewable energy sources.

However, sustainability challenges in aviation extend beyond propulsion. The structure of today's air transportation network contributes to its own inefficiency. The hub-and-spoke system, notionally depicted in Figure 1.4, that defines most commercial aviation today requires passengers and goods to route through centralized hubs that often results in travel far out of their way, all while compounding total energy use, congestion, and emissions. These inefficiencies motivate the search for a less circuitous model of flight that connects communities directly rather than through major airports. The distributed nature of AAM, therefore, carries sustainability advantages that extend beyond energy supply. By creating networks of smaller nodes instead of concentrating operations at large airports, AAM reduces the need for extensive land development and reduces the surface congestion associated with traditional airport access. When paired with renewable power generation, such networks can operate with a substantially smaller environmental footprint while expanding mobility options across regions. As a result, environmental imperatives do more than justify AAM; they shape its framework from the outset.

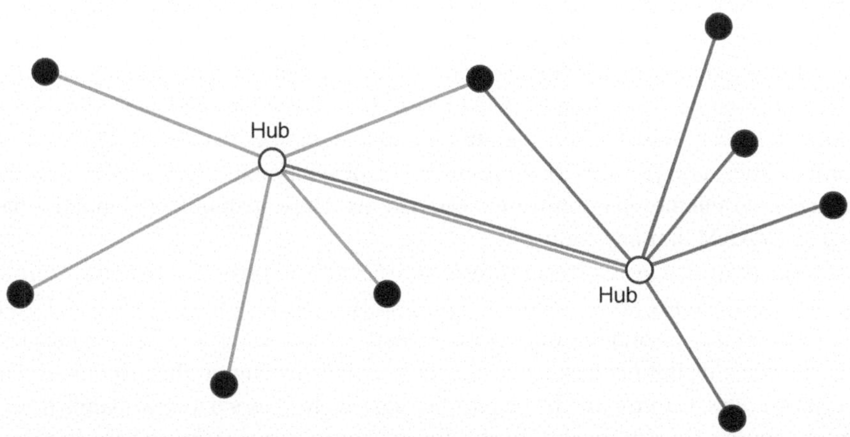

Figure 1.4 Contemporary aviation hub-and-spoke operations.

AAM is the response to *new economic conditions* that make short-range air transport marketable and scalable. Population growth, urbanization, and economic decentralization, where industries, services, and workforce populations more evenly distribute across regions, create demand for efficient movement within and between metropolitan areas. Thus, AAM offers a way to restore productivity by using underutilized airspace for local and regional transport. The global rise of e-commerce and logistics exacerbates this demand, as the pressure for rapid ground delivery finds a parallel solution in extending distribution networks into the aerial domain.

The advantages that electrification brings to improved vehicle control, efficiency, and redundancy further contribute to the economic feasibility of AAM. Electric and hybrid-electric systems promise lower maintenance costs, reduced fuel expenses, and improved operational reliability compared to conventional aircraft. These efficiencies make shorter routes, which have traditionally been uneconomical, viable for commercial operation. The combination of lower direct operating costs and reduced infrastructure requirements opens the door to new business models based on frequent, point-to-point service rather than hub-and-spoke operations. Investment dynamics in AAM further suggest enthusiasm for this potential. Concomitantly, public agencies recognize AAM's potential to stimulate economic growth, particularly in regions underserved by current transportation networks.

Beyond technology, the environment, and economics, AAM development is propelled by *social and policy motivations* that seek to create a more equitable and resilient transportation system. Many regions, both urban and rural, face persistent accessibility challenges. Traditional ground networks cannot always meet mobility demand due to congestion, geography, or infrastructure cost. AAM provides a way to connect people and deliver goods in areas where expanding roads or railways is impractical. Consequently, policymakers increasingly view AAM as a tool for meeting these challenges and advancing public welfare. More so, the same AAM systems developed to move passengers and cargo more efficiently can also be adapted for emergency response, medical transport, and disaster relief missions. AAM infrastructure is thus a dual-use asset, supporting routine transport in normal conditions and emergency logistics when surface systems are compromised. In this sense, AAM contributes to societal resilience as well as mobility.

For many nations, AAM development is also part of broader *strategic and geopolitical considerations*. Aviation has historically been both a technological frontier and a marker of national capability and innovation. Just as the jet age and space race defined previous eras of competition and collaboration, AAM now occupies a similar role as the sustainable and autonomous flight frontier. Nations recognize that leadership in AAM translates to leadership in electric propulsion, digital infrastructure, and mobility innovation. Accordingly, while public investment in AAM serves environmental and economic goals, it also champions strategic ones, such as safeguarding supply chains, modernizing the aerospace workforce, and fostering technological self-reliance. The government programs that support AAM research and development serve the dual purposes of stimulating domestic industry and maintaining global competitiveness. Collaboration among agencies, universities, and private firms promotes the development of standards and intellectual property that strengthen national innovation ecosystems. These geopolitical dynamics have accelerated progress in AAM in many countries.

Finally, AAM is propelled by a series of *cultural and behavioral shifts* that have transformed public perception and expectations about mobility. The past two decades have normalized on-demand, app-based transportation, conditioning users to expect immediate and personalized access to movement. The same digital platforms that coordinate ride-sharing on the ground are now incorporating aerial services. Culture and behavior have also evolved as aviation expands from automation toward autonomy. Once viewed with skepticism, autonomous systems have become familiar to many through driver-assistance systems in automobiles, robotic delivery platforms, and sUAS services. The COVID-19 pandemic further reshaped how people think about mobility by propelling decentralized work patterns and reinforcing the value of flexible, distributed transportation options. As a result, within this cultural environment, AAM emerges as a natural extension of broader trends in technology and lifestyle demands. The public's increasing comfort with electrification, autonomy, and shared mobility creates a receptive context for this new paradigm.

1.3 Key Stakeholders in AAM

AAM has developed through a convergence of stakeholders that extend well beyond the aerospace industry. The success of this new transportation system depends on coordination among manufacturers, operators, infrastructure developers, energy utilities, digital-network providers, regulators, financiers, academic institutions, and communities. Each of these stakeholders contributes unique expertise to help the industry progress. Understanding who these participants are and how their responsibilities intersect establishes the context for appreciating the technical, operational, and policy discussions that follow in the subsequent chapters.

At the heart of AAM are the *aircraft developers and manufacturers* advancing the design and production of electric and hybrid-electric aircraft capable of vertical or short takeoff and landing. These include established aerospace firms and new entrants alike. These *original equipment manufacturers* (OEMs) draw on expertise from electric vehicle engineering, automation and control systems, and advanced energy storage design, while also developing critical capabilities in-house to support the production of certified aircraft. However, even these innovators operate within a web of dependencies sustained by a network of *suppliers*. This network includes propulsion system suppliers, avionics firms, composite material producers, and battery manufacturers, among others. Each supplier provides specialized technologies that contribute to overall system performance.

Some OEMs anticipate also serving as the *operators* of the vehicles they are developing. Many other new and existing aviation firms plan to acquire AAM vehicles from OEMs once certified to serve as UAM and RAM operators. Separately, many firms plan to act as AAM *service providers* to support these operators. Operators and service providers are where AAM aircraft meet the market. Operators manage fleets, schedule missions, and ensure aircraft readiness for services such as passenger transport, cargo logistics, and emergency response. *Service providers* supply the operational infrastructure that supports these activities, including digital platforms, maintenance organizations, and information services, such as weather forecasting, airspace coordination, and flight data management. Responding to market demand, operators and service providers identify viable routes, service models, and operational use cases that align with existing infrastructure and regulatory frameworks. Their collective efforts shape where AAM grows and influence how these new aerial services manifest in conjunction with local communities.

Supporting these community-linked operations is the physical and energy infrastructure that enables AAM to function at scale. *Infrastructure developers* and *energy and utility providers* supply the foundation on which aerial mobility operates, extending aviation's reach into the built environment and connecting it with modern power and data networks. Vertiports, charging and refueling systems, and ground-handling facilities form the visible interface between aircraft and the community. These sites must meet regulatory requirements for safety, accessibility, and environmental compatibility while aligning with local transportation plans. City and regional planners play an important role in determining suitable locations, integrating vertiports with surface transit, and ensuring compliance with zoning and land-use rules. Energy and utility providers, meanwhile, adapt power grids, renewable generation, and storage capacity to support the high electrical demand associated with distributed aircraft operations. These ensuing efforts among developers, planners, and utilities transform AAM into a functioning component of the transportation landscape.

Surrounding these technical and operational actors are the *governance institutions* that establish the rules and procedures governing flight. Civil aviation authorities, standards organizations, and intergovernmental bodies define airworthiness criteria, pilot certification requirements, maintenance standards, and operational limitations. Their primary challenge is updating frameworks originally built for large, crewed aircraft to accommodate smaller, distributed, and increasingly autonomous systems. Beyond aviation regulators, local and regional governments play equally vital roles in zoning, land use, and community planning, determining where and how AAM infrastructure can coexist with urban and regional environments. These overlapping responsibilities make AAM governance inherently interdependent, requiring a whole-of-government approach that aligns transportation, energy, and environmental policy.

Of course, all these efforts must be funded. *Financial and insurance institutions* underpin the economic viability of AAM by providing the investment, risk management, and market confidence required for long-term stable commercial operation. Venture capital, infrastructure finance, and public–private partnerships fund aircraft production, vertiport construction, and digital-network integration, supporting the expansion of both operational and physical systems. Insurers and lenders establish the valuation and risk frameworks that govern commercial deployment, influencing how assets are financed, leased, and maintained throughout their service life. Their involvement signals a functioning market sector with economic viability and commercial maturity. Accordingly, the financial ecosystem acts in parallel with the technical one, ensuring innovation remains commercially durable and capable of scaling across regions.

Beyond the aircraft developers and manufacturers, *academia* and *workforce-development* organizations are also integral to AAM. Universities and laboratories established much of the early knowledge base that has enabled the present generation of electric and hybrid-electric aircraft, contributing to advancements in distributed propulsion, energy storage, materials science, autonomy, and noise modeling. Collaborative programs between academia, industry, and government continue to refine the technologies and operational frameworks that support certification and integration into shared, busy airspace. As AAM matures, these institutions will expand their role from research to workforce development, preparing engineers, technicians, and operators for a breadth of AAM-focused careers. In this expanded role, academic partnerships sustain AAM's innovation ecosystem by connecting education, applied research, and policy analysis on regulation and governance to advance both the technical foundation and the skilled workforce that the industry depends on.

Behind the scenes, *digital and data-network providers* form the connective architecture that enables safe and efficient AAM operations. Software, communications, and automation companies develop the digital architectures that integrate aircraft, vertiports, and airspace management into a unified network. These systems manage flight scheduling, navigation, surveillance, and traffic coordination while ensuring cybersecurity and data integrity across all nodes of operation. Advanced data analytics, cloud computing, and artificial intelligence allow real-time optimization of routes, energy use, and airspace capacity. Telecommunications and satellite providers extend connectivity beyond individual cities, linking regional corridors and supporting seamless transitions between surface and aerial transportation systems. By enabling continuous communication among vehicles, operators, infrastructure, and regulators, digital and data-network providers create the operational backbone that supports scalability, safety, and public confidence in AAM.

None of this, however, will succeed without the support of *communities* and *end users*. Ultimately, it is these stakeholders who will define the success of AAM. Public acceptance determines whether operations expand, where infrastructure is welcomed, and how frequently services are used. Local governments coordinate community engagement, evaluate environmental and noise impacts, and manage permitting processes that ensure equitable integration with existing transportation systems. Residents assess AAM through their first-hand experiences with safety, accessibility, and reliability, forming opinions that directly influence political and regulatory decision-making. As a result, public perception becomes both a performance measure and a planning constraint, shaping the pace and pattern of network development. Within this context, equitable access emerges as a central consideration, as perceived inclusivity and distribution of benefits strongly influence community support across both major population centers and underserved regions.

1.4 Summary

AAM represents the convergence of technological innovation, societal need, and systemic coordination across multiple domains of transportation. It defines a new era in aviation in which flight becomes a distributed, integrated component of how people and goods move within and between communities. AAM extends aviation's reach beyond airports by linking electrified flight, digital airspace management, and interoperable infrastructure into a single, adaptive network. This concept is defined by system-level integration that connects aircraft, energy systems, data networks, and urban design to achieve more efficient, sustainable, and equitable mobility.

AAM's emergence reflects both technological capability and institutional collaboration. Advances in propulsion, materials, autonomy, and energy systems coincide with policy commitments to sustainability, regional connectivity, and community access. Realizing this transformation depends on coordination among diverse stakeholders that include the engineers who design the vehicles, the operators and service providers who sustain them, the regulators who ensure their safety, the financiers and insurers who make them viable, the universities that develop knowledge and the workforce, and the communities that grant legitimacy. Together, these contributors define AAM as a reconfiguration of how aviation, infrastructure, and society interact in the modern era. The chapters ahead will examine these perspectives in greater depth, showing how their collective actions transform AAM as a "concept" into a functioning transportation system.

Reference

1 Atkins, E. and Ollero, A. (2016). *Unmanned Aircraft Systems*. 1st ed., Wiley.

2

AAM Vehicles

Just as lightweight piston engines enabled the birth of aviation and gas turbines ushered in the jet age, the maturation of batteries and electric propulsion is driving the next major transformation in aircraft design. Advancements in battery technology, spurred by previous investments in electric ground vehicles, are allowing new electric aircraft designs to move from concept to reality. Electric propulsion allows designers to explore new aircraft configurations through distributed propulsion architectures that provides redundancy, improve safety, reduce mechanical complexity and maintenance requirements, enable quieter flight, and support modular manufacturing and scalable production methods. These capabilities, coupled with advances in batteries and hybrid-electric energy systems, open the door to aircraft architectures and operating models that were previously impractical.

Aircraft come in many forms, and advances in machine learning (ML) and artificial intelligence (AI) are expanding the feasible design and control space, particularly for aircraft employing many independently controlled propulsive units distributed across the airframe. Although generative AI models may enable new approaches to configuration optimization, the fundamental driver of diversity in aircraft design remains the wide range of operational roles aircraft must satisfy. Aircraft are designed around missions, and those missions generate distinct design requirements. These mission-driven requirements most often produce the most evident physical differences between one configuration and another.

The variations observed among the first generation of advanced air mobility (AAM) vehicles reflect the wide variety of use cases introduced in Chapter 1, with further treatment provided in Chapter 8. Before exploring the fundamentals of AAM vehicle design, we begin with a top-level categorization of AAM aircraft types (Figure 2.1). With a focus on larger AAM vehicles, in contrast to small uncrewed aircraft systems (sUAS), we note that not all AAM vehicles take off and land vertically. These electric conventional takeoff and landing (eCTOL) aircraft, many of them electric short takeoff and landing (eSTOL) aircraft, share more characteristics with their traditional aviation counterparts. Accordingly, this chapter focuses primarily on electric vertical takeoff and landing (eVTOL) aircraft, and the reader is encouraged to reference the greater volume of literature on traditional aircraft design and its electrification.

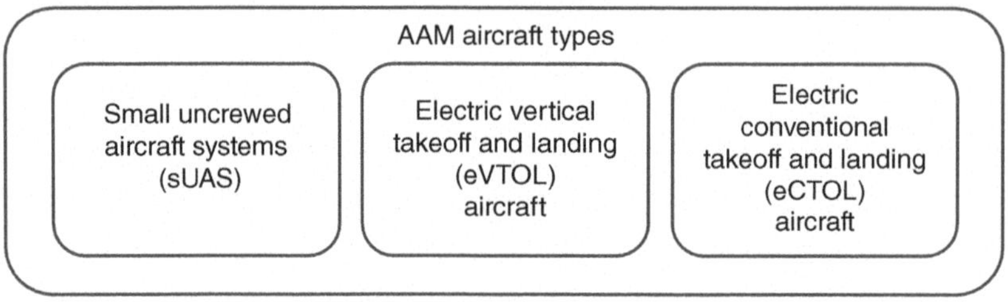

Figure 2.1 AAM aircraft types: sUAS, eVTOL, and eCTOL, which can include eSTOL.

2.1 Types of AAM Vehicles

Among the various aircraft envisioned for AAM, those capable of vertical takeoff and landing (VTOL) represent the most transformative shift from conventional aviation. Vertical flight removes the need for traditional runways and enables point-to-point operations within dense urban and regional networks, making it central to most AAM concepts. The following subsections examine the principal types of electrically powered aircraft. We begin with VTOL systems before turning our attention to electric short- and conventional-takeoff aircraft that extend AAM operations toward longer-range missions and greater reliance on existing airport infrastructure.

2.1.1 The VTOL Aircraft

Within the eVTOL category, the diversity of aircraft design is striking. Figure 2.2 shows a general categorization of eVTOL architectures, and Figure 2.3 provides representative examples. As illustrated in Figure 2.2, eVTOLs can be generally categorized into *multicopter*, *lift + cruise*, and *vectored-thrust* architectures, along with *hybrid* approaches that blur these boundaries. Vectored-thrust architectures can be further broken down into *tilt-wing*, *tilt-rotor*, *tilt-duct*, and *ducted-fan* designs.

Multicopter

Multicopter vehicles lack fixed wings and generate both lift and thrust exclusively from multiple rotors. Volocopter's VoloCity multicopter, shown in Figure 2.3, exemplifies this configuration. Rotor thrust is modulated through differential rotor speed control to achieve attitude, translation, and altitude changes while maintaining stability. This configuration offers mechanical simplicity and cost advantages because it eliminates complex tilt mechanisms and control surfaces. Multicopters are relatively efficient in hover compared to many of their AAM counterparts, and their multiple rotors can provide propulsion redundancy, fault tolerance, and precise low-speed maneuvering. However, because they do not benefit from wingborne lift, multicopters are inefficient in forward flight, resulting in lower cruise speeds and limited range. Their simplicity makes them attractive for early urban air mobility applications, although their lack of aerodynamic efficiency constrains their practicality for longer-range or higher-speed missions.

Lift + Cruise

Lift + cruise vehicles employ independent sets of propulsors for vertical lift and forward cruise. The Beta Alia 250, shown in Figure 2.4, exemplifies this architecture. In a typical configuration,

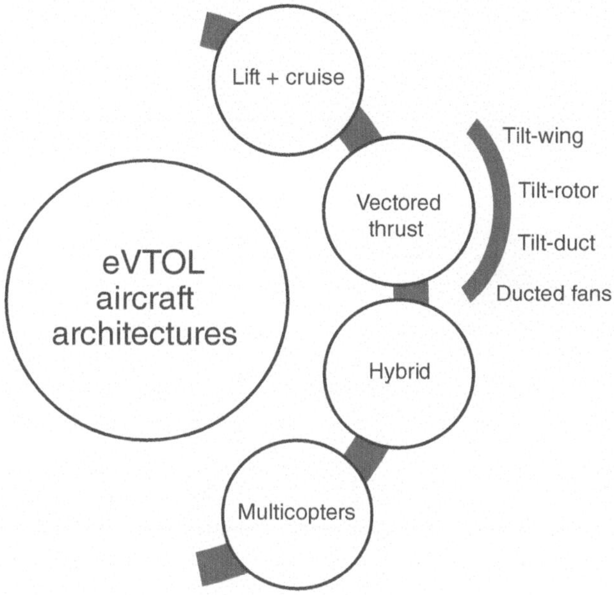

Figure 2.2 eVTOL design architectures.

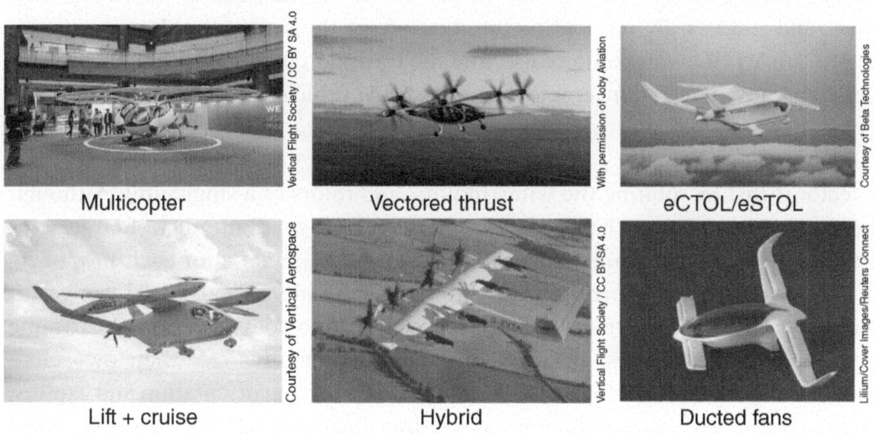

Figure 2.3 Examples of AAM aircraft.

multiple lift rotors are mounted along the wing or on booms to provide vertical thrust during takeoff and landing. Once the aircraft transitions to forward flight, these lift rotors are inactive, and a separate pusher propeller provides thrust for wingborne cruise. This arrangement eliminates the mechanical complexity and control logic associated with tilting mechanisms. It also allows the lift rotors and the cruise propeller to be individually optimized for their respective operating regimes. The design benefits from efficient cruise flight supported by aerodynamic lift from the wings. However, the inactive lift rotors impose penalties in the form of added structural mass and parasitic drag during cruise, as they contribute no thrust once they are unpowered.

Figure 2.4 Beta Alia 250 lift + cruise aircraft. *Source*: Brian Jenkins / Wikimedia Commons / CC BY-SA 3.0.

Vectored Thrust

Vectored-thrust aircraft change the direction of their propulsive flow to generate control forces and to enable both vertical and forward flight. Vectoring thrust can enhance maneuverability and augment stability and control without relying solely on conventional aerodynamic control surfaces, such as ailerons, elevators, and rudders. By reorienting the thrust from the same propulsors, vectored-thrust aircraft can perform VTOL, wingborne cruise, and the transition between them. A variety of mechanical and aerodynamic strategies are employed across AAM vehicles to accomplish this.

Tilt-wing

Tilt-wing aircraft vector thrust by rotating the wing, motors, and rotors as a single unit. Although this introduces mechanical complexity, this configuration relies on a small number of large actuators assigned to the rotating wing assembly, instead of individual actuators for each propulsor. Tilt-wings benefit from efficient wingborne cruise without the penalty of carrying inactive lift rotors. They also experience less aerodynamic interference from rotor outwash on the airframe than some other vectored-thrust designs. However, tilt-wing architectures face their own challenges. During takeoff and landing, the wing articulates to a nearly vertical orientation and exposes a large flat-plate area to the relative wind, which increases drag and complicates control. The most demanding flight phase occurs during the transition after takeoff, when the rotors must support both the aircraft's weight and accelerate it forward. As the wing rotates toward a horizontal position, it operates at high angles of attack for a period of time with partial to post-stall flow separation. An example of a tilt-wing aircraft is the Airbus A^3 Vahana (Figure 2.5).

Tilt-rotor

In contrast to tilt-wings, tilt-rotor aircraft redirect thrust by rotating their propulsors, with or without the nacelle, while keeping the wing fixed. The Joby S4 (Figure 2.6) combines both approaches: the four aft propellers tilt with their nacelles, while the two forward propellers tilt independently. Like tilt-wings, tilt-rotors use all propulsors during every phase of flight and benefit from efficient wingborne cruise. Retaining a fixed wing reduces structural loads and avoids increasing the aircraft's frontal area during vertical operations. The trade-off here is that the tilting nacelles and propulsor mechanisms introduce mechanical and control-system complexity.

Figure 2.5 Airbus A³ Vahana tilt-wing vectored-thrust aircraft. *Source*: Vertical Flight Society / CC BY-SA 4.0 / https://gallery.vtol.org/image/PHHWt.

Figure 2.6 Joby S4 tilt-rotor vectored-thrust aircraft. *Source*: Harlan Huntington / The United States Air Force / Public domain / https://commons.wikimedia.org/wiki/File:Joby_Aviation_S4_experimental_eVTOL_aircraft_at_Edwards_AFB.jpg.

Tilt-duct

Tilt-duct eVTOLs use shrouded or ducted propellers that tilt as a single unit, providing vertical lift when oriented downward and thrust for wingborne cruise when rotated forward. The Bell Nexus (Figure 2.7) is an example of this configuration. The duct can improve static thrust efficiency by reducing blade tip losses and improve safety by shielding the blades from foreign objects, along with lowering noise levels. Ducted propellers are typically smaller in diameter than open rotors, reducing their physical footprint. However, as with other tilt architectures, the ducts add structural weight and create drag during cruise. The tilt mechanism also retains significant mechanical complexity.

Ducted Fan

A ducted-fan eVTOL employs fixed ducts embedded in the wing or fuselage, with thrust vectoring achieved by redirecting airflow within the ducts. The ducts themselves remain stationary; instead,

Figure 2.7 Bell Nexus tilt-duct vectored-thrust aircraft. *Source*: Amaury Lapote / Flickr / CC BY 2.0.

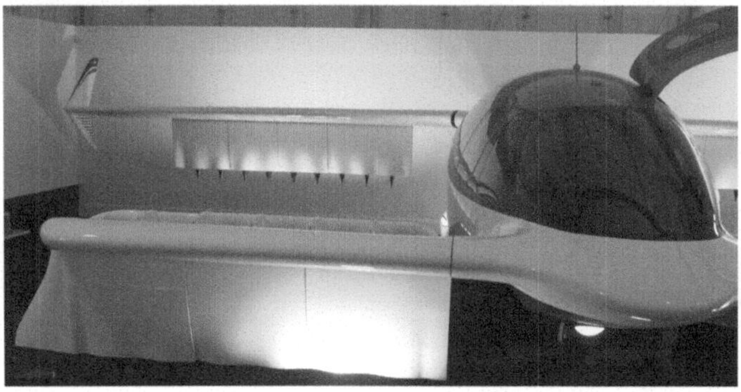

Figure 2.8 Lilium Jet ducted-fan vectored-thrust aircraft. *Source*: Leo067 / Wikimedia Commons/ CC BY 4.0/ https://commons.wikimedia.org/wiki/File:Modell_Lilium_Jet.jpg.

airflow is deflected through variable vanes or by tilting the fans within the ducts. The Lilium Jet (Figure 2.8) exemplifies this approach, integrating 36 small ducted fans into its wing flaps and canards. This distributed ducted-fan layout allows all propulsion units to contribute to both vertical and forward flight. The ducts provide potential benefits such as noise shielding, reduced tip losses, and protection from foreign object debris, while also enhancing perceived safety for urban and passenger operations. The ducts remain fixed within each fan module, while the flaps that house them pivot during transition to redirect thrust for vertical flight to forward flight. This articulation avoids the larger moving mechanisms associated with tilt-wing or tilt-rotor systems. The trade-off, however, is that these smaller fans can incur substantial induced-power and hover-efficiency penalties.

Hybrid

Hybrid eVTOLs blend features of lift + cruise and vectored-thrust architectures by enabling some vertical-lift propulsors to tilt forward for cruise. Rather than using a large number of lift rotors that are inactive during cruise, hybrid configurations reuse part of the propulsor system across all flight phases. The Vertical Aerospace VX4 (Figure 2.9) is an example of this approach, employing four fixed vertical-lift rotors aft of the wing and four tilting propellers forward of it. The tilting units support VTOL and then function as propellers during wingborne cruise, reducing drag and weight penalties compared to pure lift + cruise designs. Hybrids, therefore, occupy a middle ground in the eVTOL design spectrum by offering improved efficiency at the cost of increased mechanical complexity.

2.1.2 The eCTOL/eSTOL Aircraft

Although much of this chapter focuses on eVTOLs, eCTOL aircraft also represent an important category of AAM vehicles (Figure 2.10). Consistent with the AAM framework introduced in Chapter 1, eCTOL vehicles also employ high levels of automation and electric or hybrid-electric propulsion

Figure 2.9 Vertical Aerospace VX4 illustrating a hybrid propulsion architecture that combines vectored thrust for forward flight with dedicated lift rotors for VTOL. Courtesy of Vertical Aerospace.

Figure 2.10 Beta's eCTOL Alia CX300. *Source*: Airman 1st Class Gianluca Ciccopiedi / U.S. Air Force / Public domain / https://commons.wikimedia.org/wiki/File:231018-F-CW106-9139.jpg.

systems to move goods and people within local and regional transportation networks, particularly those that are presently underserved. Unlike eVTOLs, which promise point-to-point access within dense urban centers, eCTOL aircraft rely on existing or new runways or compact landing strips. By doing so, eCTOLs offer a complementary pathway for expanding AAM beyond city cores into peri-urban and regional environments. In contrast to eVTOLs, eCTOL aircraft rely on wingborne lift during all phases of flight, and this aerodynamic efficiency translates into improved range and payload capacity. Given current battery limitations, this makes them especially well suited for regional air mobility applications where energy efficiency is paramount. From an infrastructure perspective, eCTOL aircraft can utilize the thousands of underused municipal and regional airports already in operation, reducing the immediate need for entirely new facilities. Additionally, the regulatory certification pathway for eCTOL aircraft may be more straightforward because their architecture remains closer to conventional fixed-wing aircraft, positioning them as a lower-risk entry point into electric aviation.

Advances in electric propulsion distributed across the airframe, known as distributed electric propulsion (DEP) and discussed later in this chapter, combined with modern high-lift systems, have enabled eCTOL aircraft to operate as eSTOL designs (Figure 2.11). These aircraft are potentially capable of operating from runways on the order of 100 m and achieving takeoff and landing speeds as low as 25 knots. These capabilities create the possibility for compact community-scale operating sites, including short prepared strips, repurposed airfields, floating platforms, and certain adapted heliport facilities, to support regional eCTOL operations.

Conventional short takeoff and landing (STOL) aircraft have historically relied on large and mechanically complex high-lift devices such as flaps and slats. However, with the advent of electric propulsion, designers can now mount multiple small propulsors along the wing and actively control the airflow over high-lift surfaces. This distributed propulsion architecture is possible because electric motors can drive each propulsor directly, eliminating the previous need for heavy gearboxes and driveshafts used to mechanically distribute power from one or two central piston or turbine engines. The resulting "propeller–wing integration" helps maintain attached flow at much higher angles of attack, thereby generating significantly greater lift at very low speeds. This integration is referred to as *blown-lift*, where the accelerated propeller slipstream augments lift by increasing local airflow velocity across the wing and high-lift surfaces. The energized flow also helps maintain attached flow at high angles of attack and lower flight speeds.

Figure 2.11 Electra's nine-passenger EL9 aircraft with its ultrashort takeoff and landing capability is designed to unlock new access points that bypass today's commercial airline hub-and-spoke system. Courtesy of Electra.aero, Inc.

2.2 The eVTOL Design Considerations

Designing an eVTOL aircraft, like all aircraft, begins with a set of core requirements defined by customers, operators, and regulators. Quantitative design drivers, such as range, payload capacity, cruise speed, cost, and safety, have always shaped how aircraft are designed. What distinguishes eVTOLs are additional design considerations associated with VTOL operations, the adoption of electric propulsion, and expanded possibilities for higher levels of automation. Moreover, emerging AAM missions (passenger service, cargo logistics, emergency response, etc.) shift the relative weighting of traditional performance metrics. Collectively, these factors reshape how traditional design drivers are prioritized and highlight the enabling technologies that now make eVTOLs viable.

2.2.1 Design Drivers

Range is one of the most important design drivers for AAM because the viability of air taxi and regional missions depends on how far an aircraft can fly, whether between destinations or between charging events. The primary technological limiter is battery specific energy, measured in watt-hours per kilogram [Wh/kg].[1] Battery specific energy determines how much energy can be stored for a given battery weight. Short ranges may satisfy dense urban air taxi operations, where energy demands are modest, whereas intercity and regional missions impose greater range requirements that, under current battery limitations, demand higher cruise efficiency. Because eVTOLs must carry heavy battery packs that do not lighten during a flight, the aircraft's total weight constrains achievable range. Additional passengers or cargo add to this constraint. The resulting design space places a premium on weight minimization, where any increase in payload or other mission performance demand comes at the expense of range.

Capacity is another key driver of eVTOL design and can be considered in terms of number of seats or useful payload. Through an economic lens, carrying more passengers lowers the cost per seat-mile and reduces the carbon footprint per unit payload. Larger vehicles can be structurally more efficient, since weight does not necessarily scale linearly with passenger or payload capacity and they distribute fixed costs, such as crew, across more passengers. However, increasing capacity introduces performance penalties. Each additional passenger increases the thrust required for takeoff and landing, raising power demand and acoustic output. Smaller aircraft are often better aligned to meet strict community noise constraints, but this comes at the expense of higher per-passenger costs and less favorable emissions per seat-mile. Capacity, like range, is a balancing act between aircraft large enough to achieve economic and environmental viability and sizes small enough to be compatible with operational and community acceptance constraints.

Cruise speed is another essential design parameter. Here, a speed that is fast enough to deliver meaningful time savings is desired, yet the speed should not be so high that aerodynamic drag and power consumption impose prohibitive penalties. This balance reflects efficiency and utility, where eVTOLs cruise fast enough to justify their role in the transportation network but not so fast that they compromise their already limited range.

While time spent in hover is minimized in normal operations because of the significant power penalties it incurs, this vertical phase of flight introduces *hover efficiency* as a key design consideration. As we will see later in this chapter, hover efficiency is governed by disk loading,

1 Throughout the chapter, parameters are placed within parentheses (); units are placed within square brackets [].

defined as rotor thrust divided by total rotor disk area. Lower disk loading reduces hover power requirements, downwash velocity, and acoustic intensity, but the accompanying larger rotors introduce structural, aerodynamic, and operational challenges. Multicopters, for example, achieve low disk loading and high hover efficiency but, as we have seen, pay a penalty in cruise performance. Vectored-thrust architectures reverse this trade-off, sacrificing some hover efficiency in exchange for greater aerodynamic efficiency during forward flight.

An eVTOL must operate at costs low enough to be commercially viable. Replacing gearboxes, complex fuel systems, and numerous moving parts with simpler electric drivetrains, combined with modular components for easier maintenance and replacement, is intended to lower maintenance costs. The *direct operating cost* (DOC), which is the ultimate measure of economic viability, combines maintenance, energy consumption, and utilization rate. Minimizing DOC depends on both vehicle design and operational factors such as turnaround time, charging infrastructure, and fleet scheduling efficiency.

Finally, *aircraft dimensions* and *noise*, long-standing concerns in conventional aviation, take on heightened importance in AAM operations. Maximum vehicle dimensions are not limited by runways and gates; instead, they are limited by the compact footprints of vertiports. Noise becomes even more problematic when AAM operations are embedded directly within neighborhoods.

2.2.2 Enabling Technologies

If the design drivers set the targets, three enabling technologies make it possible for eVTOLs to achieve them. The first is *DEP*. Electric motors are compact, torque-dense, and efficient, enabling the distribution of many small propulsors across the airframe. DEP provides propulsive redundancy, increases control authority, and allows designers to distribute thrust over multiple rotors. The second enabler is advancements in *electrical energy storage and management systems*. This technology has moved eVTOLs from concept toward operational viability. High-specific-energy batteries, advanced thermal management, and, in some concepts, hybrid architectures provide sufficient onboard energy to support practical mission ranges. The third enabler is *automation and simplified vehicle operations* (*SVO*). Fly-by-wire controls, intuitive digital cockpits, and increasing levels of autonomy reduce pilot workload and training requirements. In near-term operations, SVO expands the pool of qualified aviators. Over longer time horizons, higher levels of automation are expected to remove the requirement for onboard pilots, reclaiming usable payload capacity and reducing DOCs.

The fundamentals of eVTOL design are shaped by classical aircraft engineering principles, but they are reframed by recent technological advancements and emerging use cases. Range, payload, speed, efficiency, economics, size, and noise remain at the core, yet their relative prioritization has shifted. DEP, advanced batteries, and automation are the technologies that enable these requirements to be reconciled in this new transportation ecosystem. We will now develop this reconciliation through a first-principles approach to eVTOL design.

2.2.3 Fundamentals of eVTOL Design

Vital to AAM operations is the ability to take off and land vertically. Without vertical access, operations in cities and dense suburbs are stymied, and such services would be forced back into the mold of conventional aviation, tied to airports and long runways located far from many origins and destinations. VTOL capability frees aircraft from this limitation and permits operations embedded directly within urban and suburban areas. This allows for the placement of transportation nodes close to demand centers, creates point-to-point connectivity that bypasses congested ground networks, enables distributed networks of high-frequency flights, and ultimately shortens door-to-door travel times. VTOL capability is also required for precise and safe flight in dense, obstacle-rich environments where slower vertical maneuvers are better suited than high-speed approaches. Beyond cities, this same flexibility provides access to remote or infrastructure-limited regions where runways are impractical. In these ways, VTOL is the fundamental enabler of the AAM vision.

Of course, VTOL is not a new idea. Helicopters have long demonstrated the utility of vertical flight with their ability to reach confined locations, hover, and maneuver precisely. Helicopters also highlight the limits of traditional VTOL: they are noisy, mechanically complex, expensive to operate, and not easily scaled for high-frequency passenger service. Various militaries have also employed VTOL strategies using jet thrust vectoring or lift fans. However, these approaches come at the cost of extreme mechanical complexity, high fuel consumption, prohibitive expense, and operational complexity. AAM builds on the same advantages of vertical access but seeks to deliver it in a fundamentally different way. To better understand this, we begin with the basics of vertical lift.

First-principles of Thrust, Power, and Hovering Efficiency

A great deal of insight into VTOL design can be garnered from a first-principles approach. Using this approach, we start from basic concepts and keep our reasoning transparent and grounded in fundamentals. Though simple, such an approach often yields disproportionate insight and reveals relationships and trade-offs that might otherwise be obscured by more complex analyses. We begin with the familiar form of Newton's second law of motion:

$$F_{\text{net}} = ma \tag{2.1}$$

which states that the net force, F, acting on an object of mass, m, produces an acceleration, a, in the direction of the net force.

For one-dimensional motion, the average acceleration, a, is given by Eq. (2.2) where Δv represents the change in velocity over a given time interval, Δt,

$$a = \frac{\Delta v}{\Delta t} \tag{2.2}$$

Substituting Eq. (2.2) into Eq. (2.1) yields

$$F = m\frac{\Delta v}{\Delta t} \tag{2.3}$$

With the associative property of multiplication, we rewrite Eq. (2.3) as

$$F = \left(\frac{m}{\Delta t}\right)\Delta v \tag{2.4}$$

Here, the term $\frac{m}{\Delta t}$ represents a mass flow rate. Taking the mass of interest to be the mass of air passing through the eVTOL's rotor disks per unit time, we identify the resulting force as *thrust*, T,

$$T = \left(\frac{m}{\Delta t}\right)\Delta v \tag{2.5}$$

Equation (2.5) shows that the thrust required for vertical flight can be produced by accelerating a mass of air per unit time through the rotor disk. Equation (2.5) therefore shows that the thrust of an eVTOL can be increased by either (1) increasing the amount of mass brought through the rotor disk, $\frac{m}{\Delta t}$, or (2) increasing the change in the velocity of the air, Δv. With these two options to maximize thrust available, we must determine which strategy is more efficient, recognizing that battery technology remains a limiting factor in eVTOL design. Since *power*, *P*, represents the rate at which energy is transferred to the airflow through the rotors, we begin with the basic relation:

$$P = Fv \tag{2.6}$$

Hovering flight is among the most power-intensive phases of flight, often establishing the upper bound on battery and propulsion requirements. Therefore, additional insight into the optimal strategy for producing thrust can be gained by considering the required hover power. Taking the force within Eq. (2.6) to be the thrust force, *T*, and recognizing that the relevant velocity in the power relation corresponds to the velocity of the air as it passes through the rotor disk, we express this local velocity as the average of the air speeds immediately before, v_{in}, and after the disk, v_{out}:

$$v = \frac{v_{in} + v_{out}}{2} \tag{2.7}$$

In a hover, the air far above the rotor is assumed to be nearly stationary relative to the aircraft, while below the rotor, it has been accelerated to a finite downward velocity. In practice, some of the flow is drawn back toward the rotor through localized recirculation and induced inflow, so the air approaching the rotor is not perfectly motionless. However, this inflow speed is generally small compared with the velocity imparted by the rotor and may be neglected in this first-order analysis. The rotor, therefore, imparts a total change in air velocity:

$$\Delta v = v_{out} - v_{in} \tag{2.8}$$

and the air that crosses the rotor plane does so at a representative velocity equal to roughly half of this total change. Said differently, the air approaches the rotor at a relatively low speed and leaves it moving downward at approximately twice the velocity that it had at the rotor plane. Expressing the velocity through the disk in this averaged form gives:

$$v = \frac{\Delta v}{2} \tag{2.9}$$

Substituting this relation into Eq. (2.6), along with the thrust expression, $T = \left(\frac{m}{\Delta t}\right)\Delta v$, from Eq. (2.5), yields a useful expression for hovering power:

$$P = \frac{1}{2}\left(\frac{m}{\Delta t}\right)(\Delta v)^2 \tag{2.10}$$

As in Eq. (2.5), Eq. (2.10) once again shows that we can increase thrust by increasing either $\left(\frac{m}{\Delta t}\right)$ or Δv. However, the square of the velocity in Eq. (2.10) reveals which of these strategies demands more power, thus depleting our stored battery energy more rapidly.

To illustrate this, consider two cases. If the mass flow rate $\left(\frac{m}{\Delta t}\right)$ is doubled, Eq. (2.5) shows that thrust, *T*, doubles, and Eq. (2.10) shows that hover power also doubles. In contrast, if the change in velocity, Δv, is doubled, Eq. (2.5) again shows that thrust doubles, but Eq. (2.10) indicates that power increases fourfold (2^2). Thus, it is **far more efficient to increase thrust for vertical flight by accelerating a larger mass of air more slowly than by accelerating a smaller mass of air more rapidly**. In the context of traditional gas-turbine-driven aviation, this is the rationale behind the use of the high-bypass jet engine.

Disk Loading and Rotor Solidity

Our first-principles look at thrust and power in the context of hover has shown that accelerating a large mass of air by a small amount is more efficient than accelerating a small mass of air by a large amount. When discussing hovering efficiency, another core idea is that of disk loading. *Disk loading* is defined as the ratio of an aircraft's total weight to its total rotor swept area, and it provides a measure of the aircraft's weight supported by each square meter of rotor disk area:

$$\text{Disk loading} = \frac{\text{Aircraft weight}}{\text{Total rotor swept area}} \tag{2.11}$$

Figure 2.12 illustrates the rotor swept areas of a representative helicopter and quadrotor eVTOL. For illustrative purposes, let us take the weight of each of these vehicles to be the same. Inspecting Figure 2.12, we see that, due to the helicopter's larger rotor swept area, it exhibits a lower disk loading than the eVTOL with a smaller swept area. For aircraft of equal weight, a larger total rotor swept area corresponds to a lower disk loading. Conversely, for a fixed vehicle weight, a smaller total rotor area requires each square meter of rotor disk to support more weight, resulting in higher disk loading. Lower disk loading is advantageous for hovering flight because, as established by the analysis above, it allows a rotor to generate the required vertical lift by accelerating a larger mass of air through the disk with a smaller induced velocity, reducing the power required to produce thrust. This effect distributes the lift requirement over a larger rotor area, allowing more efficient use of the surrounding air. In contrast, high disk loading forces the rotor to accelerate a smaller volume of air more forcefully, which, as established earlier, requires greater induced power to maintain lift and reduces hovering efficiency. Because helicopters typically employ very large

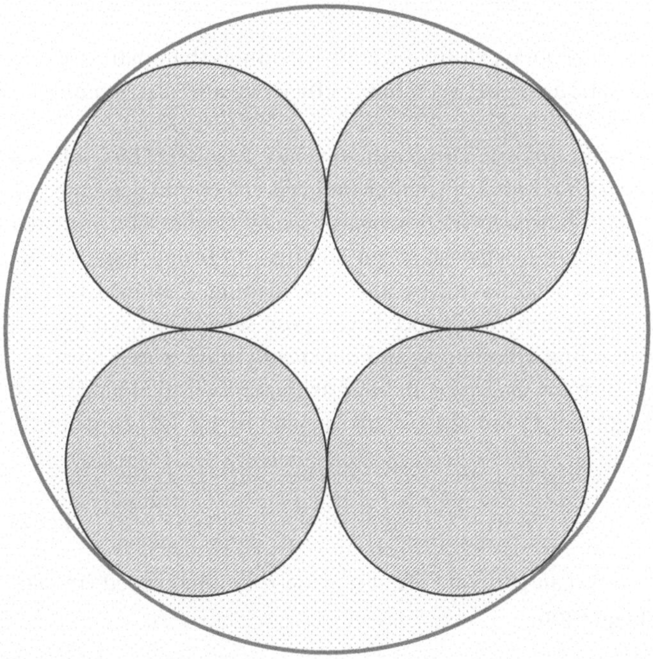

Figure 2.12 Representative helicopter (lighter shading) and quadrotor eVTOL (darker shading) rotor swept areas.

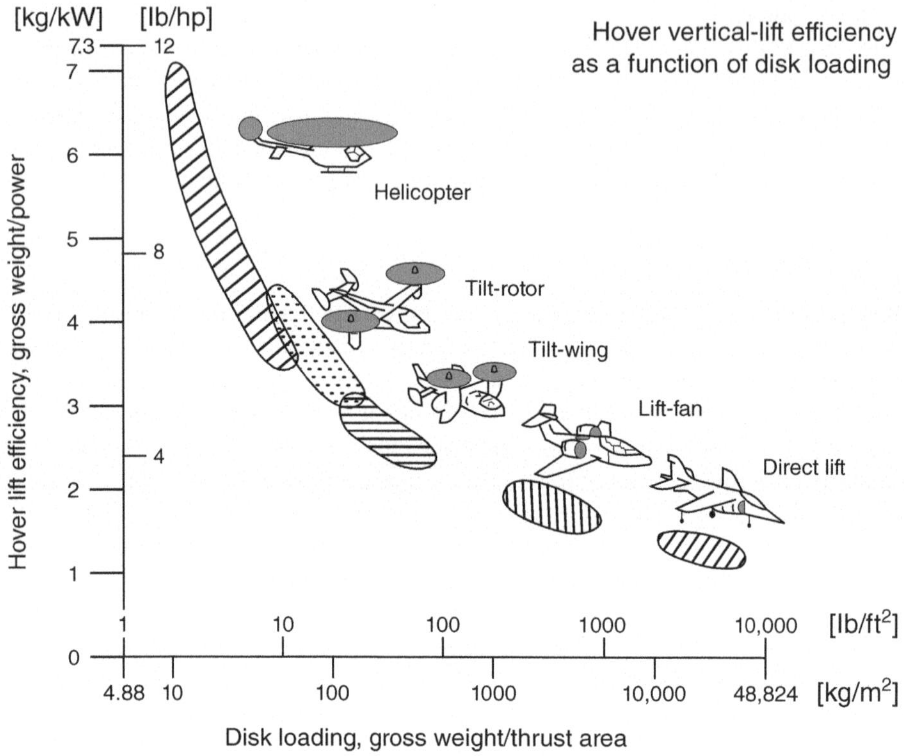

Figure 2.13 Hover efficiency versus disk loading for a variety of aircraft architectures. *Source*: NASA / Public Domain.

rotor swept areas relative to their weight, they often exhibit lower disk loadings than many eVTOL configurations. Figure 2.13 shows this relationship between hover efficiency and disk loading for various vehicle architectures.

The velocity imparted to the airflow by the rotor is termed the *downwash speed*. Thus, when a larger rotor acts on a greater mass of air and requires a smaller induced velocity to generate lift, an aircraft with a large rotor swept area will have a lower downwash speed. Conversely, a smaller rotor swept area necessarily results in a higher downwash speed. Figure 2.14 brings together the relationships between hover efficiency, downwash velocity, and disk loading. Downwash is an important design consideration because high downwash speeds can pose safety risks within vertiport operating environments and surrounding urban areas and may result in increased noise.

Another key parameter in rotor design is *rotor solidity*, σ, defined as the ratio of total blade area, a_b, to the overall swept disk area, A (Figures 2.15 and 2.16). This is given by Eq. (2.12), where N represents the number of rotor blades,

$$\sigma = \frac{Na_b}{A} \tag{2.12}$$

Solidity influences how efficiently a rotor can generate lift at a given rotational speed (revolutions per minute [RPM]). Using the lift equation,

$$L = C_L \frac{1}{2} \rho v^2 a_b \tag{2.13}$$

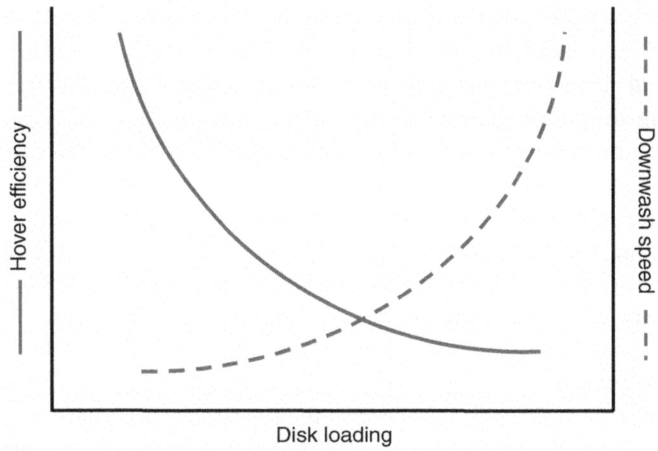

Figure 2.14 Hover efficiency (solid line) and downwash speed (dashed line) as a function of disk loading.

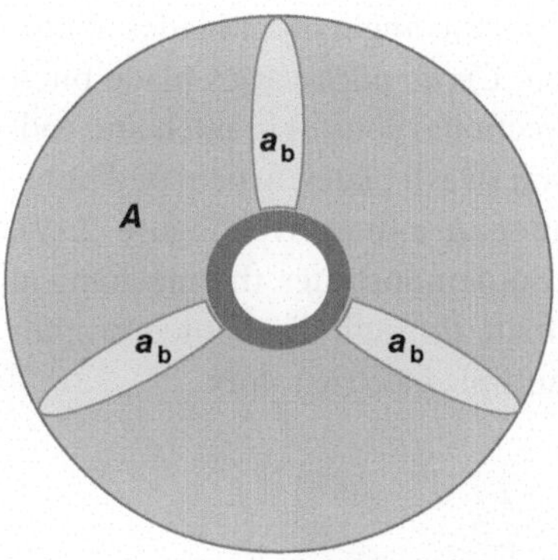

Figure 2.15 Rotor solidity is the ratio of total blade area to the swept disk area.

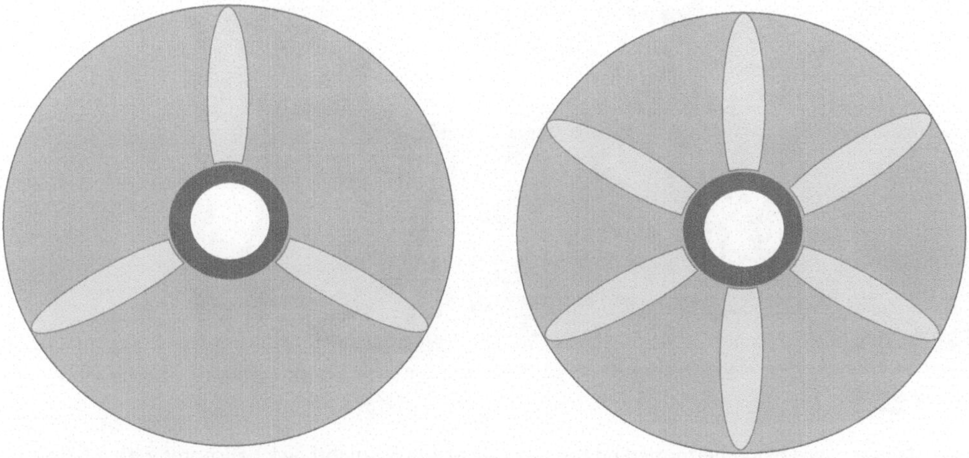

Figure 2.16 Low (left) and high (right) rotor solidity.

where C_L is the lift coefficient, ρ represents air density, and v represents the speed of the blade through the air, we see, keeping all else constant, that increasing blade area, a_b, and hence blade solidity, σ, increases the lift generated. However, this increase in lift comes at a cost. Greater blade area also increases profile drag through increased wetted surface area and can increase overall aerodynamic losses depending on rotor loading and operating condition. In practice, a higher solidity rotor allows a given lift requirement to be met at a lower rotational speed, keeping tip speeds lower. This relationship follows directly from Eq. (2.13) when used as a scaling relation. By increasing the blade area, a_b, the required velocity, v, over the blades can be reduced for the same lift, translating into lower tip speeds. This characteristic is attractive in eVTOL design because lower tip speeds reduce acoustic emissions. This concept is explored further in Chapter 7.

Control Strategies: RPM, Collective, and Cyclic

For eVTOL aircraft, understanding rotor control begins with examining the systems that have historically governed conventional helicopters. Traditional rotorcraft achieve flight control through a combination of nearly constant rotor speed and variable blade pitch. The main rotor typically spins at a fixed RPM determined by the efficiency range of the turboshaft engine and transmission system. At this fixed speed, control authority is achieved by adjusting blade pitch, thereby modifying the local angle of attack of the blade. *Collective pitch* changes the angle of all blades simultaneously, increasing or decreasing lift to govern vertical motion. *Cyclic pitch* varies blade pitch differentially during rotation, tilting the rotor disk to generate directional control in pitch and roll, enabling forward and lateral movement. These mechanisms rely on swash plates, articulated hubs, and linkage systems that are mechanically intricate and maintenance-intensive (Figure 2.17). While this control architecture has proven highly capable, it also demonstrates the mechanical and operational complexity that eVTOL designers seek to eliminate through distributed propulsion, digital flight control, and software-based coordination of multiple electric rotors.

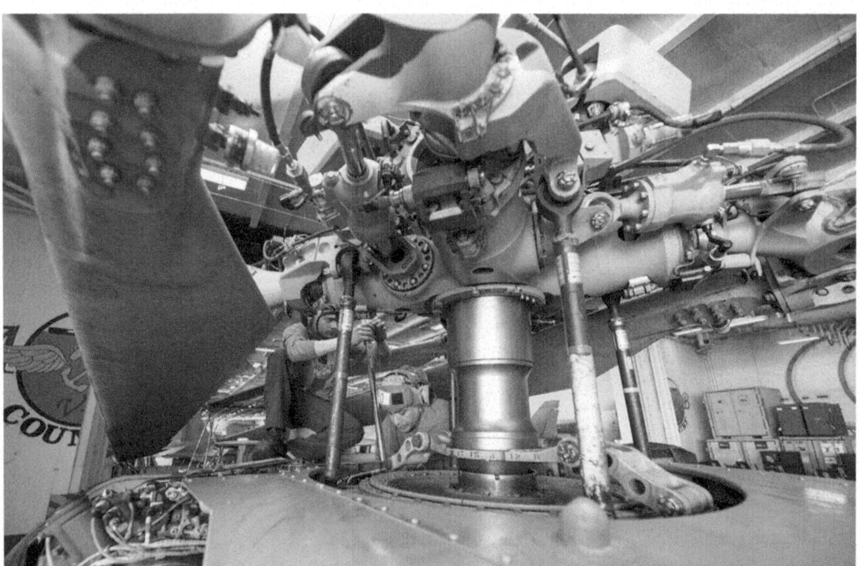

Figure 2.17 Maintenance being performed on the Sikorsky MH-60S Knighthawk helicopter with a fully articulated rotor system. Both the swashplate and linkage mechanisms are shown. *Source*: U.S. Navy photo / Flickr / CC BY 4.0.

eVTOLs expand the control space in fundamentally different ways. Multicopter configurations generally employ fixed-pitch rotors and achieve control entirely through RPM modulation. By increasing or decreasing shaft speed, thrust is varied directly. Differential thrust between rotors provides pitch, roll, and yaw control. This method is simple, robust, and scales naturally when many small rotors are available, making it especially well suited to multicopters. Other eVTOL architectures, such as tilt-rotor and lift + cruise designs, sometimes incorporate variable collective pitch, particularly for larger rotors that must operate efficiently in both hover and forward flight. Collective adjustment allows designers to balance hover power requirements with propulsive efficiency during forward flight. Cyclic pitch control, however, is generally absent from most eVTOL designs. With multiple distributed rotors, cyclic functions can be replaced by differential thrust commands across propulsors. Eliminating cyclic pitch removes the need for swash plates and articulated hubs, simplifying mechanical design and reducing weight. This marks a significant departure from the helicopter paradigm and reflects how eVTOL control tasks shift from mechanical actuation to software-based flight-control allocation and motor-controller logic.

An important aspect of distributed rotor systems is the management of torque balance (Figure 2.18). Every rotor that spins in one direction produces thrust and a reactive torque on the airframe. To prevent the aircraft from continuously yawing in the opposite direction, rotors must be arranged in counter-rotating pairs, some spinning clockwise and others counterclockwise, so that their torques cancel in steady flight. Maintaining balanced torque and lift distribution becomes more challenging following a motor failure. If a rotor suddenly stops producing thrust and its associated reactive torque, the remaining rotors must rapidly adjust their speeds to restore yaw equilibrium and maintain control. To make this possible, eVTOL motors are typically sized with extra power margin beyond nominal operating levels, so that healthy units can be commanded above their rated power for short durations to compensate for the loss. Modern flight-control laws are designed to coordinate these rapid adjustments automatically, a strategy known as *graceful degradation*. Rather than a single failure leading to catastrophic loss of control, the system reconfigures its control allocation to maintain stability, albeit with reduced performance, until the vehicle can land safely.

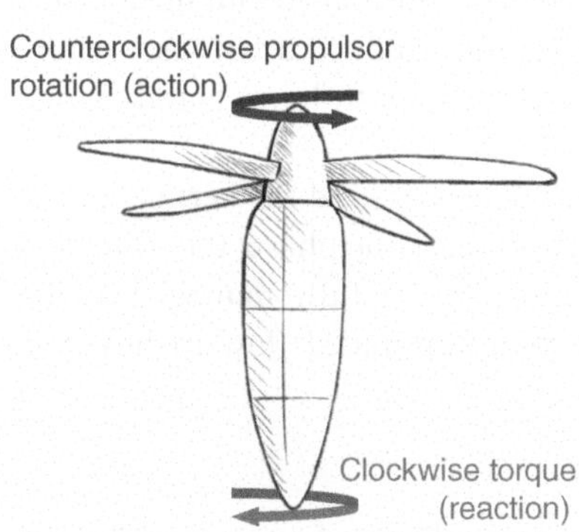

Figure 2.18 Reactive torque produced by a propulsor.

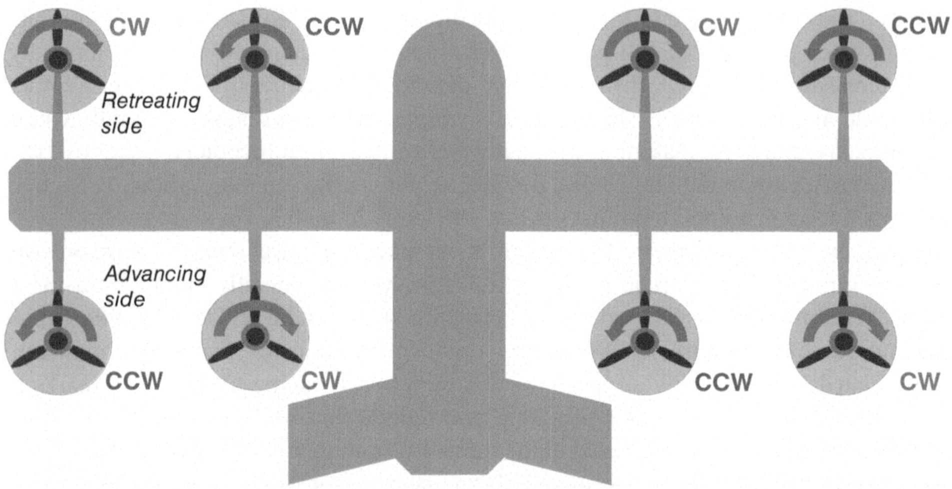

Figure 2.19 Counter-rotating rotors cancel reactive torque on the airframe. Following a motor failure, remaining rotors adjust RPM to restore torque balance. An FRRA rotation pattern assigns opposite rotation between front and rear rotors so that the rear-advancing blade aligns with the front-retreating blade, reducing wake interaction and associated noise and aerodynamic interference in forward flight. Advancing and retreating blade definitions are relative to the aircraft's forward velocity.

There are multiple ways to assign clockwise and counterclockwise rotation among a distributed set of rotors. One approach is the *front-retreating/rear-advancing* (FRRA) architecture. In this arrangement, each aft rotor spins in the opposite direction to the rotor immediately forward of it. The primary objective of this assignment is to maintain balanced rotor loading and control authority across the vehicle, ensuring that no rotor pair experiences systematically different inflow conditions, preventing asymmetric lift and control response. As illustrated in Figure 2.19, the FRRA pattern also yields a secondary aerodynamic benefit related to wake interaction: the rear rotor's advancing-blade region aligns with the front rotor's retreating-blade flow. By contrast, if front and rear rotors in a column rotate in the same direction, the rear-advancing blade encounters the accelerated wake of the forward advancing blade. The FRRA arrangement reduces the severity of this wake interaction, which can lower noise, improve efficiency, and reduce vibratory loads, thereby limiting structural fatigue over time. For eVTOL designs intended for urban use, where both controllability and community noise are critical, this combined control and aerodynamic benefit makes FRRA an attractive architecture.

Thus far, we have framed most of our discussion of control strategies around the vertical phase of flight. However, equal attention must be given to an even more demanding phase: the *transition* from rotor-borne to wingborne lift, where control authority must be carefully managed as the aircraft moves between distinct aerodynamic regimes, from rotor-dominated flow in hover to wing-dominated lift in forward flight.

Transition to Wingborne Flight

While we have discussed the effect of disk loading on power consumption in hover, regardless of whether the loading is high or low, it is generally advantageous for an eVTOL to transition to wingborne flight as quickly as possible in order to exploit the efficiency of fixed-wing aerodynamics, where lift is generated more efficiently through forward motion instead of sustained vertical thrust. The transition between rotor- and wing-borne flight is one of the most demanding and dynamic phases of eVTOL operation. In hover, the rotors provide nearly all the lift and control. In cruise, by contrast, the wings support the vehicle's weight while the propulsive system is devoted primarily to producing forward thrust. Moving between these two regimes requires careful coordination of lift distribution and control authority as the aircraft passes through an intermediate state in which neither the rotors nor the wings are fully effective in their most efficient operating condition. In this translational flight regime, stability margins shrink and aerodynamic disturbances become more consequential. Rotor wake interference with the wings and tail, reduced aerodynamic control effectiveness at low speeds, a shifting center of lift, and small imbalances in lift or thrust can all produce significant handling challenges. To address these challenges, eVTOLs rely heavily on digital flight-control systems capable of coordinating multiple actuators in real time. These systems blend thrust vectoring, RPM modulation, and aerodynamic surface deflections to maintain a smooth and controlled transition.

The nature of this transition depends strongly on vehicle configuration. *Multicopters*, which lack wings, avoid this complexity altogether, but at the cost of cruise efficiency since their rotors must support the full weight of the aircraft throughout the entirety of the mission. For other architectures, the challenge is more acute. *Tilt-rotors* and *tilt-wings* reorient their propulsors (and in the case of tilt-wings, the entire wing) to progressively vector thrust from the vertical to the horizontal. Control laws must account for rapidly changing inflow and asymmetric aerodynamic loading on both the wings and propellers. Lift + cruise designs employ separate lifting rotors and cruise propellers. For these aircraft, transition involves phasing out lift from the rotors while simultaneously transferring weight support to the wings. Timing is critical: if lift rotors are throttled back too quickly, the wing may not yet provide adequate lift; if throttled back too slowly, excess drag will penalize performance.

Aerodynamic forces inherently generate aerodynamic moments, and the changing nature of the aerodynamic forces during transition produces corresponding changes in moments about the aircraft's center of gravity (CG) (Figure 2.20). A moment, or a torque, arises when a force does not

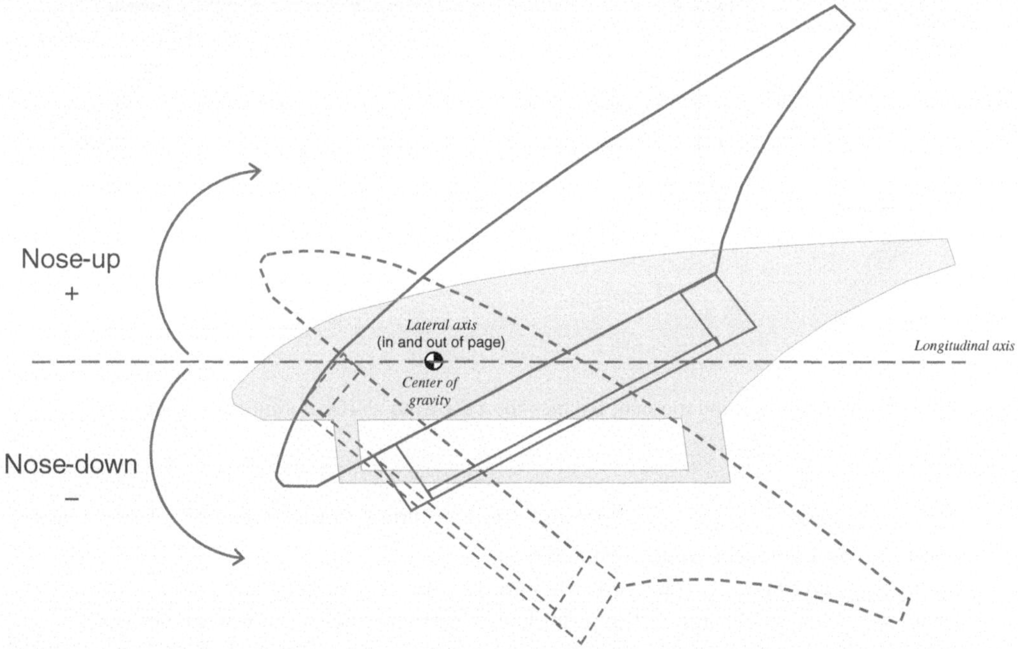

Figure 2.20 Nose-up and nose-down vehicle pitch about the lateral axis (in and out of the page) that passes through the CG.

act through the aircraft's CG. The CG of an aircraft is the point at which its weight is considered to be concentrated; this is also the point about which the aircraft rotates. If all forces act on an aircraft through the CG, the aircraft will have no tendency to rotate. With symmetry about the aircraft centerline, moments about the aircraft's lateral (pitch) axis are the most critical. Any unbalanced moment about the lateral axis determines whether the nose of the aircraft rises or falls. These *pitching moments* are fundamental to the control and stability of the vehicle. Worked Example 2.1

Worked Example 2.1:

Goal:
Gain a qualitative understanding of the effect of various forces on the overall pitching moment on a simplified tilt-wing eVTOL during transition from hover to forward flight and prior to the onset of forward velocity.

Background:
A moment, also known as a torque, is the rotational analog of force. While a force causes linear acceleration, a moment causes angular acceleration. A moment arises when a force is applied at a point that does not lie along the axis of rotation, producing a "turning effect." An aircraft rotates about its CG. Therefore any force on the aircraft that does not pass through the CG creates rotation. Forces that make the nose pitch up or down create a pitching moment (Figure 2.21).

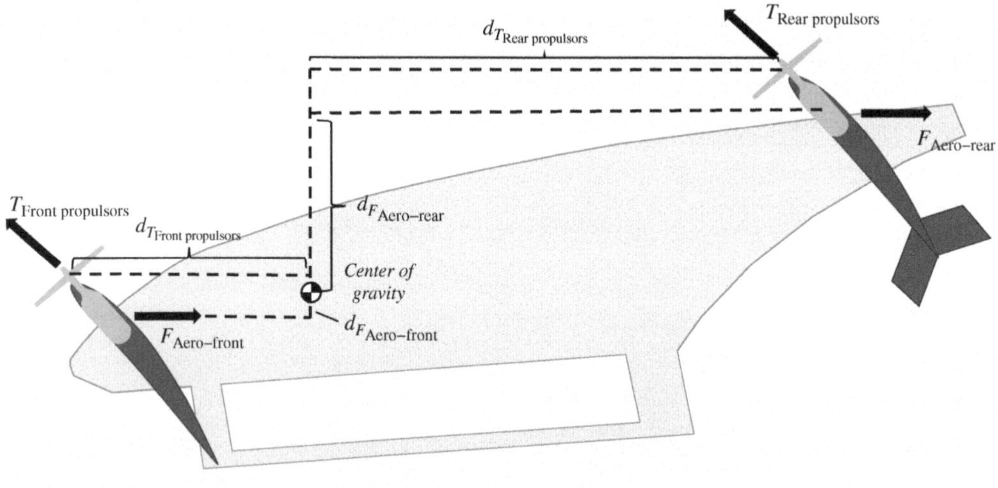

Figure 2.21 Simplified force and moment balance for a tilt-wing eVTOL during early transition.

Given:

- $T_{\text{Front propulsors}}$ and $T_{\text{Rear propulsors}}$ represent the combined thrust forces of all propulsors across the front and rear wings, respectively.
- The $F_{\text{Aero-front}}$ and $F_{\text{Aero-rear}}$ represent the aerodynamic forces generated on the front and rear wings, respectively, due to the propulsor slipstreams (with no forward velocity).
- d represents the perpendicular distance between the point of application of each respective force and the CG of the aircraft.

Find the Answer:

Define the pitching moment:

Quantitatively, a moment about the CG, M_{CG}, is defined as the product of a force and the perpendicular distance from the point of rotation (CG) to the line of action of the force, $d_{\perp}$, referred to as the moment *arm*,

$$M_{CG} = Fd_{\perp} \tag{2.14}$$

Identify qualitative contributions:

- $T_{\text{Front propulsors}}$: The upward thrust force acts forward of the CG. Because the horizontal arm is much larger than the vertical arm, this produces an overall nose-up pitching moment.
- $F_{\text{Aero-front}}$: The rearward aerodynamic force acts slightly below the CG, creating a nose-down pitching moment.
- $T_{\text{Rear propulsors}}$: The rear thrust vector acts at an approximate 45° angle and, with a greater horizontal than vertical arm, an overall nose-down pitching moment is created.
- $F_{\text{Aero-rear}}$: The rearward aerodynamic force acts significantly above the CG, producing a nose-up pitching moment.

Synthesize:

The front and rear propulsors and wings produce offsetting pitching moments at both ends of the vehicle. The net pitching moment depends on the summation of the four individual moments, each determined by the magnitude of its force and its moment arm relative to the CG.

Exposition:

As the wing tilts from a vertical to a horizontal orientation during transition, and as the vehicle begins to gain forward velocity, the orientation, magnitude, and effective arms of these forces change significantly. This example illustrates how pitching moment characteristics and required control strategies can shift rapidly during transition. Managing these moments is essential to maintain fuselage attitudes within acceptable limits, preserve passenger comfort, and ensure vehicle stability.

examines how competing aerodynamic forces on a tilt-wing eVTOL generate nose-up and nose-down pitching moments early in transition.

From a handling perspective, maintaining moderate fuselage attitudes is critical. Pilots typically prefer to limit nose-up angles for visibility, and large pitch excursions in either direction can reduce passenger comfort. Managing aerodynamic moments so that the vehicle remains trimmed without requiring extreme attitudes is a primary challenge within the transition phase. The guiding principle is to minimize the time spent in this inefficient intermediate regime. A successful transition is rapid, stable, and energy-efficient, leaving the aircraft fully established in wingborne flight. Once established in wingborne flight, performance considerations return to classical aerodynamic parameters such as aspect ratio, lift-to-drag ratio, and induced drag for determining cruise efficiency.

Composite Materials

A defining feature of modern aircraft design, and one that is critical for AAM vehicles, is the use of composite materials (Figure 2.22). For AAM aircraft, weight reduction is paramount. Because eVTOLs operate under strict energy constraints, every kilogram saved is critical and contributes directly to range and payload capability. Composite airframes and rotor systems allow eVTOL

Figure 2.22 Airbus A400M eight-bladed composite propeller. *Source*: flickr.com / myyorgda / CC BY 2.0.

designers to achieve the stiffness and strength required for flight loads without incurring the penalties of heavy metallic structures. In some circumstances, carbon-fiber composites can offer 3–5 times the specific strength of steel while only having 20% of steel's density.

Composites are engineered materials made from two or more distinct constituents: a high-strength reinforcement, usually fibers, embedded in a lighter-weight matrix phase that binds the system together (Figure 2.23). The fibers, most often carbon or glass, provide high strength and stiffness while carrying most of the structural loads. Typically, these strong fibers are embedded within a polymer matrix that transfers loads between fibers, maintains the geometric form of the structure, and protects the fibers against environmental degradation. The result is a material system that combines the best attributes of both constituents to deliver mechanical properties unattainable by either material alone.

The polymer matrix can be classified into two broad categories: thermosets and thermoplastics. While both matrix types can be used with continuous fibers, *thermosets*, such as epoxies, are processed as liquids and cure irreversibly into rigid solids. Once cured, they cannot be remelted or reshaped, but they offer excellent mechanical performance and have a long track record in aerospace structures. *Thermoplastics*, by contrast, soften when heated and harden when cooled, allowing them to be reshaped, welded, or recycled. Although they require higher processing temperatures, thermoplastics enable much faster production cycle times. This makes thermoplastics especially attractive for AAM, where vehicle production rates may need to be an order of magnitude higher compared to present conventional aircraft production. Additionally, thermoplastics offer improved recyclability, supporting AAM's sustainability objectives. The choice between using thermosets and thermoplastics is a timely issue within the AAM industry, namely the balance between certification maturity and the need for scalable manufacturing. Early AAM vehicles may rely heavily on thermosets, but scaling up to thousands of vehicles will demand broader adoption of thermoplastics and resin infusion methods that support automation.

The employment of glass or carbon fibers in aerospace applications depends on the purpose. Glass fibers are often used where cost, corrosion resistance, or dielectric properties are prioritized, such as in fairings, radomes, or interior panels. Carbon fibers, on the other hand, dominate in primary load-bearing structures where maximum stiffness and strength at minimum weight are

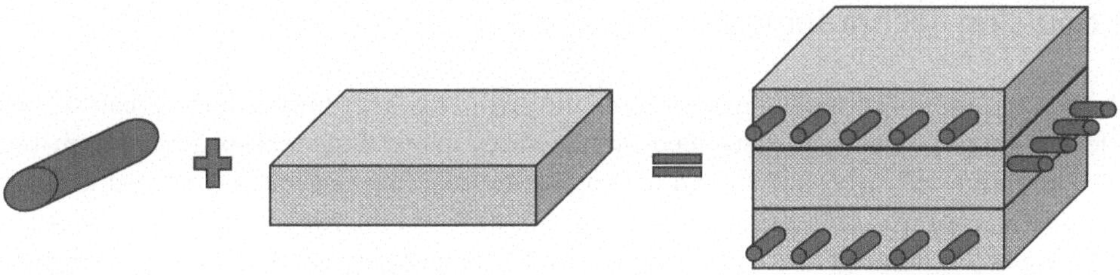

Figure 2.23 A composite material made from a reinforcing fiber and a binding matrix. A CFRP uses carbon fibers embedded in a polymer matrix.

essential, and, as such, *carbon-fiber-reinforced polymers* (CFRPs) represent the primary structural material for modern airframes. Carbon fibers are produced from precursors such as polyacrylonitrile (PAN) or pitch. PAN is a synthetic polymer that, when oxidized and carbonized, produces high-strength, fine-diameter carbon fibers, while pitch is a tar-like substance that can be derived from petroleum or coal and can be processed into extremely stiff, brittle fibers. The exceptional stiffness and strength of CFRPs make them the backbone of modern aircraft structures.

Carbon fibers are used when structural performance must be maximized and weight minimized. This superior structural performance appears in several distinct engineering attributes. The first of these attributes is a *high strength-to-weight ratio.* Carbon-fiber composites rival the strength of steel at a fraction of the mass, allowing for thinner wings and lighter fuselages without compromising load-bearing capacity. A second key attribute is a *high stiffness-to-weight ratio.* Lightweight structures must also resist aeroelastic effects such as flutter. Carbon composites provide the necessary stiffness while avoiding the weight penalties associated with metallic reinforcements. Additionally, CFRPs offer *tailorability* by orienting fibers along principal load paths, creating structures that are strong in specific directions but lightweight overall. This anisotropy enables optimization for distributed propulsion architectures where multiple mounting points introduce complex load paths. Finally, composites in general are highly suited to *aerodynamic shaping* where smooth, complex surfaces can be molded with fewer joints and fasteners, reducing both drag and maintenance requirements.

One of the most widely used product forms in aerospace is pre-impregnated fabric or tape (prepreg). *Prepregs* contain fibers that are pre-coated with a precisely formulated resin and ensure consistent fiber alignment, controlled resin content, and excellent mechanical performance. They are widely used in high-performance structures such as wings, fuselages, and rotor blades. Other fabrication methods, such as vacuum-assisted resin transfer molding (VARTM) and related infusion systems, are being developed for faster, high-rate production suitable for the scaling demands of AAM.

Composites are indispensable for AAM vehicles to achieve viable range and payload performance. Because a high percentage of a vehicle's takeoff weight is consumed by energy storage systems, primary and secondary structures must be optimized for minimum weight. Composite materials are central to meeting these weight and performance requirements. However, this is not to suggest that they do not introduce their own challenges. Manufacturing composites requires stringent quality control, repair and inspection demand specialized nondestructive evaluation techniques, and recycling remains limited. Moreover, compliance with aviation authority certification guidance adds complexity. For those working with the United States' Federal Aviation Administration (FAA), Advisory Circular 20-107B [1] and the industry-standard Composite Materials Handbook (CMH-17) [2] provide detailed best practices and qualification requirements for composite structures in aircraft design.

2.3 Propulsion Technologies

For AAM aircraft, propulsion technology presents the greatest opportunity and the greatest constraint. Electric motors and their associated architectures allow designers to depart from the mechanical limitations of turboshaft-driven helicopters by enabling variable rotor speed, distributing thrust across multiple propulsors, and integrating different energy sources within a single powertrain. These advances permit new strategies for efficiency, noise reduction, and redundancy. However, these same architectures also introduce challenges in system integration, thermal management, and energy supply. The following subsections examine how the physics of torque, speed, and power govern the behavior of electric propulsion systems, and how these relationships give rise to distributed and hybrid configurations that define AAM vehicle propulsion.

2.3.1 Power, Torque, and RPM

Having established how rotors accelerate air to generate thrust for vertical flight, we can now generalize this understanding to propulsion in all flight regimes. Thrust is the aerodynamic force produced by a propulsor, whether a rotor, propeller, or jet. In vertical flight, thrust must balance or exceed weight to hover and climb; in forward flight, thrust must counteract drag to sustain cruise. In every case, thrust defines the aerodynamic requirement that the propulsion system must satisfy. Power, in contrast, measures the rate at which mechanical work is performed to produce that thrust. This distinction between the aerodynamic demand (thrust) and the energetic cost (power) is fundamental to analyzing propulsion for AAM aircraft.

As we have seen, accelerating air to generate thrust demands power from the propulsion system. In this sense, thrust specifies the aerodynamic requirement, while power represents the energetic cost of achieving it. For eVTOL propulsion systems, this relationship is especially important because batteries store only a finite amount of energy. Motors must therefore convert stored electrical energy into mechanical power efficiently enough to sustain the required thrust throughout all phases of flight.

The mechanical power delivered at the motor shaft is expressed as:

$$P = Q\omega \tag{2.15}$$

where P is power [W], Q is torque [N·m], the rotational analogue of force, and ω is the angular velocity [rad/s] of the rotating shaft. This relationship linking torque and rotational speed to power mirrors our earlier translational form for power reflected in Eq. (2.6), $P = Fv$. Equation (2.15) shows that, for a given power requirement, torque and angular velocity are inversely related. If angular velocity decreases, torque must rise; if angular velocity increases, torque falls.

Breakout box:

Shaft torque represents the mechanical reaction to the aerodynamic forces acting on the rotor blades. As the rotor accelerates air to produce thrust, aerodynamic drag forces act on the blades in the opposite direction of rotation. The magnitude of this aerodynamic loading increases when either the air is accelerated more (a higher exit velocity) or when a larger rotor disk moves a greater mass of air at a lower rotational speed. In both cases, the motor must deliver greater torque at the hub to overcome these aerodynamic forces and sustain steady rotation.

This trade-off is fundamental to propulsion design. A large-diameter rotor operating at lower angular velocity can generate thrust efficiently because it accelerates a greater mass of air with a smaller velocity increment, but it imposes high torque demands on the motor. Smaller, faster-turning rotors act on a smaller mass flow of air and, consequently, reduce the torque required for a given shaft power at a higher rotational speed, but typically require higher shaft power to produce the same thrust because the induced velocity increment is larger. As previously established, this leads to higher tip speeds, raising the noise level and reducing hover efficiency. Hence, every eVTOL design must resolve the balance between torque, rotational speed, and total power to meet efficiency, noise, range, and payload mission requirements.

Conventional helicopters manage this trade-off in a fixed manner. Turboshaft engines operate efficiently only within a narrow band of rotational speeds, and rotor and drivetrain constraints favor a nearly constant main-rotor RPM. For example, the S-70 Black Hawk helicopter spins its main rotor at approximately 258 RPM in all flight conditions. Thrust and maneuvering are not achieved by changing the rotor speed; they are achieved by altering the aerodynamic pitch of the blades. Departures from the governed rotor speed can produce a steep degradation in turboshaft efficiency, as reflected by rising specific fuel consumption (Figure 2.24 (left)), and simultaneously introduce aeroelastic, acoustic, and handling penalties, making wide-range variable RPM operations impractical for most conventional rotor systems. SFC represents the rate of fuel mass flow divided by the useful output of the engine and provides a measure of fuel efficiency independent of engine size. Lower values indicate a higher conversion efficiency of chemical energy into useful mechanical work.

Electric motors powering eVTOLs, by contrast, are fundamentally different. Electric motors maintain high efficiency across a broad range of torques and rotational speeds, allowing rotor RPM to vary without significant efficiency loss (Figure 2.24 (right)). This flexibility permits designers to optimize operation for each phase of flight. Thus, RPM can be scheduled by flight phase, with higher RPM often used during high-thrust segments and lower RPM selected in cruise to reduce tip speeds, acoustic emissions, and drag. *The ability to modulate rotor speed efficiently in flight is a fundamental departure from the helicopter paradigm and is one of the defining enablers of AAM.*

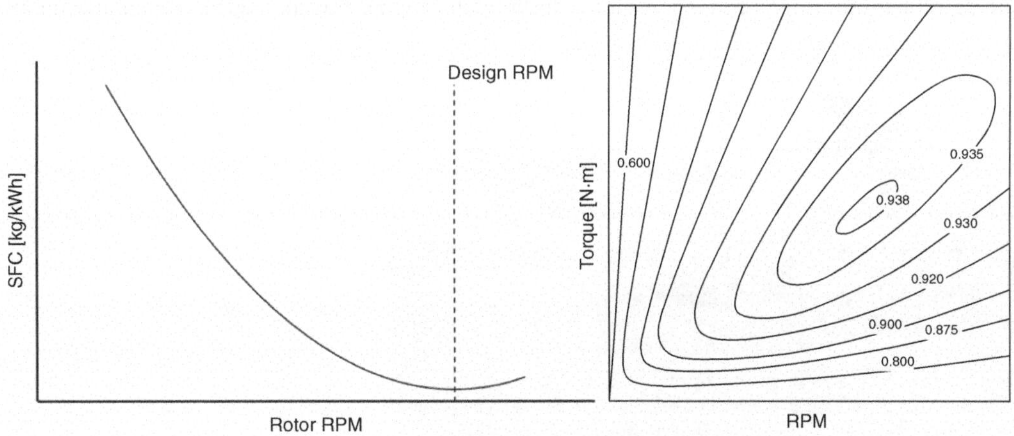

Figure 2.24 (Left) Representative gas turbine performance map (Courtesy of Kayla D. Taylo); (right) representative electric motor torque–speed efficiency map. *Source*: Duan et al., (2020) / IEEEE / CC BY 4.0.

The fundamental difference in efficiency behavior between gas turbines and electric motors is illustrated in Figure 2.24. Figure 2.24 (left) shows a representative turboshaft SFC curve, where efficiency is tightly concentrated around a single design RPM and degrades steeply as the shaft speed deviates. In contrast, the representative electric motor torque–speed efficiency map of Figure 2.24 (right) shows a broad region of nearly constant efficiency across varying RPM. Collectively, these figures demonstrate how gas turbine helicopters are restricted to fixed-RPM operation while eVTOLs can exploit the design freedom associated with variable RPM.

The ability to efficiently vary RPM in flight is transformative. It enables eVTOLs to match power, torque, and speed to each phase of flight. Depending on the mission profile, this adaptive capability can improve rotor efficiency during hover and reduce energy consumption during cruise. It also allows control strategies for acoustic signature reduction and thermal management. Still, this same flexibility introduces new challenges. Rotor systems capable of maintaining high efficiency across wide RPM ranges are still early in their development, and their design is further constrained by the weight and energy limitations of current battery technologies. Rotor aerodynamic performance and structural loads must also remain acceptable across the scheduled RPM range.

2.3.2 Distributed Electric Propulsion

The interplay of torque, rotational speed, and power captured in Eq. (2.15) drives a new propulsion strategy known as DEP. In its simplest form, DEP replaces one large rotor–motor system with multiple smaller rotors, each powered by its own electric motor (Figure 2.25). This strategy is a major departure from the helicopter paradigm of concentrating propulsion in a single large rotor and arises from the unique scaling properties of electric motors.

In turboshaft engines, weight tends to scale with total power output. Because shaft power is proportional to the mass flow of gas through the engine per unit of time, producing greater power requires larger compressors, turbines, and gearboxes, all of which add weight. For this reason, helicopters achieve the best weight efficiency by centralizing power into one or two large engines and distributing it mechanically to drive large rotors at a fixed speed. Attempting to divide power among many small turboshafts would add substantial weight and complexity, eroding efficiency.

Electric motors, however, scale differently (Figure 2.26). Motor weight scales primarily with the torque that the motor must produce. A motor designed for higher torque requires larger and more

Figure 2.25 A Joby S4 eVTOL with a DEP architecture consisting of six motors. *Source*: Harlan Huntington / the United States Air Force / Public domain.

Figure 2.26 Archer Midnight electric motor. *Source*: Matti Blume / Wikimedia Commons / CC BY-SA 4.0.

powerful magnetic material, additional copper windings, and a more robust mechanical structure to withstand the associated forces. Because these elements account for much of the motor's total mass, the motor's weight scales roughly in proportion to its torque output.

This scaling behavior brings us back to Eq. (2.15). For a given power requirement, operating at a lower rotational speed demands proportionally higher torque, leading to a heavier motor. Conversely, motors designed to spin at higher speeds can deliver the same power with less torque and, consequently, have a lower weight. This explains why high-performance electric machines achieve favorable power-to-weight ratios by spinning at high rotational speeds. These characteristics imply that a single large motor designed to deliver very high torque at moderate speed becomes disproportionately heavy. Alternatively, several smaller motors can each supply lower torque at higher rotational speeds and, collectively, meet the thrust requirement with favorable power-to-weight characteristics. Thus, DEP emerges directly from the torque–speed scaling laws of electric machines.

One of the most immediate advantages of distributing propulsion across several smaller rotors is redundancy. Because thrust is generated across many propulsors, the architecture provides a natural pathway to redundancy and, crucially, in the event of a failure, the remaining propulsors can compensate by temporarily operating above their nominal (100%) thrust rating for short durations. This capability for continued controlled flight is vital for urban operations where suitable emergency landing areas may be scarce.

DEP also transforms how maneuvering forces are generated. The independent controllability of each propulsor enables *propulsion-controlled aircraft* (PCA), where pitch, roll, and yaw motions are produced through differential thrust in place of traditional aerodynamic control surfaces. This approach can reduce the need for complex rotor hub mechanisms, discussed in Section 2.2.1, or large moveable surfaces, simplifying vehicle design and maintenance.

Vehicle acoustics also benefit from a DEP configuration. Distributing the thrust requirement among many rotors allows each rotor to operate with a lower blade tip speed for a given thrust level because the small rotor radii keep tip speeds moderate even when shaft RPM is relatively high, reducing acoustic emissions. Because the rotors in a DEP system are small and lightly loaded, they do not require the flapping and feathering degrees of freedom built into a helicopter's large, heavily loaded main rotor. This design choice enables the use of simple rigid hubs in lieu of the articulated hubs of conventional helicopters (Figure 2.17). Accordingly, DEP enhances safety, reduces noise, and simplifies mechanical structures. The implications of acoustic performance for AAM operations are explored in greater depth in Chapter 7.

Beyond these immediate benefits, DEP affords tighter integration between aerodynamics and propulsion, known as *aero-propulsive coupling*. Historically, aircraft designers have sought to position propulsion units where they minimally disturb the flow and, in doing so, have treated thrust production and airframe aerodynamics as largely independent of the other. DEP reverses this logic by placing propulsors so that they can constructively interact with the local airflow around the airframe. Among the most studied integrations is *boundary layer ingestion* (BLI). In a BLI configuration, propulsors ingest and re-energize the slower-moving air that forms along the vehicle's surface (the boundary layer). Because, as we have previously concluded, the most efficient thrust is obtained by accelerating a large mass of air by a small velocity increment, recovering this low-energy airflow improves overall propulsive efficiency. Alternatively, BLI can be considered within an energy framework. From an energy perspective, the wake behind an aircraft represents a region of slower-moving, low-energy air relative to its upstream counterpart. If a propulsor ingests this slower air and restores it toward the free-stream velocity, it recovers energy that would otherwise persist as pressure drag. The engineering challenge is that this boundary-layer flow is nonuniform and turbulent; thus, propellers must be designed to tolerate distorted inflow without excessive vibration or performance loss.

The higher number of propulsors in a DEP configuration also permits propulsor placement to influence spanwise flow near the wingtips. The flow of higher-pressure air below the wing into the low-pressure region above it generates wingtip vortices, a major source of induced drag. A propulsor positioned near the wingtip can redirect or weaken this vortex system, thereby reducing induced drag and simultaneously producing thrust.

Taken together, these arrangements illustrate the potential of aero-propulsive coupling in which propulsion units are not external devices attached to an airframe; instead, they are integral elements of the aerodynamic design itself. By aligning thrust generation with flow control, DEP configurations can reduce drag, enhance efficiency, and open new possibilities for integrated aircraft concepts in the future.

The advantages DEP offers must be balanced against the challenges that it introduces. DEP adds system complexity, including high-voltage distribution networks, thermal management for motors and associated electrical conversion systems, along with susceptibility to electromagnetic interference (EMI). Aero-propulsively coupled designs require structurally robust propellers capable of handling distorted flow fields, adding both weight and manufacturing cost. Among these considerations, the dominant constraint is that the advantages of DEP derive from the properties of electric motors and depend on the availability of sufficient onboard electrical power. This dependence motivates emerging interest in hybrid-electric propulsion architectures, discussed in the following subsection.

2.3.3 Hybrid Propulsion Systems

Battery technology defines the performance envelope of today's fully electric aircraft, constraining both range and payload. To address these limitations, developers pursue a broader set of propulsion architectures that integrate onboard combustion engines with electric motors and generators. Figure 2.27 illustrates the major propulsion architecture categories currently under consideration for electric aviation. These categories include both fully electric and hybrid-electric systems. Within the hybrid-electric category, the principal sub-architectures are parallel-hybrid and series-hybrid systems. Taken together, fully electric and hybrid-electric architectures represent the principal technological pathways under active development for AAM. In this subsection, we focus on the hybrid-electric configuration.

A representative fully electric DEP architecture is shown in Figure 2.28.

A representative turbo-electric DEP architecture is shown in Figure 2.29.

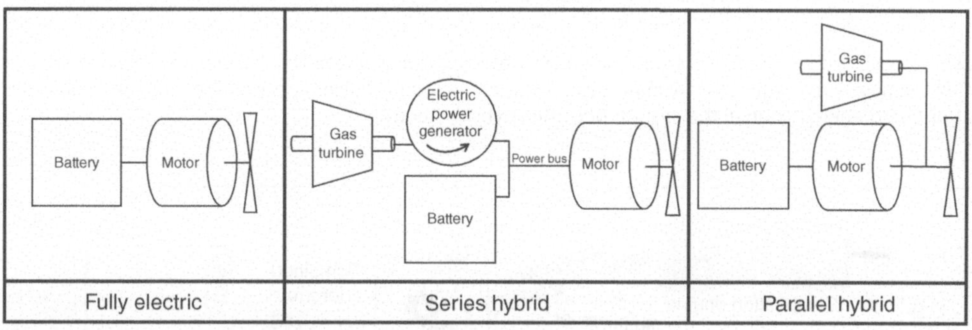

Figure 2.27 Electric aircraft propulsion architecture categories.

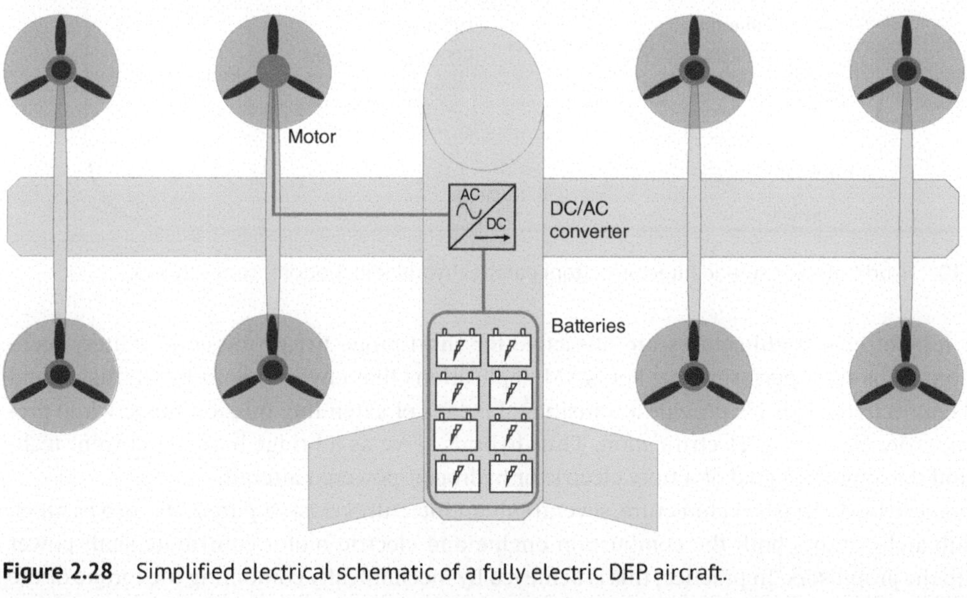

Figure 2.28 Simplified electrical schematic of a fully electric DEP aircraft.

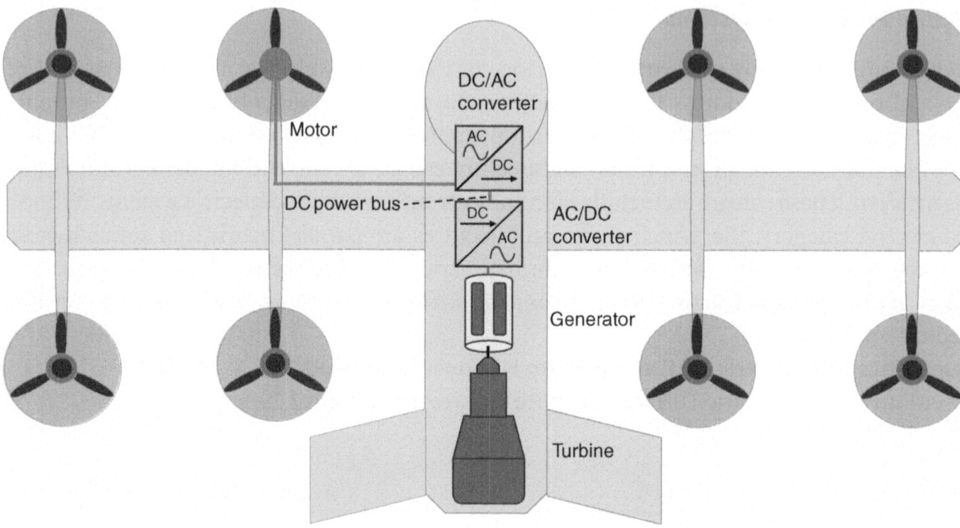

Figure 2.29 Simplified electrical schematic of a turbo-electric DEP aircraft. The DC bus architecture enables flexible power management from multiple sources, including battery packs. When batteries provide a material onboard energy source, the system becomes hybrid-electric.

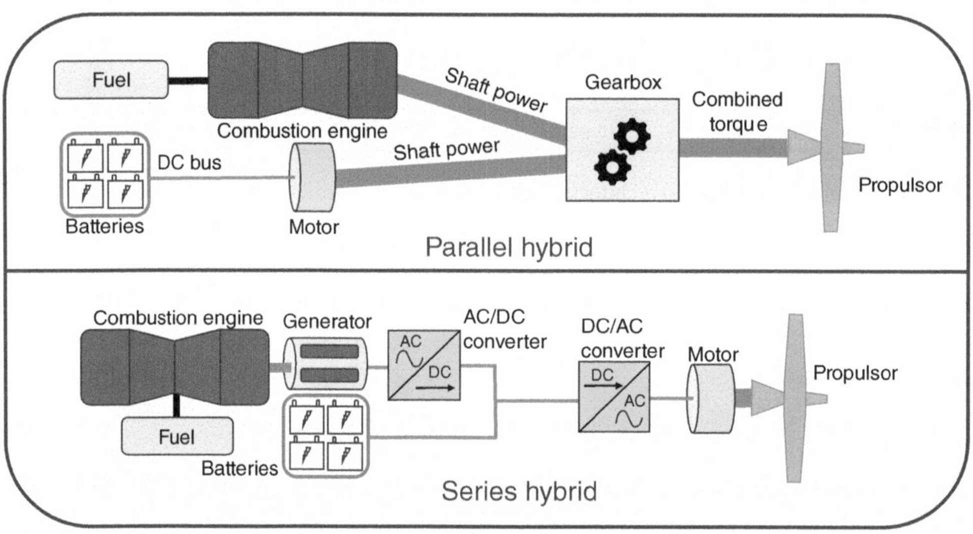

Figure 2.30 Hybrid-electric sub-architectures: (top) parallel hybrid and (bottom) series hybrid.

While fully electric architectures are attractive for short-range urban missions, battery technology constrains their performance. For AAM applications that envision both urban flights and regional connectivity, hybrids provide a transitional means of extending mission range while preserving the core benefits of electrification. Thus hybrids serve as a bridge between current technology and the long-term goal of a fully electric or hydrogen-powered aircraft.

Within the hybrid-electric architecture, several sub-architectures exist (Figure 2.30). In a *parallel-hybrid* sub-architecture, both the combustion engine and electric motor contribute shaft power directly to the propulsors. In practice, this is achieved by mechanically combining the torque of the engine and motor through a gearbox that drives the propulsors. This arrangement allows the engine to supplement the batteries during power-intensive phases such as takeoff and climb, where thrust and power demand are highest, while permitting battery-only operation in noise- and emission-sensitive areas. A *series-hybrid* sub-architecture, by contrast, does not mechanically connect the combustion engine to the rotors. Instead, the engine drives a generator whose alternating-current (AC) output is rectified, that is converted to direct current (DC), and fed into a common electrical bus alongside the batteries. An inverter then converts the DC to AC and supplies this power to the motors. This decoupling allows the engine to operate at its most efficient point, and the electric motors provide the advantages of distributed propulsion. Variations of *blended* or *series–parallel-hybrid* design concepts are presently being explored in conventional aviation, though they are less common in AAM vehicle designs. Series hybrids are particularly attractive for DEP because they preserve distributed propulsion without the complexity of mechanical power transmission.

The shaft-equivalent power demand at the propulsors, $P_{\mathrm{dem}}(t)$, can be expressed as the sum of contributions from cruise, maneuver, and vertical flight:

$$P_{\mathrm{dem}}(t) = P_{\mathrm{cruise}}(t) + P_{\mathrm{maneuver}}(t) + P_{\mathrm{vertical}}(t) \tag{2.16}$$

This total demand must be balanced by the combined shaft-equivalent power delivered by the engine–generator and battery:

$$P_{\mathrm{dem}}(t) = P_{\mathrm{engine}}(t) + P_{\mathrm{battery}}(t) \tag{2.17}$$

A common design choice is to size the engine slightly above the continuous cruise requirement so that the engine can sustain cruise power while simultaneously providing margin for battery charging during steady flight. The battery is then relied on for short-duration peaks (e.g., takeoff, climb, and gust response). This strategy is referred to as *peak shaving*. Alongside sizing, designers must also select an operational mode for power management over the course of a flight. The two most common operational modes are *charge-depleting* and *charge-sustaining*. Charge-depleting intentionally draws the battery down to extend performance. In contrast, charge-sustaining, as its name implies, maintains the battery's state of charge (SOC) at a prescribed level by modulating engine output.

A measure of power sharing is the *degree of hybridization in power*, H_{p},

$$H_{\mathrm{p}} = \frac{P_{\mathrm{batt}_{\mathrm{peak}}}}{P_{\mathrm{dem}_{\mathrm{peak}}}} \tag{2.18}$$

$P_{\mathrm{batt}_{\mathrm{peak}}}$ in Eq. (2.18) represents the peak power drawn from the battery and $P_{\mathrm{dem}_{\mathrm{peak}}}$ is the total peak shaft power demanded by the aircraft. This ratio quantifies what fraction of maximum power is electrified. Here, we use a power-based definition at peak demand, but an energy-based form can also be used. Using the power-based expression, a system with $H_{\mathrm{p}} = 0.2$ relies on the battery for 20% of its peak power, with the engine supplying the remainder. The degree of hybridization in power is a key design parameter that influences aircraft performance, fuel efficiency, emissions, and cost.

The benefits of hybrid-electric propulsion for AAM are significant. By carrying an onboard combustion source, a hybrid aircraft can achieve ranges far greater than those permitted by current battery technology alone. This makes hybrids especially attractive for intercity and regional flights, particularly where charging infrastructure may be limited. Nevertheless, these gains in range are accompanied by penalties in vertical flight. The additional weight of the engine, generator, and fuel must still be supported in hover. With thrust equivalent to weight, W, in hover, the induced power, P_{ind}, required to generate the necessary thrust grows faster than weight itself, increasing approximately with $W^{3/2}$ for a fixed rotor disk area:

$$P_{\text{ind}} = \frac{T^{3/2}}{\sqrt{2\rho A}} \propto W^{3/2} \tag{2.19}$$

where ρ represents the air density and A and T, once again represent the total rotor disk area and thrust, respectively. Consequently, even a modest increase in weight causes a disproportionate rise in hover power. Conversely, as fuel is consumed during flight, landing weight decreases and hover margins improve accordingly.

Mission flexibility is another advantage of hybrid flight. Short-range flights can be conducted entirely on battery power, enabling quiet, emission-free operation in urban cores; longer-range flights can draw on the combustion fuel source. The presence of two power sources can also enhance redundancy, although the degree of redundancy depends on the level of hybridization. In architectures where the battery or engine can independently sustain a significant portion of the propulsive load, a partial failure can be accommodated with degraded performance. In lower-hybridization designs, the battery may supply only peak power, making continued flight on battery power alone infeasible.

These benefits afforded by hybrid-electric propulsion come with notable challenges. As previously established, a hybrid aircraft must carry both an engine and a battery system. This increases aircraft weight and system complexity. Overall prowertrain efficiency, η_{overall}, is reduced by the product of the component efficiencies, η:

$$\eta_{\text{overall}} \approx \eta_{\text{engine}} \cdot \eta_{\text{generator}} \cdot \eta_{\text{rectifier}} \cdot \eta_{\text{DC bus}} \cdot \eta_{\text{inverter}} \cdot \eta_{\text{motor}} \cdot \eta_{\text{propulsor}} \tag{2.20}$$

Even with high-quality components, cumulative losses are consequential and must be considered in system-level analysis. Thermal management is also critical. Both the engine–generator and the electrical conversion systems produce heat, but limited cooling airflow in hover and approach reduces cooling effectiveness. Associated radiator sizing, the use of auxiliary cooling fans, and their combined impact on weight, drag, and noise must all be carefully balanced.

Reliability and fault tolerance pose further challenges. Series hybrids concentrate mechanical power in the engine–generator, meaning loss of that unit requires sufficient battery capacity and bus redundancy to sustain controlled flight. Parallel hybrids divide power mechanically, so an engine-out event can be mitigated by motors if the battery and inverters can supply the necessary transient power. However, motor torque and thermal limits quickly become critical in such circumstances.

Unlike purely electric aircraft, hybrid aircraft experience weight and CG changes throughout the flight. Fuel burn shifts the CG depending on tank placement, and although changes in battery SOC do not alter aircraft weight, they do affect available power and thermal margins. Consequently, dispatch planning must account for CG envelopes at both takeoff and landing, along with hover power capability at maximum takeoff weight and minimum-fuel landing weight.

Acoustic performance also requires special attention for hybrid aircraft. Hybridization supports DEP and low rotor-tip speeds, but the combustion system introduces turbine, gear, and broadband exhaust noise. One mitigation strategy is to prioritize battery-only operation during departure and arrival, reserving engine use for cruise segments where community noise exposure is lower.

Hybrid control strategies further influence battery life and reserve policies. Limiting battery duty cycles to short, high-power transients and maintaining the SOC within a mid-range band can substantially improve battery cycle life. Further, operators must reserve energy for contingencies and power for high-demand events such as missed approaches or hover diverts. Under these constraints, hybrid control must ensure an adequate power margin before committing to a VTOL landing.

The role of hybrid propulsion differs across AAM vehicle classes. For eVTOLs, any additional installed weight is particularly problematic because vertical takeoff performance is highly sensitive to weight. Depending on the mission, a hybrid system may weigh more or less than a purely electric system. For example, meeting longer-range requirements in an all-electric configuration often demands a large battery pack. What remains consistent is that added mass in any form increases disk loading and reduces hover efficiency, placing hybrid propulsion at a disadvantage for compact multicopter designs. Larger tilt-rotor or lift + cruise architectures are better positioned to accommodate this mass because they rely more heavily on wingborne lift during transition and cruise. In contrast, for eCTOL and eSTOL aircraft that take off from runways, hybrid propulsion is more practical since wingborne lift mitigates the weight penalty, and the extended range aligns with regional missions.

Within AAM, hybrid propulsion is best understood as a transitional technology. It extends the reach of AAM vehicles beyond what batteries alone presently enable while retaining the design flexibility of electric drive systems. When combined with sustainable aviation fuels, hybrids may support near-term decarbonization and provide time for advances in energy storage. We tackle a more detailed examination of energy storage and management systems next.

2.4 Energy Storage and Management Systems

In the design of an eVTOL aircraft, the propulsion system and the onboard energy storage system are tightly coupled and must be considered as an integrated whole. The battery system, in particular, strongly influences vehicle performance. The weight of the battery must be low enough to preserve range and payload capacity. It must also be capable of delivering the high power required for vertical takeoff, hover, and landing. In addition, the stored chemical energy must be sufficient to sustain cruise flight and contingency reserves. Balancing the competing requirements for high power delivery and sufficient onboard energy defines a primary design trade-off for all electric aircraft.

Lithium-ion (Li-ion) batteries are the dominant technology used in current eVTOL aircraft designs. The designation "Li-ion" encompasses several related chemistries that differ in energy density, power capability, safety characteristics, and cost. At the same time, alternative energy storage technologies, including sodium-ion (Na-ion) batteries, solid-state batteries, and fuel-cell systems, are under active investigation for AAM-scale aircraft. Regardless of the specific chemistry, every battery operates based on the same fundamental principle: converting chemical potential energy into electrical energy through electrochemical reactions.

We begin with the basic structure and operation of a battery, followed by its performance characteristics, system integration, and management strategies as relevant to AAM applications. We then look ahead to hydrogen fuel cells and their incorporation into AAM powertrains, including conceptual hybrid hydrogen battery architectures that combine the advantages of both systems.

2.4.1 Battery Basics

A *battery* is an electrochemical device that converts chemical potential energy into electrical energy through a controlled electrochemical reaction that drives a flow of electrons. It is a device composed of three basic components: an *anode, a cathode*, and *an electrolyte*. As shown in Figure 2.31, the anode is the negative electrode where oxidation (the loss of electrons) occurs, and the cathode is the positive electrode where reduction (the gaining of electrons) occurs. Together, these reactions form a *reduction–oxidation* (*redox*) process that transforms a battery's stored chemical energy into electrical energy. The electrolyte electrochemically interacts with both electrodes and serves as the medium through which ions move between them. Electrolytes are conventionally liquid, but a *solid-state battery* employs solid electrolytes that perform the same ion-transport function. Solid-state batteries remain an area of significant research interest for AAM because these systems promise higher energy storage capability per unit weight, improved intrinsic safety, and potentially reduced thermal-management requirements compared to liquid-electrolyte Li-ion cells. However, many present solid-state chemistries exhibit limited power-delivery rates, constraining their applicability to electric propulsion systems that demand rapid power availability. Ongoing advances in solid-state battery development must address both performance and manufacturability before these technologies can support the full range of AAM power requirements. Nonetheless, these fundamental electrochemical principles apply to all battery types, independent of chemistry or design.

Batteries are classified as either primary or secondary. *Primary batteries* are single-use cells whose electrochemical reactions are not reversible. *Secondary batteries* are rechargeable. All eVTOL batteries belong to this secondary category. During discharge, a redox reaction drives electrons from

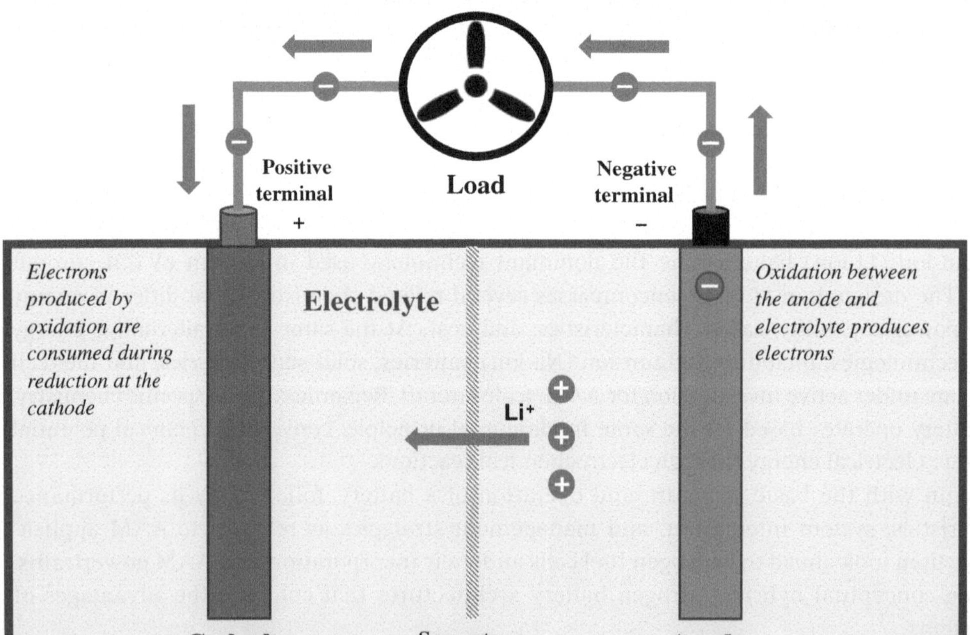

Figure 2.31 Fundamental components of a battery.

the anode through the external circuit to the cathode, delivering electrical power, while positive ions migrate through the electrolyte to maintain charge neutrality. When the battery is recharging, the reaction reverses.

The redox reaction described above is the general mechanism by which a battery's chemical potential energy is converted into electrical energy. A variety of battery chemistries can facilitate this process. As previously stated, Li-ion chemistries are currently the most prevalent choice for AAM applications, although other chemistries, such as Na-ion, are under active development. Key performance factors for comparing varying chemistries include specific energy, specific power, cycle life, and cost.

The first two parameters, specific energy and specific power, introduce the important ideas of electrical energy and power. Here, a battery paradox arises. A battery can be optimized for either high power or high energy, but not both simultaneously. To understand this limitation, and batteries in general, we must first review some foundational ideas and vocabulary regarding work, energy, and power, along with their associated units. Being attentive to units is always prudent but, in the context of analyzing battery systems, it is especially important for clear comprehension.

In physics, *work*, W, is defined as a force, F, acting through a distance, d (strictly speaking, the component of force in the direction of motion, $F_{||}$, multiplied by the displacement):

$$W = F_{||}\, d \tag{2.21}$$

The unit for work combines the unit for force, the newton [N], and the unit for distance, the meter [m], producing the newton-meter [N·m], also known as the *joule* [J]. However, this concept of work alone does not take into consideration the time required to perform the work. Here, we turn to the concept of *power* (P), which introduces the dimension of time, t. Average power, P_{avg}, is defined as the amount of work, ΔW, that is done in a time interval, Δt:

$$P_{avg} = \frac{\Delta W}{\Delta t} \tag{2.22}$$

The unit of power is the joule per second [J/s], also known as the *watt* [W].

Energy, also measured in joules, represents the capacity to perform work. Performing work on an object can transfer energy to it. If we apply the concept of work to pushing (force) a charged particle through an electric field (distance), the particle gains *electric potential energy*, ΔU, by virtue of its position in the electric field. The unit of *electric charge* (q) is the *coulomb* [C]. Considering this change in electrical potential energy per unit charge yields the concept of *electric potential*, measured in volts [V],

$$\Delta V = \frac{\Delta U}{q} \tag{2.23}$$

The potential difference between two points, commonly referred to as *voltage*, represents the work required to move a unit charge between these points. Thus, a fully charged 800 V eVTOL battery can be thought of as having a positive terminal that is 800 V higher in electric potential than its negative terminal. From Eq. (2.23) and its associated units, this means the battery provides 800 J of energy to every 1 C of charge that passes through it. For such a battery, it is useful to define the battery's minimum, nominal, and maximum voltages (V_{min}, V_{nom}, and V_{max}). While minimum and maximum voltage values are intuitive, a battery's nominal voltage is taken to be the average voltage over a standard discharge from fully charged to fully discharged.

When the ends of a conductor are connected across an electrical potential difference (voltage), charge flows from the higher potential to the lower potential. By convention, current is considered to flow in this direction, from positive to negative. However, in reality, the electrons themselves move in the opposite direction, from the negative terminal toward the positive terminal. This flow of electrical charge constitutes electric *current* (I). The rate of electrical charge flow with respect to time is measured in *amperes*, commonly called an *amp* [A],

$$I = \frac{q}{t} \tag{2.24}$$

The magnitude of this current depends on both the applied voltage and the electrical *resistance* opposing the flow of charge, measured in *ohms* [Ω]. The relationship between voltage, current, and resistance is given by *Ohm's Law*:

$$V = IR \tag{2.25}$$

Returning to the ideas of power and energy, electrical power can be calculated as the product of electrical current and voltage [watt = amperes × volts]:

$$P = IV \tag{2.26}$$

Equating work, W, and energy, E, and noting the equivalent units for both, our expression for power, P, in Eq. (2.22) can be rewritten:

$$P = \frac{\Delta E}{\Delta t} \tag{2.27}$$

From the definition of power as the rate of energy transfer, total energy can be obtained by multiplying power by the duration of time over which that power is delivered. This yields the unit watt-hour [Wh], where 1 Wh corresponds to 1 W of power sustained for 1 hour,

$$E = Pt \tag{2.28}$$

Through dimensional analysis, and noting that 1 ampere-hour [Ah] equals 3600 C (from Eq. 2.24, where $q = It$, and recalling 1 hour contains 3600 seconds), we can relate the fundamental units of voltage and charge to energy. From Eq. (2.23), one volt corresponds to one joule per coulomb (1 V = 1 J/C). Therefore, multiplying voltage by charge yields energy. Hence, an expression for the total *energy capacity* (E) of a battery can be written as the product of its nominal voltage (V_{nom}) and its charge capacity (Q), expressed in ampere-hours:

$$E = V_{nom} Q \tag{2.29}$$

Expressed in watt-hours [Wh] or joules [J], energy capacity quantifies the total energy a battery can store and deliver. Alternatively, it can be interpreted as the total work the battery can perform before requiring recharge.

With *charge capacity* (Q) indicating how much current a battery can deliver over a specified time, the *C-rate* expresses the rate at which a battery charges or discharges relative to its total charge capacity. This value is stated as a multiple or fraction of the battery's charge capacity. Thus, a C-rate of 1C means the battery would ideally fully charge or discharge in 1 hour; a C-rate of 2C corresponds to 0.5 hours. The C-rate provides a standardized measure for comparing battery performance by describing how quickly a battery can be charged or discharged relative to its total capacity. Achievable rates differ for charging and discharging due to thermal and chemical limits, but eVTOL systems are often designed so that charging can occur within a comparable or shorter time than a typical flight. Higher C-rates enable faster energy transfer but also increase heat generation and impose greater demands on thermal management, a topic addressed later in this chapter.

These ideas of energy and power for batteries are often expressed in terms of *energy density* and *power density*, respectively. Energy density is defined as the total energy a battery can store [Wh] per unit mass [kg] and is expressed in watt-hours per kilogram [Wh/kg],

$$\text{Energy density}\left[\frac{\text{Wh}}{\text{kg}}\right] = \frac{\text{Total energy stored [Wh]}}{\text{Mass of battery [kg]}} \tag{2.30}$$

Likewise, power density quantifies the peak power capability of the battery [W] per unit mass [kg] and is expressed in watts per kilogram [W/kg],

$$\text{Power density}\left[\frac{\text{W}}{\text{kg}}\right] = \frac{\text{Peak power capability [W]}}{\text{Mass of battery [kg]}} \tag{2.31}$$

Power density reflects how rapidly energy can be delivered relative to weight. Energy density and power density normalize their respective quantities by mass. It is reasonable to anticipate that well over one-quarter of an eVTOL's mass consists of its battery system.

A Ragone plot graphically characterizes an energy storage technology by its specific energy and specific power. This plot thus provides insight into the total energy that an energy-storage system can store per unit mass relative to the rate at which that energy can be delivered. The Ragone plot provides a concise visual framework for evaluating the trade-offs between these two fundamental performance metrics. Figure 2.32 illustrates how, among electrochemical systems, high-specific-energy systems such as fuel cells and hydrocarbon-fueled engines enable longer endurance. Fuel-cell systems, however, generally respond more slowly to transient power demands than

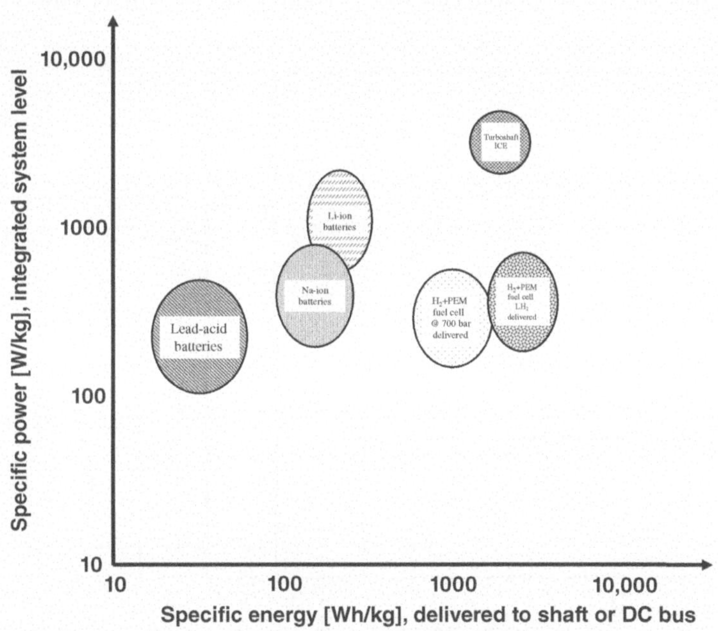

Figure 2.32 A Ragone plot on a log-log scale comparing aviation energy sources on a delivered-energy basis. The *x*-axis represents specific energy density [Wh/kg] delivered to the shaft or DC bus, using pack-level values for batteries, system-level values for hydrogen fuel cells, and fuel plus engine values for internal combustion engines (ICEs) with an assumed 30–40% engine efficiency for delivered energy. The *y*-axis represents integrated system-specific power [W/kg], using pack-level values for batteries, complete system values for fuel cells, and engine-level values for turboshaft systems.

batteries. Meanwhile, high-power systems, such as Li-ion batteries, can deliver rapid bursts of electrical power for high-thrust flight conditions but store less total energy. Notably, turboshaft engines achieve both high specific energy and high specific power, placing them in the upper-right region of the plot. This distinction underscores why battery-electric eVTOL aircraft face such stringent energy-to-weight constraints compared to conventional combustion-powered aircraft. By mapping technologies on this plane, designers can assess which energy storage solution, or combination of solutions, best satisfies a given mission profile's energy and power requirements.

2.4.2 eVTOL Battery Modules and Packs

The design of a battery pack for an eVTOL is based on meeting a defined mission profile and that mission's power and energy requirements. Figure 2.33 illustrates how power and energy requirements vary over a typical eVTOL flight profile. Power represents the instantaneous rate at which energy is drawn from the battery and peaks during the most power-intensive phases of flight when the propulsion system must generate high thrust to overcome both the aircraft's weight and the additional aerodynamic losses associated with lift generation. These power-intensive phases include vertical takeoff, low-speed hover or transition operations near takeoff or landing, and the climb to cruise altitude. As the aircraft transitions to forward flight, aerodynamic efficiency improves and the required power decreases, reaching its minimum during steady cruise flight. During descent, power demand remains comparatively low until a second sharp increase occurs during hover and landing. In contrast, the energy trace depicts the cumulative energy expended over time and increases monotonically throughout the mission. The slope of the energy curve at any point corresponds to the instantaneous power demand, becoming steeper where power

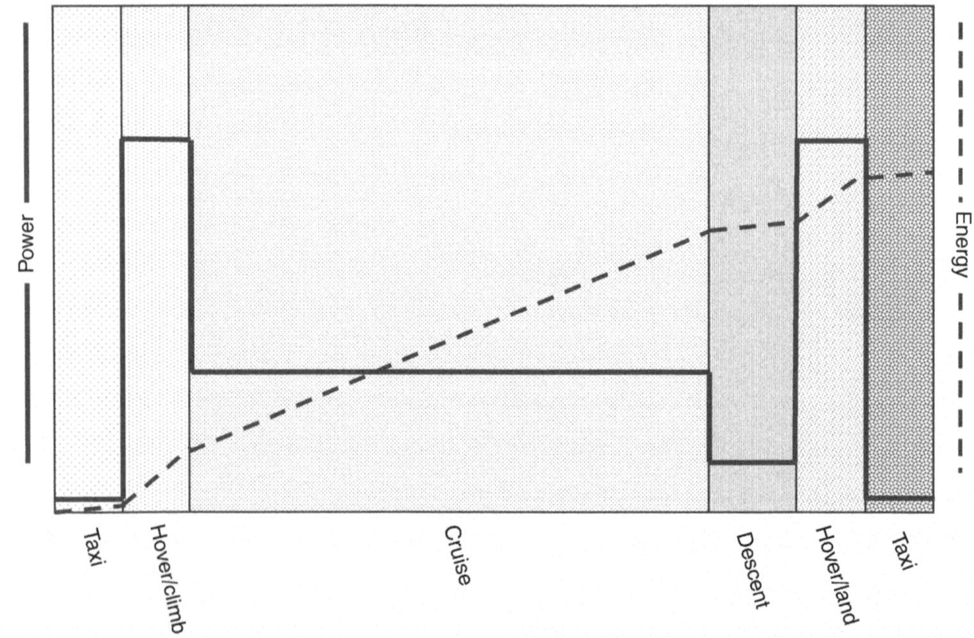

Figure 2.33 Power and cumulative energy requirements for a representative eVTOL flight. Power demand varies across flight phases, peaking during takeoff, hover, and landing. The cumulative energy curve increases continuously, with its slope proportional to the instantaneous power draw.

demand is high and shallower where power is low. The final value of the energy curve at the end of the flight represents the total electrical energy consumed for the entire flight, including reserves.

Ultimately, energy capacity dictates flight duration. However, as shown, high power output is necessary during the takeoff and landing phases, particularly during hover, where power requirements can be 2–4 times greater than during cruise. Meeting these high power demands within a flight, without carrying excessive stored energy, is a significant challenge in eVTOL battery design. The next subsection examines how cell configuration and pack architecture are designed to satisfy both power and energy requirements effectively.

Modern energy storage systems are composed of three key hierarchical components: *cells*, *modules*, and *packs* (Figure 2.34). *Cells* are the fundamental electrochemical units that store and release energy. Their arrangement (specifically, the number of cells connected in series or in parallel) defines the system's electrical characteristics, including its total voltage, maximum current, and energy capacity. In the strictest sense, a "battery" consists of two or more connected cells, although oftentimes the terms battery and cell are used interchangeably.

Higher system voltage improves overall electrical system efficiency by reducing current for a given power level and thereby reducing resistive losses, whereas higher current capability supports the instantaneous power demands associated with vertical flight. To achieve the voltage and current required for propulsion, cells are combined in two distinct ways: parallel and series (Figure 2.35). Cells connected in parallel form a cell *group*, and the number of cells in parallel determines the system's total capacity and current capability. Cell groups are typically connected in series to form a *module*, and the number of series connections determines the module's voltage (Figure 2.36). This is the primary level at which electrical safety can be introduced as each module is fused and limited in voltage to allow safe handling and maintenance.

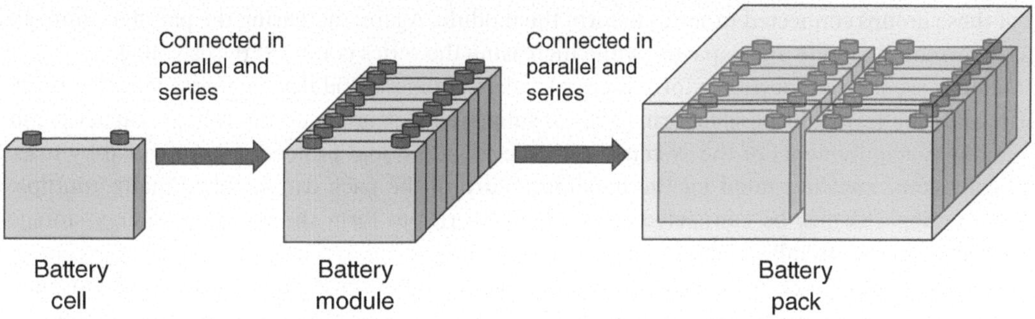

Figure 2.34 Modern energy storage systems are composed of cells, modules, and packs. Cells are combined in series and parallel to form modules, and modules are interconnected to form packs using series and parallel connections as required to meet system-level voltage, capacity, and redundancy requirements.

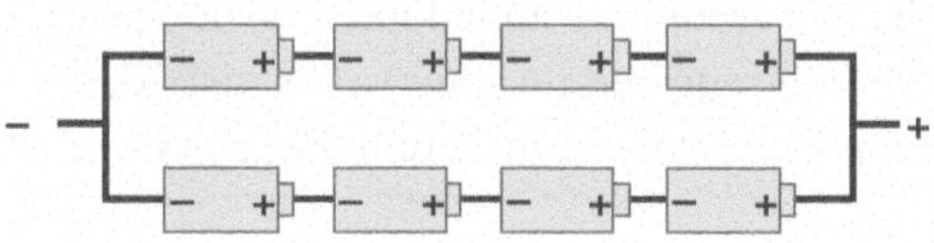

Figure 2.35 A simple battery module consisting of two parallel-connected strings, each string containing four cells in series (4S2P configuration). The series connection within each string increases the total voltage, and the parallel connection between strings increases total capacity and current-handling capability.

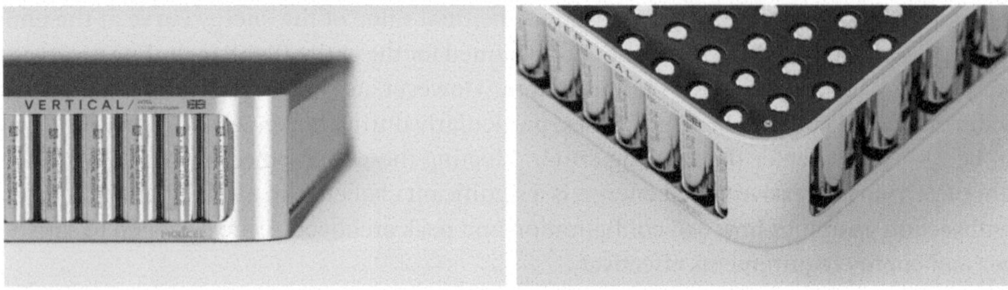

Figure 2.36 Vertical Aerospace VX4 battery module. Courtesy of Vertical Aerospace.

Increasing system voltage provides a primary benefit by enabling the high power levels required for eVTOL propulsion without excessive current draw. Increasing system voltage also offers a secondary advantage during aircraft-level integration. Reexamining Eq. (2.26), we observe that for a given propulsor power requirement, raising the bus voltage decreases the electrical current that must be carried through cables, connectors, inverters, and protection devices. Because electrical conductor weight scales with current capacity through cross-sectional area, higher voltage architectures permit smaller gauge wiring with lower total weight. Reducing current also decreases resistive heating and associated thermal-management burden. These considerations further motivate the use of elevated pack voltages within eVTOL powertrains, provided the accompanying insulation and spacing requirements are satisfied.

The series and parallel arrangement of cells within a module is described by the S–P convention, where S indicates the number of series connections and P the number of parallel connections. For example, a 12S8P configuration denotes 8 cells connected in parallel to form each cell group, with 12 of these groups connected in series to form the module. Again, increasing the parallel count (P) raises current capability and capacity, while increasing the series count (S) raises voltage.

Modules are then connected to form higher-level assemblies. Modules are interconnected using series and parallel connections to form packs or subpacks, depending on the voltage, capacity, and redundancy requirements of the system. At this level, additional protection is provided by fuses and contactors, switches rated for the maximum current the pack can supply. Finally, multiple packs or subpacks may be connected in parallel or series to form the complete energy storage system used in the aircraft.

This hierarchical structure of cells forming modules, modules forming packs, and packs forming the full energy storage system allows engineers to tailor both power and energy characteristics to the demands of a given eVTOL mission. Designing for high power requires delivering large currents quickly to satisfy peak demands during takeoff, hover, and landing phases. This is achieved through higher parallel cell counts (larger P) and the use of high-discharge-rate cells with low internal resistance. Comparatively, designing for high energy requires storing sufficient total electrical energy to sustain flight for longer durations without excessive current draw. This is accomplished by increasing the series count (S) and selecting high-energy-density cells that minimize added mass. When power and energy requirements for a given mission profile (including required reserves) are satisfied, several additional factors must also be considered in the design and integration of battery packs.

Voltage Matching

The pack's output voltage, including its minimum, nominal, and maximum limits (V_{min}, V_{nom}, and V_{max}), must remain within the allowable input range of the power inverter that converts DC power from the battery into AC power for the electric motors.

Cycle Life

The *cycle life* of a cell is the number of charge–discharge cycles it can undergo while maintaining acceptable performance. Although Li-ion chemistries can exceed thousands of cycles under ideal conditions, *cycle fade*, a gradual reduction in capacity, occurs naturally. If high depths of discharge and rapid charge–discharge rates take place in the context of eVTOL operations, this degradation might be greatly accelerated. Certification and operational requirements will likely mandate battery replacement well before a cell reaches its theoretical end-of-life, making cycle life a key driver of operating cost.

Weight

Weight sensitivity is especially pronounced for eVTOL aircraft. The total weight of the energy storage system, as well as its distribution within the airframe, directly affects payload capacity, hover power requirements, and overall flight efficiency. Because battery weight does not decrease during flight, unlike fuel, battery weight penalties persist throughout all phases of operation, reinforcing the importance of careful sizing and placement.

Form Factor

Modular designs allow custom form factors for battery packs that optimize space utilization within the aircraft. A useful metric in this context is *volumetric energy density*. This represents the amount of electrical energy stored per unit volume of the battery and is commonly expressed in watt-hours per liter [Wh/L]. Volumetric energy density serves as a useful measure of how effectively available space within the aircraft is utilized for energy storage.

$$\text{Volumetric energy density}\left[\frac{\text{Wh}}{\text{L}}\right] = \frac{\text{Stored electrical energy [Wh]}}{\text{Battery volume [L]}} \tag{2.32}$$

Cost

High initial and replacement costs for battery packs strongly influence operational strategies. Operators may elect to recharge battery packs between flights or to swap depleted packs for charged ones to maximize utilization. Improvements in cell chemistry, manufacturing scale, and achievable cycle life will be critical in reducing cost over time.

Safety

Safety considerations, including thermal management, isolation, and containment, will be discussed in detail in the next subsection.

2.4.3 Battery Safety

Failures within the high-energy-density battery systems required for an eVTOL can lead to catastrophic consequences, such as thermal runaway, fire, and explosion. Proactive measures to mitigate these risks are implemented in the battery management system (BMS), discussed in the next subsection. However, before outlining prevention strategies, it is first important to understand the possible failure modes of battery cells and packs and the risks that they present.

One of the most significant hazards associated with Li-ion batteries is *thermal runaway*. Thermal runaway occurs when excessive heat generation within a cell initiates an uncontrollable self-sustaining chemical reaction that generates more heat than the battery cell can dissipate (Figure 2.37). The resulting rapid rise in the temperature of the cell can lead to the venting of

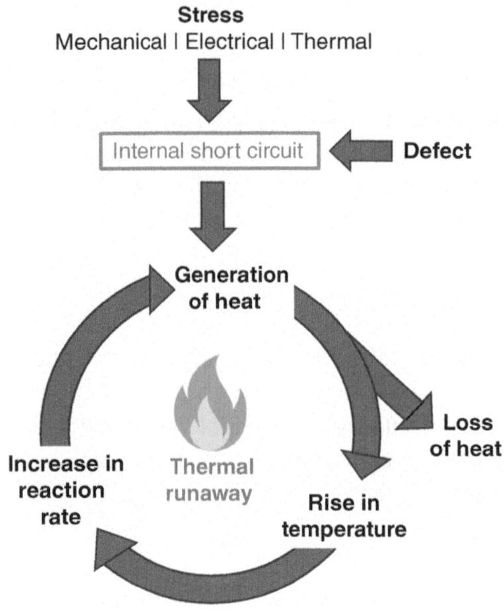

Figure 2.37 Primary initiators and cycle of thermal runaway.

flammable gases and, in severe cases, fire or explosion. The primary initiators of thermal runaway can be broadly categorized as mechanical, electrical, and thermal. Mechanical damage, such as crushing or puncturing a cell, can cause an internal short circuit by damaging the separator that isolates the anode and cathode. Internal shorts, external short circuits, overcharging, overdischarging, and aggressive high-C-rate charging or discharging can generate excessive internal heating and degradation that contribute to thermal runaway. Thermal factors such as high ambient temperatures, failure of a cooling system, or exposure to external fire can also raise cell temperatures to a point where exothermic reactions become self-sustaining.

Propagation occurs when thermal runaway in one cell triggers thermal runaway in neighboring cells, resulting in a chain reaction. Unchecked, propagation can lead to catastrophic failure of the entire battery pack. Therefore, propagation control is critical. It can take the form of thermal barriers, internal cell arrangement with thoughtful spacing, orienting cells such that venting is directed away from adjacent cells, and implementing active cooling or fire suppression systems.

The high-energy battery systems required for an eVTOL operate at high voltage, as introduced earlier in the battery fundamentals discussion. Therefore, *arcing* is a concern for high-voltage battery packs and occurs when current jumps across a gap between two conductors due to a high potential difference. Two key mitigation strategies are used to prevent arcing: clearance and creepage. *Clearance* refers to the shortest distance between two conductive parts as measured through the air. *Creepage* refers to the shortest path between two conductive parts measured along the surface of the insulation. Maintaining appropriate clearance and creepage helps prevent this type of electrical discharge. The occurrence of arcing can produce intense localized heating and pressure waves strong enough to damage nearby components and ignite fire.

Beyond propagation control, partitioning and containment play essential roles in managing potential failures. *Partitioning* divides the battery pack into smaller, independent subpacks so that the failure of a single cell or module does not compromise the entire system. This approach limits the impact of localized failures and promotes redundancy. *Containment*, in turn, aims to confine the effects of a failure through structural and protective design measures. Strategies include crash-worthy enclosures and fire-resistant casings that maintain integrity under impact or thermal stress.

2.4.4 Battery Management Systems

To mitigate these risks, an eVTOL employs a BMS. The BMS is designed around the overarching principle that no single-point failure, such as the loss of a single module, should lead to a catastrophic condition [3]. In an aircraft that depends entirely on electric propulsion, a well-designed BMS ensures safety and the reliable and efficient operation of the propulsion and energy storage systems. Composed of both hardware and software elements, the BMS continuously monitors, protects, and optimizes the performance of the battery pack. Key hardware and software components of a BMS include the following.

Hardware Architecture

BMS hardware includes a network of *battery sensors* that measure individual-cell voltage, current, and temperature in real time. *Microcontrollers and processors* interpret these signals and execute the control algorithms that govern charging, discharging, and protection functions. *Power-management circuits* regulate charge and discharge rates to prevent overcurrent conditions. *Integrated cooling and protection* devices, such as thermal-management channels, fuses, and relays, guard against overheating and electrical faults. Because individual cells age at different rates, *cell-balancing circuits* maintain uniform voltage across all cells, either by redistributing charge or dissipating excess energy. In addition, *communication interfaces* link the BMS to higher-level vehicle systems, enabling the exchange of diagnostic data and control commands between the battery and the aircraft's flight computer.

Software Architecture

Software within the BMS governs estimation, protection, and optimization. *State-estimation algorithms* calculate the SOC and state of health (SOH), two of the most important diagnostic parameters. As detailed in the next subsection, SOC expresses the remaining capacity as a percentage of the total, and SOH estimates the overall condition of the battery based on capacity fade and internal resistance. *Fault-detection and diagnostic* routines identify anomalies such as short circuits, overcharge, or abnormal thermal events and initiate protective responses. *Charge-control algorithms* manage fast-charging operations to minimize degradation, and *thermal-management algorithms* regulate cell temperature. Li-ion batteries perform best near room temperature; cold environments increase internal impedance and reduce capacity, whereas high temperatures accelerate structural degradation. Finally, *data-logging* and *predictive-maintenance* routines record operational history, monitor parameter exceedances, and use accumulated data to forecast maintenance needs and extend battery life. An excellent reference for additional information and guidance on the design and development of BMS architectures for rechargeable lithium batteries used in aerospace applications is SAEAIR6897 [4].

2.4.5 Battery State Indication and Display

The information that a BMS provides must ultimately be translated into data meaningful to the pilot. Conventional fuel gauges provide a pilot with a quantity that allows them to reasonably predict remaining aircraft capability, such as its range and endurance. Electric propulsion does not provide such a direct mapping because a battery's usable energy and power depend on a confluence of factors like temperature, discharge rate, SOC, and SOH. Therefore, displaying only a "percentage of charge remaining" (like the percentage indicators familiar from consumer electronics) provides an incomplete and potentially misleading picture of the aircraft's actual energy capability. A single value such as "percentage of charge" does not indicate whether the energy can be delivered at the power levels required by the mission, especially for vertical takeoff, hover, and landing.

In technical terms, "percentage of charge" is better referred to as SOC and represents the ratio of remaining charge, Q, to the total usable charge of the battery, Q_{max},

$$SOC(t) = \frac{Q(t)}{Q_{max}} \tag{2.33}$$

While SOC conveys how much charge remains, it does not represent how much usable energy or available power can still be extracted under real operating conditions. These performance measures are governed by the *state of energy* (SOE) and *state of power* (SOP), respectively. As the battery discharges, its voltage decreases and internal resistance increases, reducing the maximum power that can be delivered. Two batteries with identical SOC values may, therefore, differ significantly in their ability to sustain the power levels required for vertical flight. Thus, a simple percentage display provides no insight into whether adequate power margin exists for hover, transition, or landing.

Effective state indication integrates SOC, SOE, and SOP into a single, operationally meaningful output (Table 2.1). It communicates both how much usable energy remains and how much power can safely be delivered under current and anticipated conditions. In this way, the pilot sees the remaining charge and the aircraft's remaining capability to perform each phase of flight.

Effective state indication must combine real-time measurement with predictive modeling. The system must determine how much energy and power are available at the present moment and

Table 2.1 Comparison of battery state parameters and their operational significance.

Parameter	Definition	Level	Represents	Operational significance
SOC	Ratio of remaining charge to total usable charge of the battery	Base parameter	How much charge remains relative to full capacity	Necessary but inadequate alone for flight decision-making
SOE	Fraction of usable energy remaining relative to total available energy	Derived physical parameter	Usable electrical energy that can be extracted given present conditions	Technically meaningful, but not directly intuitive to a pilot
SOP	Fraction of maximum power output currently available relative to nominal design power	Derived physical parameter	Maximum instantaneous power that can be safely delivered	Technically meaningful, but not directly intuitive to a pilot
Effective state indication	Model-based integration of SOC, SOE, and SOP into a single operational display	Operational output	Remaining usable energy and power capability under current and anticipated conditions	Translates electrochemical parameters into an intuitive measure of remaining flight capability, incorporating mission reserves and contingencies such as go-around

whether sufficient capability remains to complete all planned and contingency phases of flight, including an extended hover or potential go-around. This approach converts the continuous data produced by the BMS into a performance-based indication of remaining flight capability that moves beyond a static measure of stored charge.

In practice, battery state indications should express information in a form directly relevant to the pilot's decision-making. For example, instead of showing a single percentage, an indicator might depict remaining endurance in minutes, remaining range in nautical miles, or available energy relative to the mission profile. Such representations communicate whether the aircraft can complete the planned flight and still satisfy required reserve energy margins. The overarching goal is not to merely indicate how much charge remains; the goal is to indicate how much mission capability remains, given the battery's present condition and the expected power demands of eVTOL operation.

2.4.6 Fuel Cells

While a battery stores electrical energy, a fuel cell continuously generates electrical energy as long as a fuel and an oxidizer are supplied. This distinction between stored and generated energy has important implications for AAM operations. As we have discovered, once a battery's stored chemical energy is depleted, the battery must be recharged from an external source. This imposes downtime between flights. A fuel cell, by contrast, converts the chemical potential energy of its fuel directly into electrical power at the point of use. This operating principle eliminates the need for recharging and requires only refueling, a process that can be completed within minutes. The ability to refuel rapidly provides a decisive operational advantage for future high-utilization AAM fleets (Figure 2.38). Several types of fuel cells exist, distinguished primarily by their electrolytes, operating temperatures, and capability of using various fuels, including hydrogen, methanol, and ammonia. Among these, hydrogen-based proton-exchange membrane (PEM) fuel cells are the most widely adopted for aerospace applications and serve as the focus of our discussion.

Figure 2.38 Joby's hydrogen-electric technology demonstrator. *Source*: With permission of Joby Aviation.

A hydrogen fuel cell converts hydrogen (H_2) and oxygen (O_2) into electricity and water (H_2O) through an electrochemical reaction (Eq. 2.34). The electrochemical process that takes place within a fuel cell is similar in principle to the redox reactions described earlier for batteries. The same fundamental roles of the anode, cathode, and electrolyte are present; however, the overall system differs because the hydrogen and oxygen reactants are continuously supplied versus stored internally within the electrodes. In a hydrogen PEM fuel cell, hydrogen gas is introduced at the anode and dissociated on a catalytic surface into protons and electrons. These electrons then travel through an external circuit, providing useful electrical power, while the protons migrate through the electrolyte to the cathode. At the cathode, oxygen from ambient air combines with the returning electrons and protons to form water vapor as the only by-product:

$$2H_2 + O_2 \rightarrow 2H_2O + \text{Electrical energy} + \text{Heat} \tag{2.34}$$

Each individual fuel cell produces a very small voltage, so multiple cells are connected in series and compressed together between bipolar plates. These bipolar plates are conductive plates that electrically connect adjacent cells, distribute reactant gases, remove heat, and maintain structural compression. This assembly of plates and cells is called a *fuel-cell stack*, and it produces DC electrical power. Following conversion by inverters to AC, this power is used to drive the propulsion motors. Accordingly, this electrical architecture parallels that of the battery-electric aircraft schematic illustrated in Figure 2.28, differing only in the source of DC power.

The performance of a fuel cell and battery can be best compared by returning to our consideration of *specific energy*, measured in watt-hours per kilogram, and *specific power*, measured in watts per kilogram. Typical commercial Li-ion batteries provide about 150–300 Wh/kg, whereas hydrogen fuel alone provides approximately 30–40 kWh/kg. When the mass of the complete fuel cell system is considered, including storage tanks, structural hardware, cooling equipment, and associated support components, the system-level specific energy is typically on the order of 1–3 kWh/kg, which remains significantly higher than that of present Li-ion technology. However, the specific power for fuel cells is generally lower because of the slower electrochemical reaction rates and supporting components. Thus, as we saw on our Ragone plot (Figure 2.32), hydrogen fuel cells exhibit high specific energy but more limited specific power. Therefore, fuel cells are well suited for the cruise and endurance portions of flight and are less capable of supplying the short-duration power required for VTOL.

A modern PEM fuel cell operates with an efficiency of approximately 50 %, meaning that about half of the hydrogen's chemical potential energy becomes electrical work, and the remainder appears as waste heat. This heat must be dissipated to maintain a stable operating temperature. Consequently, effective thermal management is integral to the design of the power system. Liquid, air, and evaporative cooling methods are all used for this purpose, but liquid cooling is the most common for AAM applications because it employs loops similar to those used in battery systems and offers simplicity, reliability, and well-understood performance.

Hydrogen Storage Considerations

Hydrogen possesses the highest energy per unit mass of any fuel, but its usable energy per unit volume (volumetric energy density) within practical storage systems, including compressed gas tanks and cryogenic liquid tanks, is extremely low. Therefore, a central engineering challenge posed by using hydrogen is storing enough of it in a compact, lightweight, and safe form suitable for flight. This limitation stems from the storage system itself instead of from the electrochemical performance of the fuel cell. AAM vehicles, particularly eVTOLs, exacerbate this challenge because their slender and compact airframes limit the available volume for either compressed or cryogenic storage systems.

One common method is to store hydrogen as a compressed gas, typically in composite vessels operating at pressures between 350 and 700 bar. This requires heavy tanks with significant inert mass, directly reducing payload capability. Furthermore, compressing hydrogen consumes approximately 15% of its stored energy and introduces additional safety and regulatory challenges. Hydrogen can also be stored as a cryogenic liquid. Cryogenic storage increases volumetric energy density by cooling the gas to −253 °C. This approach is used in some space and long-range aviation programs, but it is less suitable for AAM-scale aircraft. The liquefaction process itself consumes nearly 40% of the hydrogen's usable energy. Concurrently, continuous heat infiltration into the tanks causes gradual fuel loss through evaporation, known as boil-off. The associated insulation, venting, and plumbing systems add mass and complexity, reducing the practicality of liquid-hydrogen storage for small vehicles such as eVTOLs.

The most promising configuration for near-term AAM employs hybrid propulsion architectures that combine the sustained energy capability of a hydrogen fuel cell with the transient power capability of a Li-ion battery. In this arrangement, the fuel cell provides continuous electrical power during cruise, while the battery supplies the high-power bursts required for VTOL, hover, climb, and rapid maneuvering. The instantaneous power balance at the propulsors can once again be expressed by Eq. (2.17), where $P_{\text{dem}}(t)$ denotes the total shaft-equivalent power required by the propulsors. This hybrid architecture exploits the complementary strengths of both technologies: the high specific energy of hydrogen for endurance and the high specific power of batteries for vertical flight. As fuel-cell stack power density and hydrogen storage efficiency improve, the balance between these two energy sources may shift toward increasingly fuel-cell-dominant systems. Fully fuel-cell-electric aircraft could ultimately achieve endurance and turnaround times approaching those of turbine-powered vehicles while producing no direct emissions. The integration of continuous chemical-to-electric conversion, efficient hydrogen storage, and coordinated hybrid power management defines a clear technological pathway toward sustainable, high-performance air mobility.

2.5 Summary

AAM vehicles introduce a new level of integration between configuration, propulsion, and energy storage, driven by the adoption of electric power and distributed propulsion systems. Vehicle architecture determines how lift and thrust are generated, and this choice governs aerodynamic efficiency, control strategy, and redundancy in ways unique to electrically powered flight. These interdependencies create exceptional sensitivity to mass and power demand. Electric propulsion replaces mechanical complexity with electrically mediated control and power distribution, yet its capability is inseparable from the energy source that drives it. The design of an eVTOL battery pack is governed by its mission profile, balancing energy density for endurance with power density for takeoff, hover, and landing. The BMS maintains safety and performance by monitoring cell conditions, regulating temperature, and ensuring that pilots receive information reflecting remaining flight capability in contrast to simple charge level.

References

1 https://www.faa.gov/regulations_policies/advisory_circulars/index.cfm/go/document.information/documentID/99693
2 https://www.cmh17.org/
3 Federal Aviation Administration (FAA). (2011). AC 23.1309-1E: System Safety Analysis and Assessment for Part 23 Airplanes. Washington, DC: U.S. Department of Transportation.
4 SAE International. (2020). AIR6897: Battery Management Systems for Rechargeable Lithium Batteries Used in Aerospace Applications. Warrendale, PA: SAE International.

3

Airspace Management and Integration

As established in the previous chapter, advanced air mobility (AAM) involves aircraft with fundamentally different operational characteristics from the vehicles for which the existing airspace system was built. These vehicles are intended to operate at high density within the lowest kilometer of the atmosphere, execute frequent point-to-point missions, and rely on automation to maintain predictable behavior. In contrast, the present air traffic management (ATM) system evolved around an environment defined by modest traffic counts, wide maneuvering volumes, and human-centered tactical control. AAM challenges these assumptions by concentrating operations in constrained airspace where spatial margins, temporal margins, and surveillance performance differ markedly from those of mid-altitude operations.

A clear understanding of how AAM vehicles integrate into shared airspace begins with the principles that govern contemporary ATM. While the existing system provides a framework for separation, surveillance, communication, and navigation, its design reflects the expectations and constraints of an earlier era. AAM operations demand a thorough examination of this framework to identify which elements are still suitable for the low-altitude environment and which elements require adaptation. Accordingly, this chapter examines AAM airspace management and integration by examining the present-day airspace structure, the role of automation in future operations, the challenges inherent to urban environments, and the communication and sensing architectures that enable predictable behavior across dense low-altitude networks.

3.1 How Air Traffic Is Managed Today

Today, ATM represents the accumulation of decades of engineering development, operational experience, and accompanying regulatory changes. The current system provides a reliable method for organizing aircraft that vary significantly in performance, mission, and equipage. Although this system has matured through history, many of its structural principles originated in an operating environment shaped by legacy traffic routing and navigation capabilities plus a level of controller involvement that reflected the technology of earlier decades. Understanding these foundational elements establishes the necessary context for assessing how AAM vehicles will integrate into a future architecture of shared airspace.

3.1.1 The Structure of Controlled and Uncontrolled Airspace

Airspace classification stipulates the level of separation service that aircraft receive and the responsibilities placed on operators. Controlled airspace provides continuous traffic management through communication, navigation, and surveillance (CNS) services that support predictable aircraft flows as vehicles transition between terminal departure, en route, and terminal arrival phases of flight. Uncontrolled airspace relies on pilot-based see-and-avoid responsibilities and voluntary cooperation, in contrast to structured separation services, an approach historically suited to low traffic densities, limited interaction rates, and comparatively simple traffic geometries.

Conventional operations within both controlled and uncontrolled airspace are also governed by visual flight rules (VFR) and instrument flight rules (IFR). These rule sets define how aircraft navigate, separate, and interact with air traffic services. VFR operations depend primarily on pilot visual acquisition and self-separation, with limited equipment requirements and discretionary access to surveillance or advisory services. IFR operations rely on certified navigation performance, procedural separation, and continuous air traffic services to maintain safe spacing in higher-density or reduced-visibility environments. Many of the navigation and surveillance capabilities discussed later in this chapter originate from IFR-based operations. However, neither VFR nor IFR frameworks were developed to support the traffic densities, interaction rates, or automation-dependent separation concepts anticipated for AAM. AAM, instead, adapts instrument-based capabilities as part of a broader shift toward cooperative, automation-mediated traffic management.

Beyond controlled and uncontrolled designations, airspace is further broken down into multiple classes. The structure of Classes A–E airspace, depicted in Figure 3.1, defines altitude-based and geographically constrained regions optimized for conventional operations. Class A supports high-altitude flight traffic through standardized instrument procedures and continuous surveillance. Classes B–D apply increasing levels of separation and communication requirements around airports with varying traffic densities. Class E remains controlled airspace but with fewer communication burdens. Finally, Class G functions as an uncontrolled environment in which hazard avoidance depends entirely on pilot situational awareness.

This classification framework balances controller workload, surveillance fidelity, and predictable flight behavior. However, this airspace system was *not* designed for high-density, low-altitude

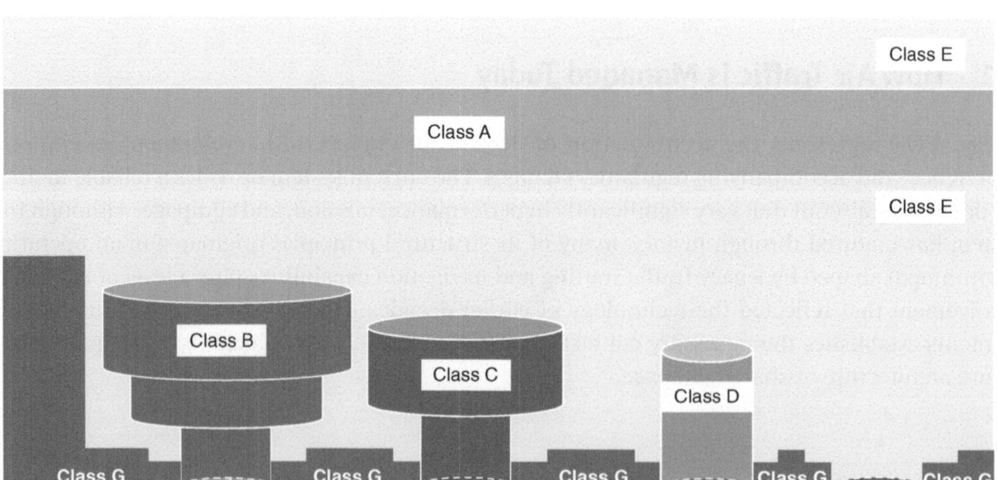

Figure 3.1 Representative classification of controlled and uncontrolled airspace. The diagram illustrates the commonly used Classes A–G airspace structure. Not all classes are implemented uniformly across national airspace systems; for example, Class F is defined in some international frameworks.

operations, including within metropolitan regions. Consequently, a high density of aircraft with short mission durations, operating within a concentrated spatial footprint and relying minimally on pilot-driven tactical decisions, will significantly strain today's ATM system.

3.1.2 Separation Minima, Surveillance, and Communication

The spatial separation structure of contemporary aviation is based on deterministic minima empirically derived from wake turbulence dynamics, human visual-acquisition performance, and surveillance accuracy. Radar and automatic dependent surveillance-broadcast (ADS-B) systems provide cooperative surveillance for most, but not necessarily all, aircraft operating in controlled airspace, with equipage varying across general aviation (GA), rotorcraft, agricultural operations, and legacy fleets (Figure 3.2). Controllers maintain lateral and vertical separation through procedural control or direct vectoring, and pilots comply through continuous two-way voice communication with air traffic control (ATC).

This method of separation management depends on three factors that support conventional air traffic operations but limit scalability: the predictability of aircraft performance that enables controllers to anticipate future positions; human cognitive bandwidth that limits the number of aircraft that can be managed within a given controlling sector; and reliable communication that ensures instructions are received and executed with acceptable latency. Together, these constraints define system throughput limits, particularly as traffic density increases. The requirement that every tactical maneuver pass through a human controller or pilot further constrains entry rates, reducing the capacity of the airspace system to absorb additional operations. As AAM introduces short-duration missions and rapid trajectory changes, these constraints are exacerbated.

3.1.3 Performance-based Navigation and Trajectory Conformance

Airspace classes and separation responsibilities define how air traffic is organized. However, the ability of controllers to manage this structure depends critically on how precisely each aircraft follows its assigned path. This reliance on predictable route following leads directly to the role of performance-based navigation (PBN).

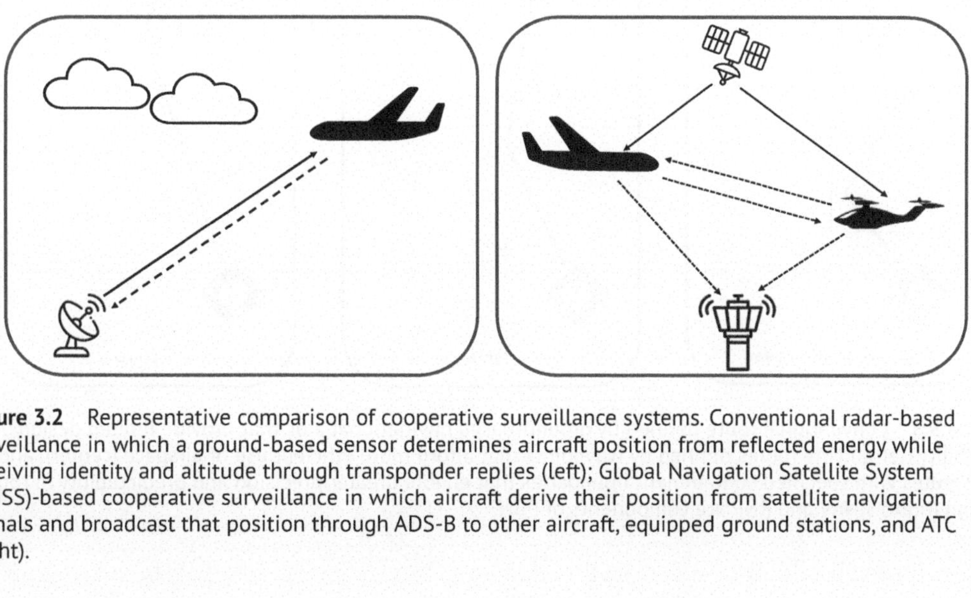

Figure 3.2 Representative comparison of cooperative surveillance systems. Conventional radar-based surveillance in which a ground-based sensor determines aircraft position from reflected energy while receiving identity and altitude through transponder replies (left); Global Navigation Satellite System (GNSS)-based cooperative surveillance in which aircraft derive their position from satellite navigation signals and broadcast that position through ADS-B to other aircraft, equipped ground stations, and ATC (right).

Traditionally, navigation has been defined by ground-based infrastructure, such as conventional radio navigation beacons, but PBN is now recognized as a more effective means of ensuring that aircraft remain within specified lateral and vertical limits throughout a flight. PBN provides the precision necessary for tighter route spacing and higher achievable traffic density in en route and terminal (airport) environments. It is based on the demonstrated (or performance-based) accuracy, integrity, continuity, and functionality of an aircraft's internal navigation system. The adoption of area navigation (RNAV), a navigation specification within the PBN framework, allows more customized point-to-point routing between any two geographically defined points through onboard navigation systems in lieu of ground-based radio beacons. Required navigation performance (RNP), another navigation specification within the PBN framework, is distinguished by onboard performance monitoring and alerting, ensuring aircraft remain within defined containment boundaries (Figure 3.3).

PBN provides the precision necessary for tighter route spacing, improved flow regularity, and reduced controller workload when paired with sufficient planning structure and automation support. Within this framework, trajectory conformance becomes a key requirement for the modern airspace system. Accurately executing assigned trajectories allows controllers to forecast interactions between aircraft and to maintain separation through strategic planning instead of continuous tactical intervention. This reliance on precise trajectory conformance aligns directly with the needs of AAM, where high-density operations require much tighter conformance tolerances than traditional aircraft due to smaller maneuvering volumes and higher numbers of simultaneously operating vehicles.

These advances in navigation reduce uncertainty in aircraft trajectories at the individual aircraft level. However, increased routing flexibility can also increase system-level complexity if not supported by automation that restores predictability across the traffic flow. Even with improved conformance, the system remains centered on human interpretation, voice communication, and

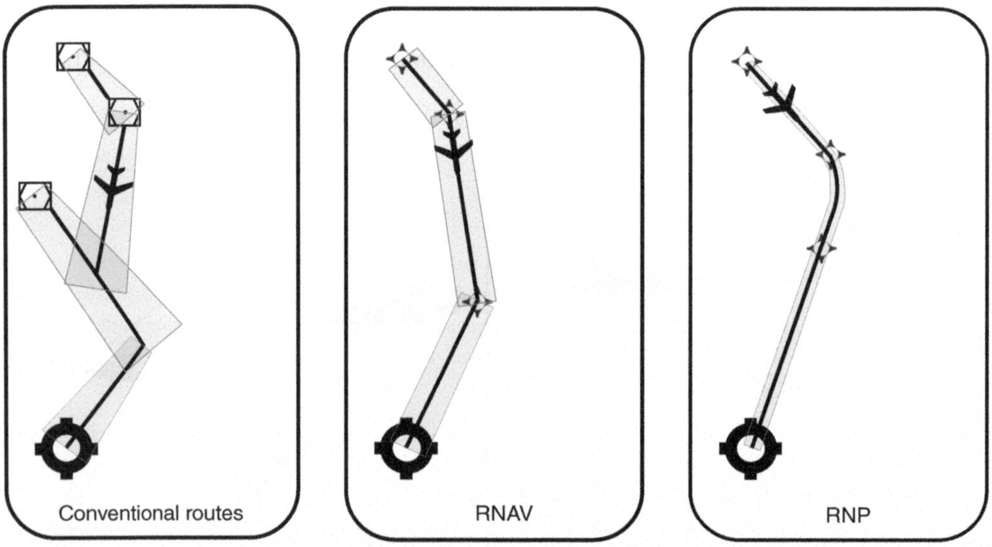

Figure 3.3 Conventional route defined by ground-based navigational aids that provide limited design flexibility (left); RNAV routes defined by waypoints that afford more efficient use of airspace (center); and RNP routes with specified containment boundaries that enable greater precision and predictability in route design (right). RNAV and RNP are components of PBN.

sector-based oversight. Without additional automated coordination and decision support, these characteristics impose scaling limits that become increasingly apparent when the traffic environment shifts toward high-density, low-altitude operations.

3.1.4 Limitations of Legacy ATC for Low-altitude High-density Operations

Today's air traffic system does not accommodate many of the operational characteristics of AAM. In addition to higher mission frequency, shortened flight durations, and reduced pilot involvement, low-altitude urban airspace brings together a more heterogeneous mix of traffic, including rotorcraft, GA aircraft, and small uncrewed aircraft systems (sUAS). These operations occur within an obstacle-laden environment characterized by localized variations in wind, temperature, and turbulence over short spatial scales, commonly referred to as microclimates and examined in greater detail in Chapter 7, as well as by structures that obstruct surveillance sensors. These conditions create an environment in which human-centric tactical deconfliction becomes increasingly inefficient and, at higher densities, operationally untenable.

The present system's reliance on voice-based communication and discrete controller instructions limits throughput and does not scale effectively as aircraft counts increase. The geometry of existing airspace control sectors and the workload placed on controllers impose limits on how much activity can be safely coordinated. In uncontrolled airspace, reliance on visual acquisition becomes infeasible due to inherent limitations in human visual detection, particularly for small, low-contrast, or slowly moving aircraft, and this limitation is exacerbated as aircraft concentrations increase. These constraints motivate the transition toward automated coordination, cooperative separation logic, and structured routing that reduces the need for continuous tactical input. Addressing these limitations requires a reconfiguration of airspace management principles to support scalable, high-density operations.

3.2 Integrating AAM into the Skies: Automation, Predictability, and Flow Management

Today's airspace systems are insufficient for high-volume AAM operations. At the same time, AAM must operate in harmony with existing airspace users. This condition defines the concept of shared airspace, an important consideration for all AAM operators. Integrating AAM vehicles into shared airspace requires a framework that extends the principles of contemporary ATM and supports higher operational tempo, tight trajectory-tracking requirements, and the increased traffic density expected in low-altitude environments. The foundational elements of the legacy system must evolve to accommodate routine operations in which separation assurance and flow coordination rely on automation instead of continuous tactical control. The following subsections examine how these elements might adapt to support scalable AAM operations through predictable aircraft behavior, cooperative information exchange, and strategically organized airspace structures.

3.2.1 AAM Operational Density and the Need for Predictable Behavior

Attempting to scale AAM traffic quickly reveals structural limitations in the legacy air traffic system. Human-mediated separation, voice communication, and sector-based oversight that partitions airspace into fixed control regions constrain the number of aircraft that can be managed safely, particularly for highly concentrated flows in low-altitude airspace. Legacy ATM functions effectively when moderate traffic volumes permit sufficient time for pilots and controllers to evaluate and respond to evolving situations. As density increases, these temporal and spatial margins shrink, reducing the feasibility of human-centered tactical deconfliction.

Operationally, many AAM missions are characterized by rapid turnaround, point-to-point movement, and short dwell times within obstacle-laden environments. Under these conditions, reliance on voice communication results in an accumulation of tactical instructions that exceeds human processing bandwidth because every course adjustment requires a controller–pilot exchange. As traffic density grows, the cumulative number of required instructions grows and, consequently, the time available for each instruction decreases. This dynamic impairs the controller's ability to sustain situational awareness and renders tactical deconfliction increasingly unfeasible. Simply put, controllers can only handle so much at one time.

Scalability challenges extend beyond controller workload and also arise from the maneuvering volume itself. AAM operations occur within smaller maneuvering volumes created by complex low-altitude environments and with the need to coordinate large fleets operating along intersecting routes. This requires highly predictable flight behavior and consistent adherence to planned trajectories in order to increase the capacity of the airspace.

Scalable integration, therefore, requires an airspace framework that reduces dependence on continuous human oversight and increases reliance on predictable aircraft behavior, structured information exchange, and automated support tools. This framework maintains controllers and enables them to manage larger volumes of traffic through higher-level monitoring functions rather than individual tactical intervention. These principles form the basis for scalable AAM integration and demonstrate the important role of automation in coordinating dense low-altitude operations.

3.2.2 Automation as a Prerequisite for Airspace Integration

Contemporary aviation already depends heavily on automation to sustain precision, reliability, and workload management across a wide range of operations. Modern aircraft depend on automated systems to manage navigation, stability, and energy states with levels of precision that exceed the capability of manual control and allow crews to supervise operations instead of manually performing each task. Flight management systems (FMSs) generate and execute lateral and vertical profiles in conformance with instrument procedures (Figure 3.4), autopilots maintain attitude, altitude, and track, and autothrottle systems regulate thrust to preserve energy balance. Many transport-category aircraft are even equipped with a highly reliable autoland feature.

Automation also underpins existing long-endurance uncrewed operations. High-altitude aircraft routinely conduct transoceanic and persistent surveillance missions with only indirect human monitoring and oversight (Figure 3.5). These aircraft follow planned trajectories, execute contingency procedures, and maintain stable flight behavior despite communication latency or extended mission duration. These operations showcase how aviation already relies on automation for predictable aircraft performance.

Beyond automation systems that stabilize and guide a single vehicle, AAM requires automation that supports cooperative behavior among many aircraft sharing constrained airspace. AAM vehicles are designed to operate within confined maneuvering volumes where traffic density is expected to be high and trajectories evolve rapidly due to frequent climbs, descents, and turns. Under these conditions, conflict geometry changes on timescales too short for human detection and response. To ensure safety, these operating conditions necessitate automated systems that continuously acquire, interpret, and use information to describe current and predicted future behavior of surrounding traffic.

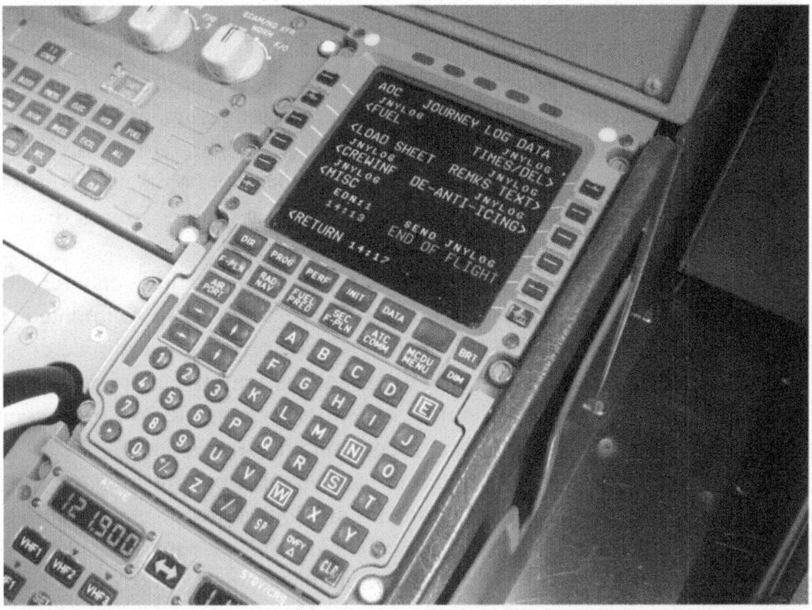

Figure 3.4 This multifunction control and display unit (MCDU) on a transport-category aircraft provides the flight crew with an interface to the FMS which supports precise trajectory conformance by automating lateral and vertical navigation along the programmed flight plan. *Source*: Christoph Paulus / Wikimedia Commons / Public domain.

Figure 3.5 *NASA's Global Hawk uncrewed aircraft* supports Earth science missions and operates with automated guidance that maintains stable trajectories for flights exceeding 24 hours at altitudes above 16 kilometers. *Source*: NASA Photo / Tony Landis / Public domain.

Cooperative coordination begins with the exchange of aircraft *state information* (position, velocity, attitude, climb or descent rates, and operational status indicators) and *intent information* (expected trajectory segments, upcoming turns, altitude changes, and waypoint arrival times). Operational status information includes flight-mode state, automation engagement, performance limitations, and contingency flags that affect maneuvering authority. With both forms of information, automated systems can project future interactions, identify conflict points, and adjust trajectories before violations occur. This capability transforms separation assurance from a reactive, human-centered process into a predictive, automation-mediated function.

The availability of shared state and intent information further enables higher-order flow-coordination tasks that extend beyond simple pairwise separation. Automated systems can maintain spacing within corridors, manage merging sequences, and coordinate crossing movements without heavy tactical input (Figure 3.6). When each aircraft behaves predictably relative to the others, the system shifts from tactical conflict resolution to strategic flow management.

Cooperative automation supports a distributed separation assurance model: aircraft and ground-based systems share state and intent information, and each aircraft uses onboard processing to conduct deconfliction and ensure safe spacing among them. This distributed process elevates human oversight to a higher supervisory level. With predictable, coordinated behavior, supervisors focus on anomaly detection, system-level monitoring, and managing off-nominal events instead of continuously issuing tactical instructions. This shift is essential to supporting AAM traffic densities and operational tempo.

3.2.3 Remote Supervision and the Ground-based Human Role

AAM introduces an evolved *supervised autonomy* operational model. In a supervised autonomy model, automated systems handle routine flight tasks, and human operators relinquish direct aircraft control and assume supervisory roles. This transition is driven by the need to maintain predictable system behavior at traffic densities that exceed the capacity of conventional pilot-based or controller-based tactical control. To achieve this, automation manages trajectory conformance,

Figure 3.6 A conceptual depiction of structured low-altitude corridors within an urban environment. Automated coordination enables multiple AAM aircraft to merge into shared routes, maintain consistent spacing, and navigate crossing or converging flows while operating among buildings and other airspace users. Courtesy of NASA / Public domain.

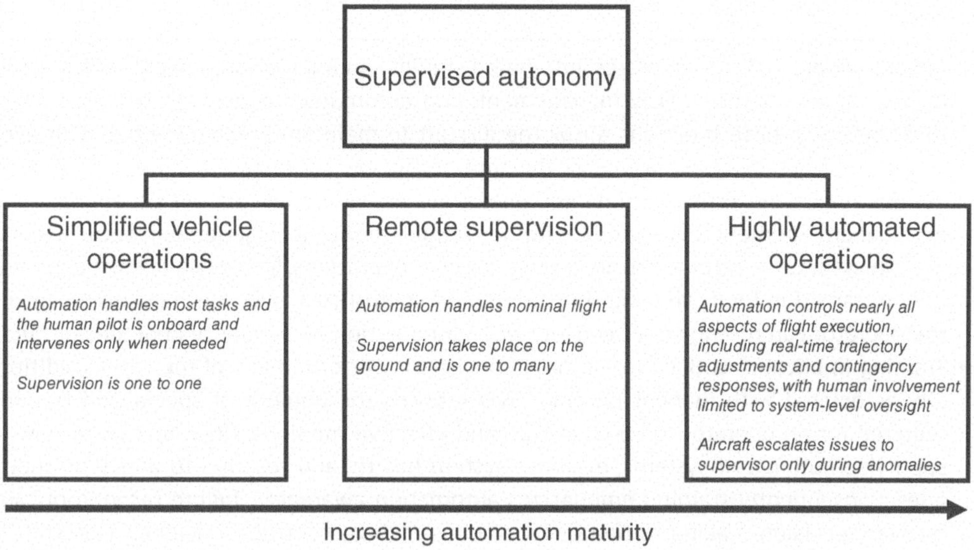

Figure 3.7 Configurations of supervised autonomy.

stability, and cooperative separation, leaving human supervisors to retain authority for oversight, authorization, and management of off-nominal events.

Supervised autonomy can be organized into three operational configurations: (1) simplified vehicle operations (SVO), (2) remote supervision, and (3) highly automated operations (Figure 3.7). Early deployment of AAM aircraft relies on onboard pilots executing under an *SVO* framework. The tasking of onboard pilots for AAM vehicles looks different than conventional crew actions under an SVO framework. Pilot workload is reduced by delegating routine tasks to automation. Under this framework, the pilot verifies conformance, authorizes transitions between major flight phases, and addresses anomalies that fall outside of normal operating parameters. Meanwhile, automated systems maintain stability, energy state, and adherence to planned trajectories (Figure 3.8). The breakout box explores the driving factors behind this early strategy.

What Is SVO?

SVO is a framework in which automated systems assume responsibility for routine flight tasks so that the human operator monitors the aircraft rather than manually managing every aspect of the flight. SVO does not reduce safety or capability. Instead, it shifts how aircraft are flown, certified, and integrated into the airspace system. The objective of SVO is to reduce pilot workload, improve operational consistency across the fleet, and create a predictable and stable foundation for future remote and autonomous operations.

- **Purpose of SVO:** Conventional aircraft require pilots to manage navigation, energy state, control inputs, and contingency procedures throughout the flight. AAM aircraft incorporate higher levels of automation to manage these tasks, reducing cognitive workload and allowing operators to focus on supervision, decision-making, and mission-level authorization. This shift supports safer operations in complex low-altitude environments and enables operational scalability as traffic density increases.

(Continued)

(Continued)

- **Core elements:** SVO incorporates automated stability augmentation, automated trajectory following, automated energy management, and automated sequencing between major flight phases. These functions allow the aircraft to maintain predictable behavior even in the presence of disturbances or dynamic traffic environments. The human operator verifies aircraft performance and intervenes only when abnormal aircraft situations or off-nominal environmental or traffic conditions require higher-level judgment. Because automation assumes control authority during nominal operations, SVO designs must explicitly define transition logic, alerting thresholds, and control-handover mechanisms to ensure the operator can re-engage effectively when automation performance degrades or fails.
- **Training implications:** Because automated systems perform many of the tasks traditionally associated with manual piloting, SVO reduces the amount of specialized training required for an operator to conduct safe and effective missions. Operators learn how to supervise automated systems, evaluate system health, and respond to alerts or anomalies. Consequently, training emphasizes automation awareness, failure recognition, and recovery decision-making in lieu of continuous manual control proficiency. This shift in responsibility supports the development of a broader operational workforce and aligns with the long-term transition toward remote supervision.
- **Pathway to remote and autonomous operations:** SVO creates a structured method for introducing automation into crewed aircraft and provides the operational data, safety evidence, and human–system interaction experience necessary for later phases of remote and autonomous flight. As automated functions mature and demonstrate reliable performance, the role of the human operator shifts from an onboard pilot to a remote supervisor. SVO is an incremental step toward scalable AAM operations in which many aircraft are overseen by a smaller number of trained personnel. This progression requires careful management of operator trust, ensuring reliance on automation remains appropriate under both nominal and degraded conditions.
- **Airspace integration benefits:** Predictable automated behavior supports strategic flow management in dense low-altitude environments. When aircraft adhere to automated flight profiles, trajectory conformance improves, merging and spacing become more consistent, and the burden on controllers and supervisors is reduced. SVO enhances both individual aircraft reliability and the collective performance of the airspace system. At the same time, system-level procedures must account for rare but consequential automation failures so that individual recovery actions do not propagate instability into the surrounding traffic flow.

Why Pilots for Early AAM Deployments?

Early AAM operations include an onboard pilot. This is not because of limitations in automated capability. Rather, it is driven by risk mitigation during early operational testing, certification and operational frameworks that govern the initial introduction of AAM aircraft, and societal considerations surrounding automated flight. An onboard pilot provides an immediate safety backstop during system maturation, allowing operators and regulators to observe real-world performance, validate automation behavior, and manage unforeseen conditions during early service entry. These factors guide how autonomy enters service and determine the role of the human operator during early phases of AAM integration.

- **Regulatory assumptions and near-term operational viability:** Existing airworthiness and operational requirements for crewed aircraft assume a human pilot will be in the cockpit. Removing the pilot requires new certification pathways for autonomous functions, automated detect-and-avoid capability, and remote supervision configurations, among others, that are still under development. As a result, piloted aircraft can be certified and operated under existing frameworks. This accelerates deployment while providing a controlled environment for evaluating automation performance under operational conditions and gathering the data required for future approval of increasingly autonomous operations.
- **Safety demonstration requirements:** Autonomous flight introduces new failure modes that must be addressed through verified automation logic, demonstrated fallback behaviors, and reliability evidence that satisfies the safety margins historically associated with human pilots. During early deployments, the onboard pilot mitigates operational risk by managing off-nominal conditions, responding to unexpected system behaviors, and ensuring safe continuation or termination of flight during automation anomalies. Establishing confidence in these systems requires substantial operational data that is most effectively obtained through initial piloted operations.
- **Integration with current airspace systems:** Contemporary air traffic procedures rely on pilot participation in communication, compliance, and tactical response. Early AAM deployments align with these expectations by including an onboard pilot. This approach simplifies near-term integration into established airspace workflows and reduces the need for immediate changes to air traffic procedures or supervisory roles during initial service introduction.
- **Public acceptance and responsibility structures:** Introducing a new category of autonomous aircraft requires public confidence and clearly defined responsibility for decision-making. Early piloted operations help build acceptance, demonstrate reliability, and reduce barriers to initial service. This approach supports gradual trust-building as automation assumes greater authority over time.

Remote supervision expands the SVO model by removing the pilot from the cockpit and empowering a ground-based operator to oversee multiple aircraft through an interface presenting aircraft state, trajectory adherence, and system health, along with airspace and environmental compatibility. Ground-based supervisors rely on detailed situational displays generated from shared state and intent information (Figure 3.9). These displays present each aircraft's planned trajectory, conformance status, environmental conditions, and expected interactions with traffic flows. For this supervisory model to function effectively, operators must maintain calibrated trust in the automation. This trust is derived from confidence that automated systems perform reliably within their defined authority and from operator awareness of the limits of that authority. Trust is reinforced through consistent system behavior, transparent alerting logic, and predictable escalation of off-nominal conditions.

Automated systems maintain separation, evaluate interactions with surrounding traffic, and execute planned trajectories. They also provide alerts when predicted trajectories diverge from

Figure 3.8 Archer Aviation Midnight eVTOL mockup cockpit. The streamlined interface showcases how AAM aircraft rely on automated systems to reduce operator workload in complex low-altitude environments. Courtesy of the Vertical Flight Society.

expected behavior or when aircraft encounter conditions that require supervisory attention. Clear delineation between automated responsibilities and supervisory intervention thresholds ensures that operators remain engaged at the appropriate level without overreliance on automation or unnecessary manual involvement.

The role of the remote supervisor is to confirm mission readiness, approve major flight transitions, and monitor overall system performance. This role includes verifying that each aircraft remains within its intended operational envelope, intervening when off-nominal conditions arise, and coordinating with airspace service providers (ASPs). This *human-on-the-loop* configuration,

Figure 3.9 Wisk Aero's fleet operations center (FOC) is the nerve center for its autonomous vehicles and where aircraft are monitored by multi-vehicle supervisors (MVSors). Courtesy of Wisk Aero.

where the operator monitors and authorizes versus continuously commanding, ensures that supervisors maintain awareness of system status without needing to issue continuous tactical instructions.

In advanced remote supervision AAM operations, the human role becomes increasingly strategic. Supervisors oversee fleet-level behavior (e.g., monitoring aggregate conformance to planned trajectories, spacing performance across corridors, and adherence to scheduled departure and arrival slots), manage contingency procedures, and coordinate with ASPs to maintain compatibility with flow-management strategies. Individual aircraft maintain safe separation and trajectory conformance through automated functions, with the supervisor focused on identifying exceptions that require human judgment. This model enables a single supervisor to manage several aircraft simultaneously without degrading safety margins because each aircraft provides predictable, automated behavior within a shared and coordinated airspace environment. Highly automated operations, where human intervention becomes infrequent, will be discussed in Chapter 9 as part of long-term AAM evolution.

The transition from onboard piloting to remote supervision does not diminish the importance of human oversight. Instead, it reallocates human attention from continuous control to system-level decision-making. Remote supervision provides scalability by allowing a smaller number of trained personnel to oversee many aircraft, provided that those aircraft maintain predictable automated behavior and adhere to structured information exchange standards. This supervisory model is fundamental to scaling AAM integration and supports the traffic density and operational tempo characteristics of urban and regional environments.

3.2.4 Strategic Flow Management for AAM Operations

AAM introduces high concentrations of low-altitude traffic with frequent departures and arrivals, short trip lengths, and highly distributed origins and destinations. Under these conditions, tactical separation alone is insufficient for maintaining efficiency and safety under these conditions. Instead, the system must rely on *strategic flow management*, where aircraft behavior is organized in advance to prevent conflicts.

This type of strategic coordination is supported by ASPs. ASPs manage flows, implement operational constraints within established frameworks, and coordinate trajectories across the airspace network to maintain compatibility well before flights commence. In the AAM ecosystem, ASPs may include a traditional air navigation service provider (ANSP) as well as new service entities.

Regardless of the specific service provider, the effectiveness of strategic flow management depends on precise trajectory generation and consistent adherence to planned routes. Automated FMSs maintain lateral and vertical profiles that align with shared airspace procedures. When each aircraft maintains consistent conformance to its assigned trajectory, the aggregate system becomes more manageable, and ASPs can coordinate flows across structured corridors and their intersections without real-time controller intervention. However, strategic flow management depends on the availability of shared state and intent information over extended planning horizons, not solely on real-time separation logic. When aircraft provide consistent trajectory intent and conformance data, automated systems can evaluate interactions across time horizons that extend beyond immediate separation distances. By comparing planned trajectories, automated systems can detect future conflicts and adjust trajectories proactively. This predictive approach transforms airspace behavior from reactive deconfliction to coordinated flow management and is essential for supporting the operational cadence of AAM.

Strategic flow management employs mechanisms such as time-based metering, trajectory-based scheduling, and corridor access management (Figure 3.10). Automated systems maintain spacing along shared routes by assigning required times of arrival (RTA), adjusting speed profiles, and synchronizing climb or descent rates with upstream schedules. When aircraft merge into a common corridor, algorithms determine sequence order and assign conflict-free trajectories through time-deconflicted slot allocations.

Environmental conditions also influence strategic planning. As we will explore in greater depth in Chapter 7, the low-altitude operational environment is composed of microclimates where environmental characteristics, such as localized wind fields, can change sharply across short spatial

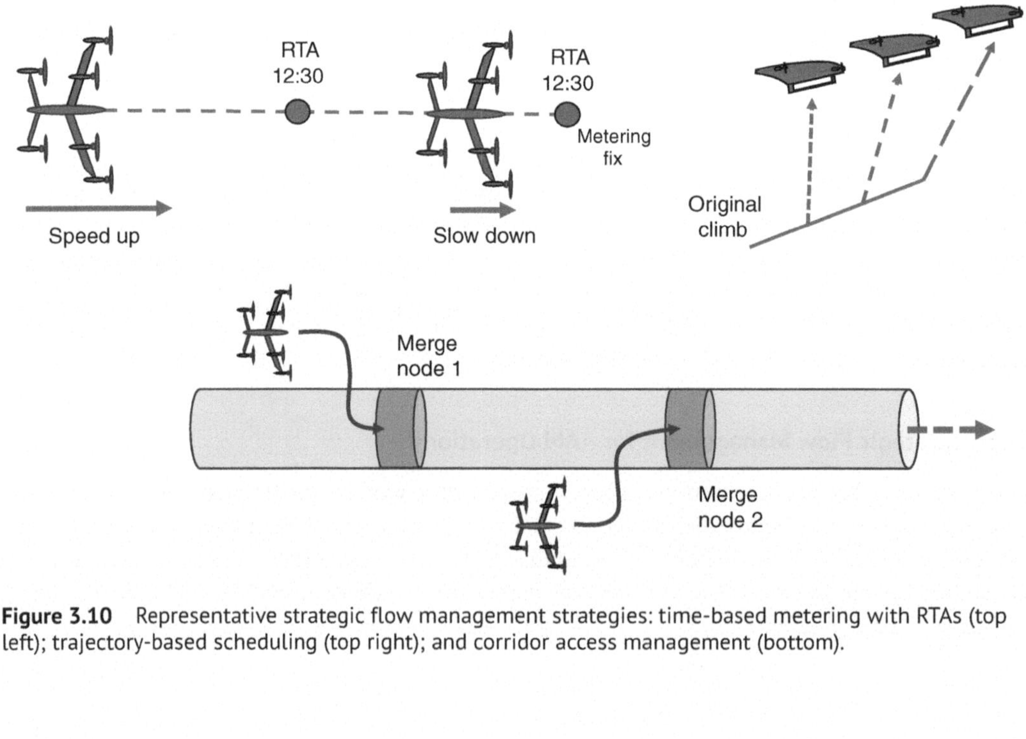

Figure 3.10 Representative strategic flow management strategies: time-based metering with RTAs (top left); trajectory-based scheduling (top right); and corridor access management (bottom).

and temporal scales. However, strategic planning can incorporate forecast and statistical representations of these environmental factors and account for them during route selection, altitude assignment, and timing coordination within the maneuvering volume. This integration improves safety margins and operational regularity and reduces the likelihood of degraded ride quality during low-altitude flight.

Strategic flow management improves efficiency by significantly reducing the frequency with which tactical adjustments are required. Automated conformance, cooperative information exchange, and structured routing allow most conflicts to be resolved before they arise. Tactical adjustments become necessary when aircraft deviate from expected behavior, encounter off-nominal conditions, or experience onboard anomalies. Such events are escalated to supervisory attention. Under a strategic flow-management architecture, most routine flow behaviors are managed by automation. This balanced approach allows the airspace system to benefit from the scalability that automation provides while still preserving the judgment of remote supervisors and ASPs.

3.3 Urban Airspace: Challenges and Engineering-based Solutions

Urban environments present a fundamentally different operating environment from the mid-altitude and en route airspace that contemporary aviation was designed to function within. In these settings, AAM aircraft operate at altitudes where surface features, built structures, and environmental variability exert a direct influence on flight behavior. This is especially the case in urban low-altitude airspace. The physical geometry of the urban environment, the performance of surveillance and sensing systems, and localized atmospheric effects collectively constrain and complicate the maneuvering volume available to the aircraft. Because these factors shape how airspace is structured and managed, they impose constraints that must be addressed explicitly in the design of urban AAM operations.

3.3.1 The Physical and Operational Characteristics of Urban Low-altitude Airspace

Urban airspace contains numerous obstacles that can include both static and dynamic features, such as buildings, towers, temporary structures like cranes, bridges, and elevated transportation infrastructure. These structures can fashion narrow corridors, limit the space available for climbs or descents, and, in general, confine maneuvering space. High obstacle density reduces the feasibility of large avoidance maneuvers, limiting diversion and contingency options. This places an increased emphasis on reliability and tightly bound emergency procedures.

Low-altitude urban operations take place in an environment where sensing and surveillance systems experience degraded performance due to signal occlusion, multipath reflections, and lighting and weather conditions that degrade perception and navigation accuracy (Figure 3.11). GNSS accuracy can degrade when signals reflect from building surfaces (multipath effects). Further, line-of-sight blockages (signal occlusion) associated with dense infrastructure can interrupt GNSS reception, ADS-B message reception, or degrade ground-based surveillance and tracking system coverage. Vision-based sensing, Light Detection and Ranging (LiDAR), and other range- and image-based onboard systems may also experience reduced range or reliability within dense metropolitan settings due to cluttered backgrounds, reduced sightlines, and highly reflective surfaces that interfere with feature extraction and ranging performance. Table 3.1 breaks down currently implemented and feasible sensing modalities for AAM airspace integration. Recent advances in vision-based perception and trajectory optimization improve local awareness and path feasibility.

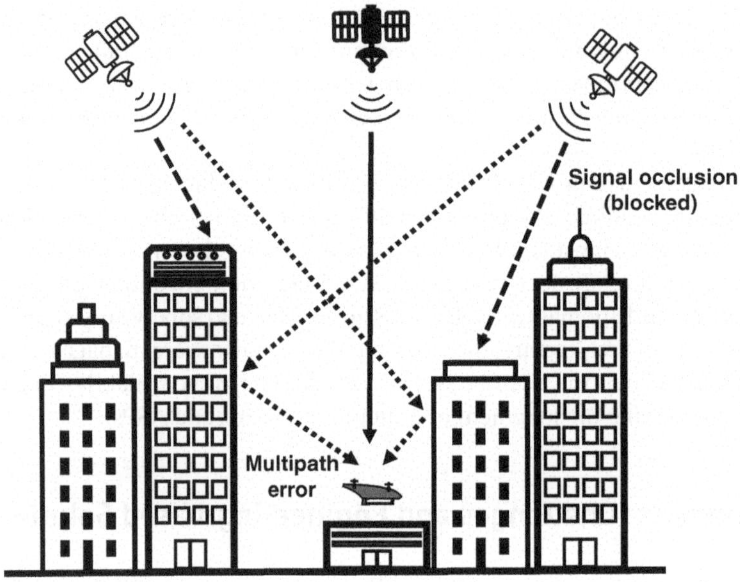

Figure 3.11 Signal occlusion and multipath effects.

Table 3.1 Onboard sensing modalities for AAM airspace integration.

Sensor type	Primary function in AAM airspace integration	Strengths	Limitation in low-altitude urban airspace
Vision-based cameras (monocular or stereo)	Supports obstacle detection, detect-and-avoid functions, and relative positioning near infrastructure	High spatial resolution, low mass, and useful for classification of objects and structures	Performance degrades in low light, glare, precipitation, or environments with significant visual clutter
LiDAR	Provide high-precision range measurements for obstacle detection, vertiport approach, and surface hazard detection	Accurate depth information and strong performance during approach and landing, where fine resolution is needed	Range and reliability degrade in fog, heavy precipitation, and complex reflective urban surfaces
Compact airborne radar (millimeter-wave or small phased array)	Supports detect-and-avoid functions, provides situational awareness in degraded visual environments, and enables monitoring of weather cells	Robust performance across lighting and weather conditions with reliable detection capability in degraded environments	Urban structures introduce clutter and multipath reflections that increase processing complexity; mass and power requirements are higher than those of optical sensors
Integrated GNSS and IMU	Provide precise state estimation needed for trajectory conformance, guidance, and automated sequencing	Integration provides high update rates and short-term inertial bridging during GNSS degradation or signal loss	GNSS multipath reflections and occlusion occur near buildings; IMU errors drift unless periodically corrected with external updates

Sensor type	Primary function in AAM airspace integration	Strengths	Limitation in low-altitude urban airspace
Infrared or multispectral cameras	Enhance object detection and situational awareness during low-visibility or nighttime operations	Improved performance in low-light environments and complementary to visible-spectrum cameras	Certification and operational pathways are still evolving; high cost and integration complexity limit adoption
Ultrawideband (UWB) ranging systems	Support precise localization for vertiport operations, gate docking, or approach corridors	High accuracy at short range and strong resistance to multipath	Limited range restricts usefulness outside terminal environments; limited standardization and few aviation-grade implementations to date
Acoustic sensing arrays	Provide supplementary detection of rotorcraft and sUAS in close proximity	Low mass and potential for passive detection without active emission	Strongly degraded by urban noise; limited directional accuracy in complex urban acoustic environments

These methods operate most effectively when embedded within cooperative surveillance architectures and structured routing frameworks that bound uncertainty and support shared situational awareness. Unfortunately, no single sensing modality provides completely reliable performance across all lighting, weather, and geometric conditions. Moreover, urban settings amplify several of these effects. Cooperative surveillance architectures thus become essential for maintaining continuous situational awareness. Although onboard sensing provides critical perception, urban AAM operations increasingly depend on distributed sensing infrastructure that supplements onboard observations with external surveillance, environmental monitoring, and precision navigation data. The associated limitations of these sensors within the low-altitude urban environment increase reliance on cooperative surveillance architectures, where aircraft share state and intent information and ground systems furnish supplemental situational data.

Localized wind fields created by building-induced turbulence, channeling, and thermal gradients can influence AAM climb profiles, approach paths, and hover performance (Figure 3.12). These effects reduce available performance margins and require automation capable of maintaining stable aircraft behavior in the presence of small-scale disturbances. The physics behind these microclimate effects is examined comprehensively in Chapter 7.

3.3.2 Structured Routing and Corridors

Structured routing provides a method for organizing AAM traffic throughout low-altitude airspace (Figure 3.13). This becomes especially important within the confined geometry of urban environments. By defining bounded lateral and vertical pathways, aircraft movements become constrained to predictable volumes, and the likelihood of uncoordinated trajectories intersecting in complex airspace is reduced. These routing structures support predictable vehicle behavior, enable the use of cooperative automation for spacing and sequencing, and limit the number of tactical interactions that would otherwise arise from free-maneuvering operations. As traffic density increases, structured routing continues to optimize the throughput and efficiency of low-altitude airspace by enabling network-level forecasting, sequencing, and systematic deconfliction across the network.

Figure 3.12 An urban canyon viewed across the East River from Roosevelt Island in New York City. *Source*: Jim.henderson / Wikimedia Commons / Public domain.

Corridor design requires balancing airspace efficiency with safety margins that account for navigation accuracy, obstacle clearance, and the need to support varying flow directions. Route placement is also influenced by land-use considerations, including the avoidance of critical infrastructure, sensitive facilities, and dense population concentrations that can impose lateral offsets or altitude constraints on corridor geometry. Corridor width is informed by navigation performance requirements, including the containment tolerances associated with RNP values, as well as obstacle-clearance criteria and lateral buffers necessary to accommodate opposing or multilane traffic. Vertical separation introduces an additional design degree of freedom by assigning distinct altitude layers to different directions of travel or to flows with varying performance characteristics. This vertical structuring also enables intentional vertical or lateral offsets where obstacle environments, built structures, or land-use constraints limit available horizontal space. Collectively, lateral and vertical structures can help create predictable and high-capacity pathways within the confines of low-altitude airspace.

Figure 3.13 Corridors define bounded lateral and vertical pathways that reduce the likelihood of uncoordinated trajectories intersecting and increase the efficiency of low-altitude airspace. *Source*: Federal Aviation Administration / Public domain.

Energy efficiency must also be considered in corridor design. Battery-powered vehicles experience significant power demands when not in wingborne flight, particularly during climb, hover, and low-speed maneuvering, and available performance margins couple directly to state of charge. Therefore, corridor altitudes, gradients, and merging geometries need to minimize extended low-altitude holding and reduce unnecessary altitude changes. Strategic flow management supports these objectives by metering traffic so that aircraft remain in energy-efficient segments of flight for as long as practicable instead of accumulating in power-intensive terminal stacks.

Corridor architectures can include one-way, bidirectional, and multilane route configurations, along with vertically separated layers depending on expected demand and local constraints. A one-way corridor simplifies merging and spacing because all aircraft simply follow the same direction of travel (Figure 3.14). A bidirectional corridor reduces the collective airspace footprint by consolidating traffic into a shared structure, but it requires larger lateral buffers between opposing flows to preserve safe separation. Multilane or vertically tiered geometries allow additional throughput where demand warrants it, provided that each lane satisfies containment and obstacle-clearance requirements and that lane-changing or merging maneuvers occur in defined transition zones.

Integrating structured routing with time-based flow management further enhances capacity. Strategic metering, sequence assignment, and slot scheduling allow aircraft to enter corridors at predetermined intervals that maintain stable spacing throughout each route segment. Trajectory-compatibility assessments verify that each planned lateral and vertical profile fits within corridor constraints and does not disrupt upstream or downstream flows. When merging into a corridor, automated systems compute sequence order, timing separation, and trajectory alignment before the merge point, reducing the need for tactical adjustments once aircraft enter the stream.

Corridor Merging Design Principles:

- *Sequence ordering* assigns each aircraft a specific position in the merge queue.
- *Time-based slot scheduling* provides aircraft with RTAs at the merge point to ensure temporal separation.
- *Trajectory-compatibility assessment* verifies that each aircraft's planned lateral and vertical flight profile fits within corridor constraints as it approaches and enters the merge point, without disturbing upstream or downstream flows.

Corridor availability may vary as a function of demand, weather, or local constraints. Dynamic activation states allow certain corridors or altitude layers to be enabled or disabled in response to changing traffic loads, environmental conditions, or temporary urban restrictions. This flexibility supports adaptive flow management and prevents unnecessary reservation of airspace that would otherwise reduce capacity for other users. Therefore, dynamic corridor management complements strategic planning by ensuring that routing structures align with real-time operational needs. Through this blend of geometric structure, performance-based containment, time-based sequencing, and demand function, corridors provide an organizing framework that allows AAM traffic to scale while maintaining precise and bounded trajectory behavior in complex urban airspace.

Finally, corridor design requires explicit provisions for contingency scenarios. Degraded performance, onboard anomalies, or deviations from expected behavior must not disrupt the primary flow. Dedicated escape volumes, diversion paths, or holding positions that account for local obstacle environments and available emergency landing options provide aircraft with safe alternatives during off-nominal conditions without compromising the stability of the main corridor network.

AAM corridor design requires a structured engineering approach. The heuristics below summarize the principles that support safe, predictable, and scalable corridor networks in low-altitude urban environments.

Corridor design heuristic	Design basis
Navigation performance defines corridor containment	RNP values and obstacle-clearance margins determine the lateral and vertical boundaries required for predictable automated tracking
Obstacle geometry constrains corridor orientation	Building height, spacing, and layout limit feasible alignments and influence minimum safe corridor altitudes
Multilane architectures require high-precision automation	Tight conformance reduces cross-lane interference and increases throughput under automated separation
Vertical segmentation expands capacity	Layered corridors support independent flow streams while maintaining safe vertical separation
Speed harmonization reduces conflict frequency	Stable speed envelopes minimize closing rates, decreasing the need for tactical adjustments
Time-structured merging ensures orderly integration	Sequence ordering and slot scheduling prevent conflicts at corridor entry points
Trajectory-based scheduling supports spacing regularity	RTAs and synchronized climb/descent profiles maintain consistent longitudinal spacing
Dynamic activation preserves safety and efficiency	Corridor status adapts to microclimate behavior, GNSS integrity, and demand patterns
Environmental characteristics influence corridor placement	Microclimates influence locations where stable and predictable aircraft performance can be maintained
Vertiport-linked geometry governs endpoints	Departure and arrival paths require alignment with corridor structure and local obstacle constraints
Contingency management requires defined escape volumes	Diversion volumes ensure that anomalies do not destabilize primary flows

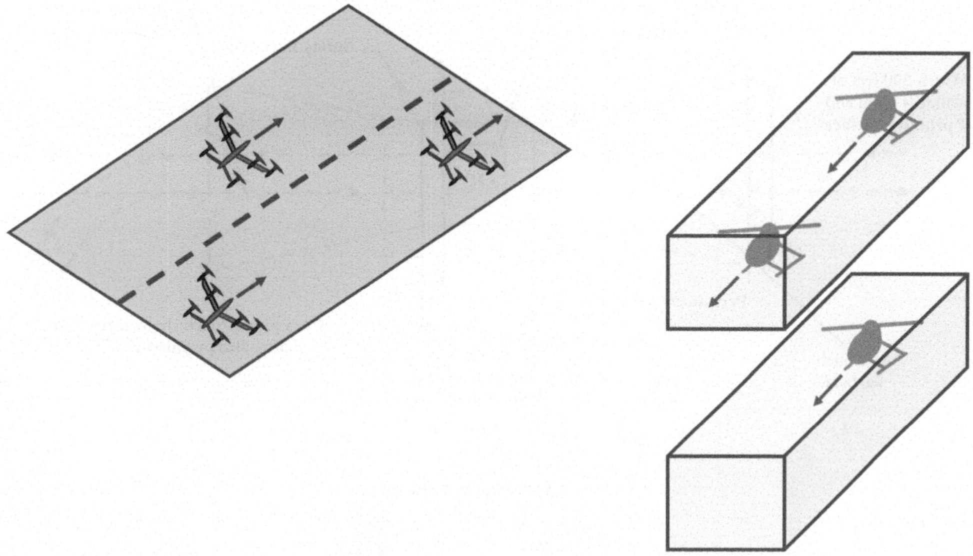

Figure 3.14 Multilane and vertically segmented corridor configurations.

3.3.3 Vertiport-linked Airspace Geometry

Vertiports function as the primary interface between surface transportation systems and the structured routing network that supports AAM operations. The airspace surrounding vertiports must facilitate reliable transitions between ground operations, vertical maneuvering, and entry into or exit from low-altitude corridors. To enable this, the terminal-area environment must allow dense AAM traffic to climb, descend, and sequence safely while near terrain, buildings, and other airspace users. Time-based sequencing or slot allocation becomes essential at vertiports with frequent departures and arrivals because both flows compete for limited maneuvering volume immediately above the site.

The geometric relationship between a vertiport and its surrounding airspace determines how aircraft enter, transit, and exit the local environment. As a result, route placement, climb and descent profiles, turn radii, and obstacle-clearance margins are constrained in the ways that link terminal operations to the nearest corridor nodes. AAM vehicles can transition between hover, low-speed maneuvering, and wingborne flight over comparatively short distances, and this enables the use of highly structured arrival and departure paths. However, the confined geometry around many vertiports still limits where these transitions can occur and how sharply a path can turn or climb in order to maintain obstacle clearance. Consequently, feasible path shapes must emerge from the intersection of vehicle performance and the spatial constraints imposed by the built environment. Because of this, vertiport siting and airspace structure must be coupled in the design process.

A central element of vertiport-linked geometry is a protected maneuvering volume that extends outward and upward from the touchdown and lift-off (TLOF) area. This volume ensures obstacle clearance via protective surfaces that bound expected trajectories and allowable deviations during arrival and departure. These protected areas resemble the obstacle-limitation or imaginary surfaces historically defined around helicopter facilities. Although terminology varies across regulatory frameworks, these surfaces share a common purpose of limiting obstacle penetration

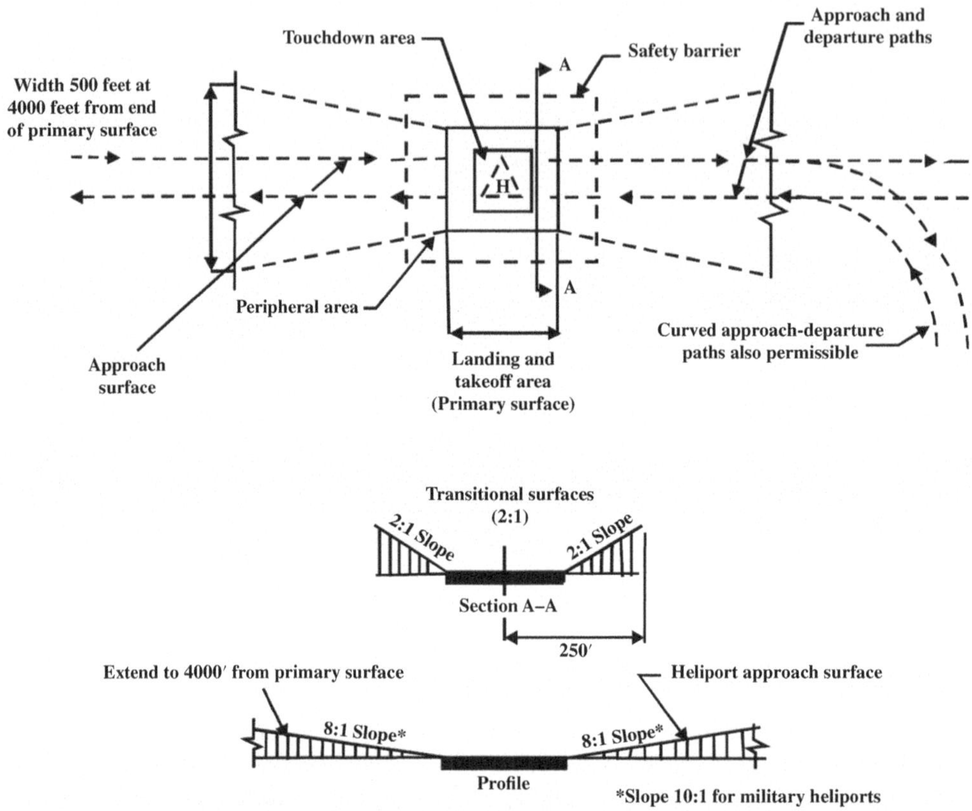

Figure 3.15 Representative imaginary surfaces defining obstacle-clearance volumes for heliport arrival and departure operations. The depicted horizontal and sloping surfaces illustrate a commonly used geometric approach for limiting obstacle penetration near vertical flight facilities. *Source*: Federal Aviation Administration / Public domain.

into approach and departure volumes. A representative example of a heliport imaginary surface is shown in Figure 3.15, illustrating horizontal extents that constrain obstacle height and sloping (or conical) boundaries that preserve a minimum vertical clearance margin. Such surfaces prevent physical intrusion into the terminal maneuvering volume. AAM operations, however, require more precisely defined three-dimensional flight paths so that automated systems can sequence, meter, and deconflict traffic at scale and do not rely on a broad protected region alone.

High-density automated operations depend on well-organized flows and predictable trajectories. Departure and arrival paths must account for the acceleration, deceleration, and mode-transition behavior of AAM vehicles within short distances, while connecting to the corridor network in a manner that maintains required spacing, minimizes conflict points, and avoids unnecessary altitude changes (Figures 3.16 and 3.17). The available maneuvering space around each vertiport determines the steepness of the initial climb, the radius of any required turns, and the altitude at which an aircraft can safely merge with a corridor. In dense metropolitan settings, nearby buildings can narrow these protected volumes and may require that an ideal departure (or arrival) path instead be vertically stacked or laterally offset to maintain a safe, continuous linkage to the en route structure.

Arrivals follow a complementary geometry. Aircraft must enter the terminal region along a stabilized descent path that maintains sufficient clearance from surrounding obstacles. These

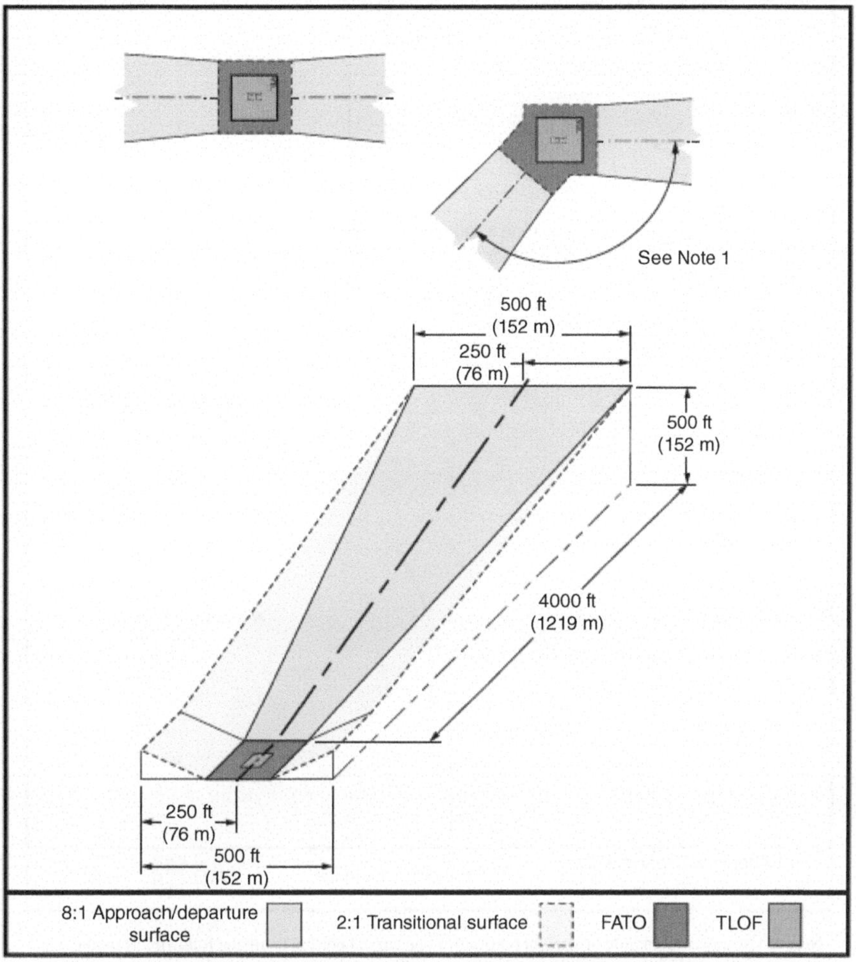

Note 1: The preferred approach/departure surface is based on the predominant wind direction. Where a reciprocal approach/departure surface is not possible in the opposite direction, use a minimum 135° angle between the two surfaces.

Figure 3.16 A representative visual approach and departure surface from FAA EB 105A [1]. The final approach and takeoff area (FATO) and TLOF are discussed in Chapter 4. *Source*: Federal Aviation Administration / Public domain.

requirements affect the placement of arrival gates and the lateral and vertical spacing between approach paths when a vertiport supports multiple directions of entry. The approach geometry must also support consistent trajectory tracking so that sequence order and timing can be maintained without relying on last-moment tactical interventions.

Vertiport-linked airspace may need to function dynamically rather than through a fixed, always-active configuration. Activation of specific arrival or departure paths may depend on time-of-day demand, corridor loading, local environmental conditions, or temporary local restrictions. This adaptability reduces the need for large, permanently reserved airspace volumes and supports efficient flow management during peak periods.

Representative engineering guidance for vertiport arrival and departure geometry is available in design circulars and technical specifications from the United States, Europe, the United Kingdom, Australia, Japan, and India [1–6].

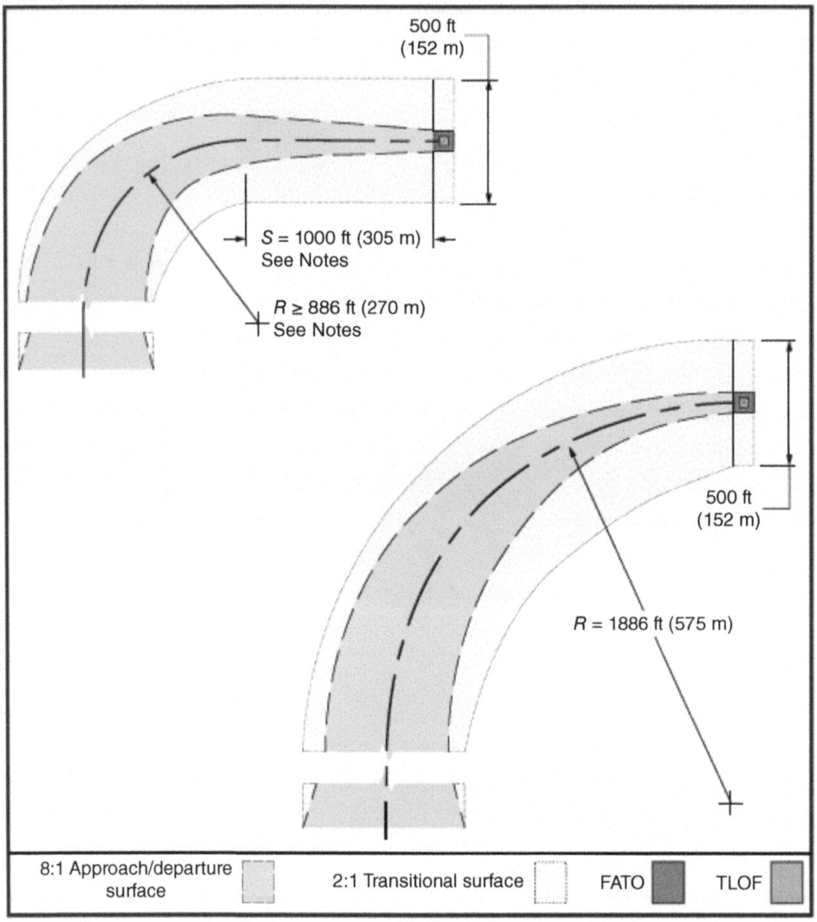

Note 1: Use any combination of straight portions of one curved portion using the following formula: $S + R \geq 1886$ ft (575 m) and R 886 ft (270 m), where S is the length of the straight portion(s) and R is the radius of the turn. Note that any combination > 1886 ft (575 m) will work.

Note 2: The minimum total length of the centerline of the straight and curved portion is 4000 ft (1219 m).

Note 3: VTOL takeoff performance may be reduced in a curve. Consider a straight portion along the takeoff climb surface prior to the start of the curve to allow for acceleration.

Figure 3.17 Representative curved visual approach/departure and transitional surfaces from FAA EB 105A [1]. *Source*: Federal Aviation Administration / Public domain.

3.4 Communication, Navigation, Surveillance, and Separation Assurance

Sustaining high-density AAM operations within constrained low-altitude environments requires a tightly coupled CNS architecture. Unlike mid-altitude traffic management, where long line-of-sight distances, broad maneuvering volumes, and moderate aircraft counts allow tactical control to absorb variability, AAM requires continuous trajectory predictability and cooperative separation behavior among large numbers of vehicles. This compels separation assurance to function as a distributed engineering process as opposed to a controller-centric one.

3.4.1 Communication and Information Exchange

AAM places stringent demands on communication links because separation, sequencing, and conformance monitoring depend on the reliable exchange of both state and intent information (Figure 3.18). Together, these elements form the basis for cooperative decision-making among aircraft and for supervisory oversight by ground personnel. Reliable communication must support continuous updates at predictable intervals. This allows onboard automation to evaluate conflict geometry, assess conformance, and adjust speed or timing with minimal latency. Any loss of continuity increases uncertainty in separation calculations and degrades the stability of the broader flow-management architecture, particularly in dense low-altitude environments where interaction timescales are short. For this reason, communication performance establishes the practical limits on achievable traffic density and the degree to which separation assurance can function as a distributed process that does not depend on controller-centric intervention.

Many environments require hybrid communication strategies that combine direct vehicle-to-vehicle (V2V) links with vehicle-to-ground (V2G) paths routed through ASPs. Because urban operations experience signal occlusion from buildings and transportation infrastructure, a robust CNS design incorporates redundant pathways, including frequency diversity, heterogeneous link types, and cooperative V2V and V2G channels that sustain continuity of critical information when a primary channel is degraded. This robustness is essential as the communication segment of the CNS architecture forms the backbone for both onboard autonomy and airspace-wide coordination.

3.4.2 Navigation Performance and Trajectory Conformance

AAM routing depends on navigation systems that sustain precise lateral and vertical performance, even when GNSS measurements degrade due to multipath, occlusion, or atmospheric effects. Navigation performance must satisfy the containment requirements of structured corridors and terminal-area approach and departure geometries. To achieve this, AAM vehicles rely on

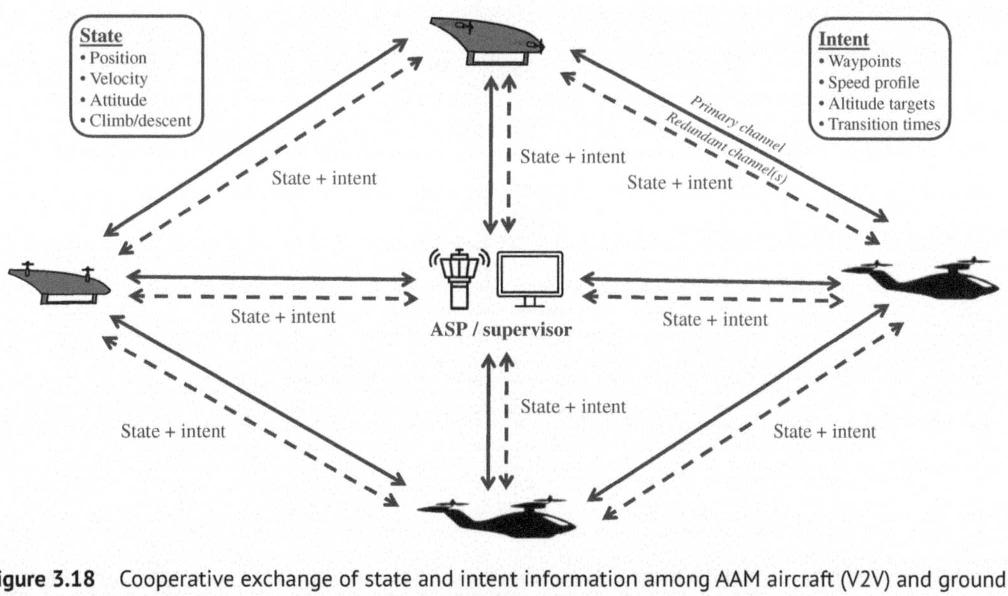

Figure 3.18 Cooperative exchange of state and intent information among AAM aircraft (V2V) and ground services (V2G) supporting communication redundancy and separation assurance.

tightly integrated navigation solutions that fuse GNSS with inertial measurement units (IMU). This integrated solution can then be augmented with LiDAR, radar altimetry, and vision-based estimation when available. These heterogeneous measurements are combined through advanced state-estimation filters, such as extended Kalman filtering [7, 8], to maintain a stable and accurate navigation solution even when individual sensors experience temporary degradation (Figure 3.19).

Trajectory conformance is foundational for scalable separation assurance. When each aircraft maintains its assigned path, altitude, and time profile within narrow tolerances, conflict geometry becomes predictable across the entire flow network. This predictability enables automated scheduling functions, such as time-based metering and merging, and allows supervisors to manage system performance without issuing continuous tactical inputs. Conversely, deviations beyond containment limits propagate uncertainty downstream and destabilize merging sequences, spacing strategies, or arrival slot allocations.

3.4.3 Surveillance and Cooperative Sensing

Surveillance in an AAM environment cannot rely on a single sensor modality. Conventional ADS-B and radar experience degradation in dense metropolitan regions due to occlusion, reflective surfaces, and limited sightlines. Vision-based sensing, LiDAR, compact radar, and multispectral systems each provide partial coverage under specific environmental conditions but suffer performance degradation under others. As a result, surveillance is achieved through a cooperative architecture where aircraft share their own state and intent data, and ground-based infrastructure provides independent observation pathways, including ground ADS-B receivers, multilateration networks, limited-range radar, and fixed optical or radio frequency (RF) sensors. These independent inputs supply additional geometry, redundancy, and continuity when onboard sensing or communication channels degrade, creating a more resilient surveillance picture for separation assurance.

Cooperative surveillance reduces the system's dependence on any individual sensing modality and creates a more stable traffic picture in environments where onboard sensors experience

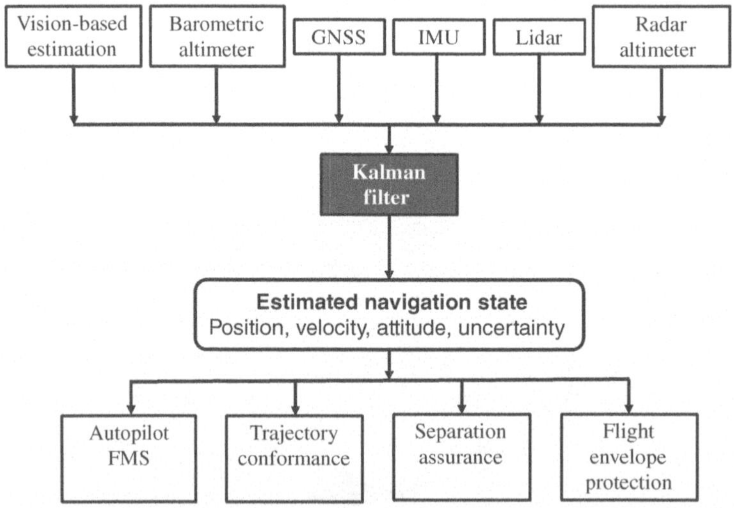

Figure 3.19 Sensor-fusion architecture for AAM navigation. A Kalman filter fuses heterogeneous sensor inputs to produce a stable navigation state used for trajectory conformance, separation assurance, and automated flight control.

intermittent degradation. By fusing independent measurements from both the air and the ground, state-estimation error is reduced. This enables automated separation algorithms to evaluate future conflicts with greater confidence.

Surveillance performance also directly influences the minimum allowable spacing between aircraft. Higher uncertainty in state estimation increases the protective buffers required for safe separation and, in turn, reduces airspace capacity. Higher-fidelity surveillance achieved through cooperative architectures and redundant sensing pathways reduces required separation margins and increases throughput.

3.4.4 Distributed Separation Assurance

Separation assurance in AAM operations is maintained through a distributed framework where each aircraft contributes to conflict detection and resolution using locally processed state and intent information. The aircraft's onboard automation evaluates predicted interactions with surrounding vehicles by propagating trajectories forward in time and assessing whether separation minima will be violated (Figure 3.20). If potential conflicts are detected, automated logic adjusts speed, climb rate, or sequencing timing to maintain spacing. Spacing, in this context, is treated as a dynamically managed buffer that reflects navigation accuracy, surveillance uncertainty, vehicle performance limits, and corridor containment assumptions. Supervisory oversight focuses on monitoring aggregate system behavior and managing anticipated contingencies, such as resolving schedule disruptions, reallocating arrival or departure slots, or adjusting corridor access when aircraft deviate from planned timing or performance envelopes.

This distributed model allows separation assurance to scale with fleet size. Instead of a single controller managing multiple aircraft, each aircraft performs local computations consistent with shared rules for spacing, right-of-way, and conflict-avoidance maneuvers. Supervisors intervene at the system level when coordinated adjustments are required across multiple aircraft or when predefined operational limits restrict automated resolution authority.

Figure 3.20 Three Beta CX300 aircraft flying within close proximity. Courtesy of Beta Technologies.

Uncertainty modeling is essential for this process. Conflict detection algorithms propagate prediction envelopes that reflect expected navigation error, communication latency, and environmental disturbances. Resolution strategies are selected to ensure that the envelopes of conflicting trajectories no longer intersect. When uncertainty grows beyond allowable bounds due to sensor degradation, communication outages, or unexpected atmospheric effects, fallback behaviors initiate predetermined maneuvers that maintain separation until normal operations can resume or supervisory intervention occurs. This architecture demonstrates how separation assurance is the integrated output of communication reliability, navigation accuracy, surveillance fidelity, uncertainty quantification, and onboard automation. The performance of this separation assurance framework determines the achievable density of AAM operations and the required structure of corridors, merging points, and terminal approach geometries.

3.5 Summary

Integrating AAM vehicles into busy shared airspace requires an airspace system that extends the core principles of contemporary ATM while supporting high operational tempos, dense traffic concentrations, and tight trajectory-tracking requirements, particularly within low-altitude environments. Present-day ATM provides deterministic separation standards, cooperative surveillance, and PBN, yet its reliance on human-mediated tactical control establishes throughput limits when aircraft operate in large numbers and within compressed spatial and temporal margins. Scalable AAM operations will depend on increased automation that maintains stable flight dynamics, preserves energy state during power-intensive segments, enforces lateral and vertical conformance, and supports cooperative separation through the continuous exchange of state and intent information. This level of trajectory predictability enables strategic flow management to coordinate merging sequences, spacing requirements, and corridor entry and exit timing through preplanned trajectory-based scheduling and time-based metering without dependence on continuous tactical intervention.

Urban low-altitude airspace introduces additional engineering constraints arising from obstacle density, sensor degradation from occlusion and multipath, localized atmospheric variability, and reduced maneuvering volume. These conditions increase dependence on structured routing, bounded corridor geometries, and protected terminal-area volumes that link vertiports to the surrounding airspace network while preserving obstacle clearance, sequencing regularity, and compatibility with automated flow-management strategies. Together, these principles establish the foundations that enable AAM aircraft to be integrated safely, predictably, and efficiently within shared airspace.

References

1 Federal Aviation Administration. (2024). Engineering Brief 105A: Vertiport Design, Supplemental Guidance to Advisory Circular 150/5390-2D, Heliport Design. https://www.faa.gov/airports/engineering/engineering_briefs/eb_105a_vertiports

2 European Union Aviation Safety Agency. (2022). Prototype Technical Design Specifications for Vertiports (PTS-VPT-DSN). https://www.easa.europa.eu/en/document-library/general-publications/prototype-technical-design-specifications-vertiports

3 UK Civil Aviation Authority. (2023). CAP2538: Considerations for Aerodromes and Vertiports Planning to Operate Vertical Take-off and Landing Aircraft (VTOL). https://www.caa.co.uk/publication/download/20387

4 Australian Government Civil Aviation Safety Authority. (2023). Advisory Circular 139.V-01v1.0: Guidelines for Vertiport Design. https://vast.aero/wp-content/uploads/2025/01/advisory-circular-139.v-01-guidance-vertiport-design.pdf

5 Japan Civil Aviation Bureau. (2023).Vertiport Design Guidelines. https://www.mlit.go.jp/koku/content/VPDesignGuidelines_en.pdf

6 Government of India Office of Director General of Civil Aviation. (2024). AC No.01: Aerodrome Advisory Circular on Vertiport. https://vast.aero/wp-content/uploads/2024/09/Advisory-Circular-India-Design_-Operations-and-Authorization-for-Vertiports.pdf

7 Brown, R.G. and Hwang, P.Y.C. (2012). *Introduction to Random Signals and Applied Kalman Filtering*. 4th ed., John Wiley & Sons.

8 Groves, P.D. (2013). *Principles of GNSS, Inertial, and Multisensor Integrated Navigation Systems*. 2nd ed., Artech House.

4

Infrastructure Considerations

As established in Chapter 1, advanced air mobility (AAM) operates as an integrated system. Within this framework, the physical infrastructure is just as essential to system performance as the aircraft themselves. Although public attention and investment often focus on aircraft development, the real-world viability of urban and regional AAM operations depends on robust, well-integrated ground and aerial infrastructure. In this chapter, we turn our attention to the surface and digital infrastructure necessary to support AAM. These systems must enable safe and efficient operations while seamlessly weaving AAM into existing urban and regional transportation systems.

Unlike conventional aviation, which concentrates operations at traditional runway-based airports that are typically separated from dense urban activity, AAM introduces a more distributed operational model where flights originate and terminate closer to points of demand. This shift brings aerial access nearer to where people live, work, and travel. However, such proximity requires explicit consideration of urban compatibility, environmental responsibility, and community acceptance. This shift creates both opportunity and constraint. The opportunity lies in the ability to drastically shorten travel times and improve connectivity; the challenge lies in integrating frequent aerial operations into an already complex, congested, and, in some cases, overcrowded urban framework.

AAM infrastructure extends well beyond merely landing and takeoff areas. AAM facilities must integrate with electrical grids, data networks, and ground transportation systems that permit rapid turnaround of vehicles and passengers in high-frequency service environments. Planning and design must also anticipate growth. Vertiports designed for initial operations should be scalable in both capacity and technology. Configurations that accommodate current aircraft geometries and charging requirements must remain adaptable as propulsion systems, energy storage, and operational concepts evolve. Ultimately, infrastructure design decisions frame how the public intersects with and perceives AAM, thereby establishing the foundation for long-term system efficiency and viability.

In this chapter, we examine four principal dimensions of AAM infrastructure. We start with the physical and operational characteristics of vertiports and then turn to the integration of these facilities into urban planning and land-use frameworks. Next, we address the charging and refueling systems that will power next-generation aircraft and, finally, the role of multimodal hubs in ensuring seamless connectivity between air mobility and other forms of transport. Collectively, these elements form the backbone of the AAM ecosystem.

4.1 Vertiports

The term "vertiport" is used broadly in this text to describe infrastructure that supports vertical or near-vertical flight operations associated with AAM. In many regulatory and planning frameworks, such facilities are treated as a specialized subset of heliports, reflecting their shared reliance on vertical takeoff and landing operations. Within the AAM context, however, vertiports are distinguished by the integration of passenger-processing, energy replenishment, digital coordination, and high-frequency operations at a scale and intensity that exceed those of conventional rotorcraft facilities.

Generally, a *vertiport* refers to a facility where passengers embark and disembark, cargo is transferred, energy systems are replenished, and limited aircraft servicing or maintenance activities are supported (Figure 4.1). Facilities that provide only a landing surface and associated safety areas, without passenger amenities or energy infrastructure, are more appropriately described as a *vertipad*, by direct analogy to helipads.

Differences in vertiport scale, throughput, and network role are described in terms of operational capacity and functional characteristics, without the introduction of distinct facility names Figure 4.2). A vertiport may support low-frequency local service or operate as a high-capacity node within a regional network, but the underlying infrastructure classification remains the same.

A properly designed vertiport is conceived as a modular and scalable infrastructure system that can adapt as technologies mature, operations expand, and patterns of urban use evolve. This adaptability is not unconstrained. While passenger-processing, energy, and ground-support systems can often be expanded or reconfigured over time, the surrounding airspace and protected operational volumes establish hard upper bounds on achievable capacity, particularly in dense urban settings.

Within these constraints, the vertiport functions as a configurable framework that can be expanded, contracted, or reconfigured in response to capacity requirements or changes in vehicle geometry associated with evolving propulsion and energy systems. Modularity allows structural

Figure 4.1 Artist's rendering of a conceptual vertiport configuration, depicting a small, single-pad facility with minimal passenger amenities intended for low-throughput AAM operations. Courtesy of Joby Aviation.

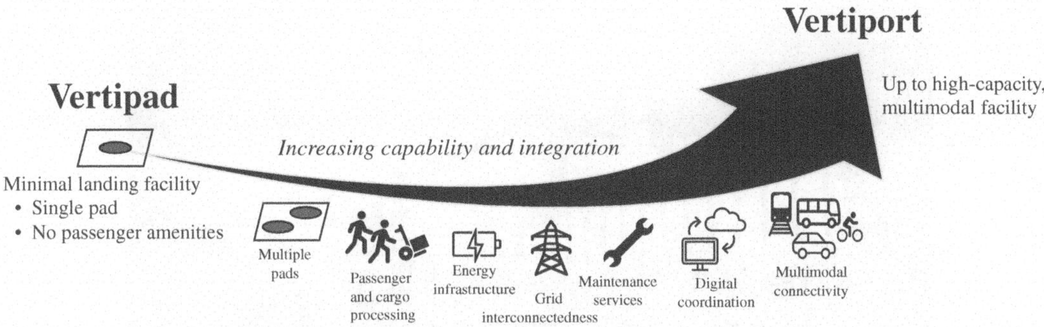

Figure 4.2 Conceptual continuum of vertiport infrastructure capability from vertipad facilities to high-capacity, multimodal vertiports.

elements, electrical systems, and passenger-handling components to be fabricated as standardized units and assembled to suit site-specific conditions. This flexibility applies to ground-based systems only; airspace availability remains fixed and constrains overall capacity. Standardization also supports consistency in safety margins, operational logic, and interface design across facilities.

During the planning phase, vertiport performance is evaluated using digital simulation models. These models quantify aircraft throughput (i.e., the number of aircraft passing through the vertiport), turnaround time, and passenger flow through the facility. Simulation enables designers to assess how the spatial arrangement of pads, taxi lanes, charging bays, and pedestrian pathways, service-vehicle routes, and emergency-access corridors influence efficiency, noise exposure, and scheduling at both site and network scales. A modular layout informed by a priori performance modeling supports regulatory compliance and long-term scalability, ensuring that the vertiport maintains operational performance as aircraft technologies and service concepts evolve.

4.1.1 Core Functions of a Vertiport

Vertiports provide the bridge between aerial and surface access. Their efficacy is measured by (1) how efficiently aircraft can arrive and depart and (2) how seamlessly passengers and goods can transfer between air and ground modalities. The ability of AAM to reduce travel time depends on this interface functioning reliably and efficiently. Curb access plays a decisive role in determining if the time saved in the air is maintained after the traveler reaches the ground (Figure 4.3). Curb access includes areas designed to accommodate concentrated activity from transportation network companies (TNCs), such as app-based ride-hailing services (think Uber or Lyft), proximate public transit (think trains, buses, and subways), clearly defined pedestrian routing, and dedicated micromobility corridors for bicycles, e-bikes, and scooters (Figure 4.4).

A poorly designed vertiport lacking integration with these other popular forms of transportation may negate the advantages AAM strives to provide. A well-planned and well-designed vertiport can anchor a multimodal network, with short aerial segments served by AAM aircraft linking other mobility options and transforming travel from a sequence of disconnected (and often frustrating) steps into a continuous journey.

A vertiport serves as a node (i.e., the connecting point) for three interconnected systems: transportation, energy, and information (Figure 4.5). The *transportation* domain governs the movement of passengers, goods, and aircraft into and through the facility. The *energy* domain governs the delivery, management, and protection of electrical power (or alternative propulsion energy)

Figure 4.3 Artist's rendering of curb access outside a vertiport. Curb access plays a decisive role in determining if the time saved in the air is maintained after the traveler reaches the ground. Courtesy of Joby Aviation.

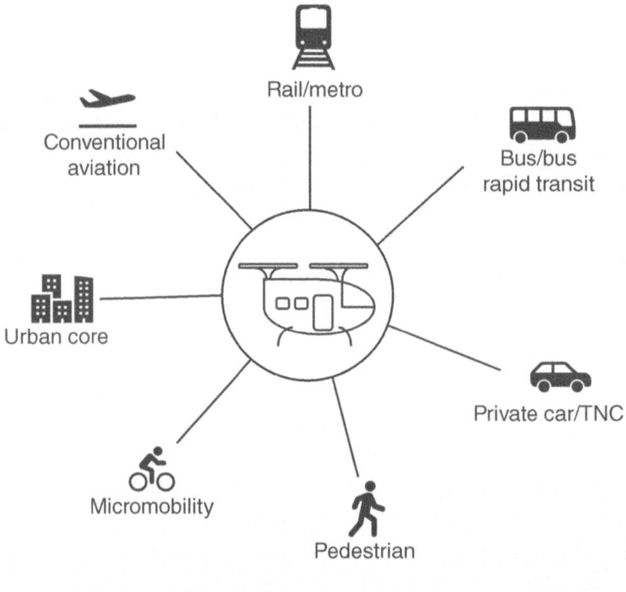

Figure 4.4 The vertiport functions as a hub linking aerial and surface transport modes in a continuous mobility network.

required to support AAM vehicles. Finally, the *information* domain enables digital communication, automation, scheduling, and data exchange that coordinate operations across the site and connect the vertiport to the broader aviation and airspace operational environment supporting both AAM and conventional aircraft. A vertiport's operational quality thus depends on how well the transportation, energy, and information domains work in parallel to support reliable, efficient, and safe movement through urban and regional airspace.

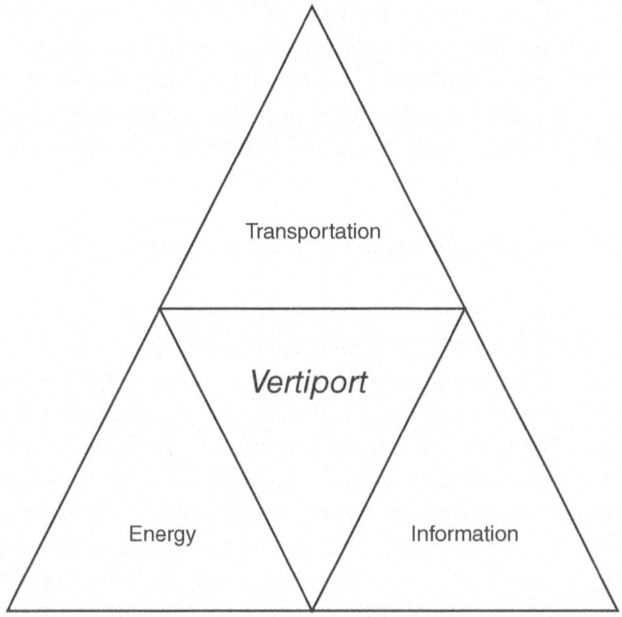

Figure 4.5 A vertiport serves as a node for three interconnected systems: transportation, energy, and information.

4.1.2 Functional and Spatial Architecture of a Vertiport

Vertiports rely on these three functional domains (transportation, energy, and information) to support safe and reliable operations. Each functional domain spans the full physical layout of the facility. The layout's configuration determines how aircraft approach, land, recharge, and depart. The layout's configuration also determines how passenger movement, energy-delivery systems, and ground services all operate within the same space.

The physical environment of a vertiport is organized into three spatial zones: the airside zone, the interface zone, and the landside zone. First, the *airside zone* encompasses the operational envelope used for approach, landing, takeoff, and immediate ground handling. This zone includes the *touchdown and liftoff*, or TLOF, area; the *final approach and takeoff*, or FATO, area; the associated approach and departure surfaces that protect the aircraft flight path; and the surrounding *safety area* that provides appropriate spatial separation and obstacle clearance.

The airside zone also encompasses the *downwash caution area* (DCA) which accounts for rotor- or propeller-induced flow effects that extend beyond the immediate landing surface. Although defined with respect to aircraft operations, the DCA can influence design requirements across adjacent interface and landside areas by constraining allowable proximity, pedestrian and vehicle circulation, and occupancy.

The size of each area is determined by the controlling dimension (D) of a reference aircraft. The *reference aircraft* represents the largest or most operationally demanding vehicle expected to use the facility under normal operations. The resulting D is generally assumed to be the diameter of the smallest circle that encloses the aircraft's planform in a takeoff or landing configuration, including rotors or propellers turning. This approach parallels established aircraft-based scaling logic used in conventional airport planning, adapted here for AAM vehicle geometries and operating envelopes.

Figure 4.6 illustrates the relationship between D, which encloses the entire aircraft projection, and the rotor diameter (RD), which encloses the propulsion swept areas. The geometric dimensions of the TLOF, FATO, and safety area scale parametrically with the geometry of the aircraft class being served. This scalable logic provides a consistent spatial framework that can be applied to a variety of vehicle types without altering the underlying operational principles. Such a framework allows vertiports to accommodate multiple aircraft classes and adapt to evolving configurations while maintaining established safety margins. In current common practice, the TLOF and FATO are sized based on the RD, with the surrounding safety area sized based on D.

The TLOF is the load-bearing pad where an aircraft physically lands or departs. The FATO surrounds the TLOF and defines the area used for the transition between hover and forward flight. Figure 4.7 illustrates the nested relationship between the TLOF, FATO, and safety area. In this representative configuration, the TLOF is defined as one times the RD (1RD), the FATO as two times the RD (2RD), and the safety area as two-and-one-half times the D (2.5D). The safety area extends beyond the FATO to reduce the risk associated with minor deviations in aircraft position or trajectory during normal operations, whereas rotor- and propeller-induced wake effects are managed through the DCA. These ratios represent a generalized geometric relationship. Complementary reference material is available from the FAA [1], EASA [2], CASA [3], and GCAA [4].

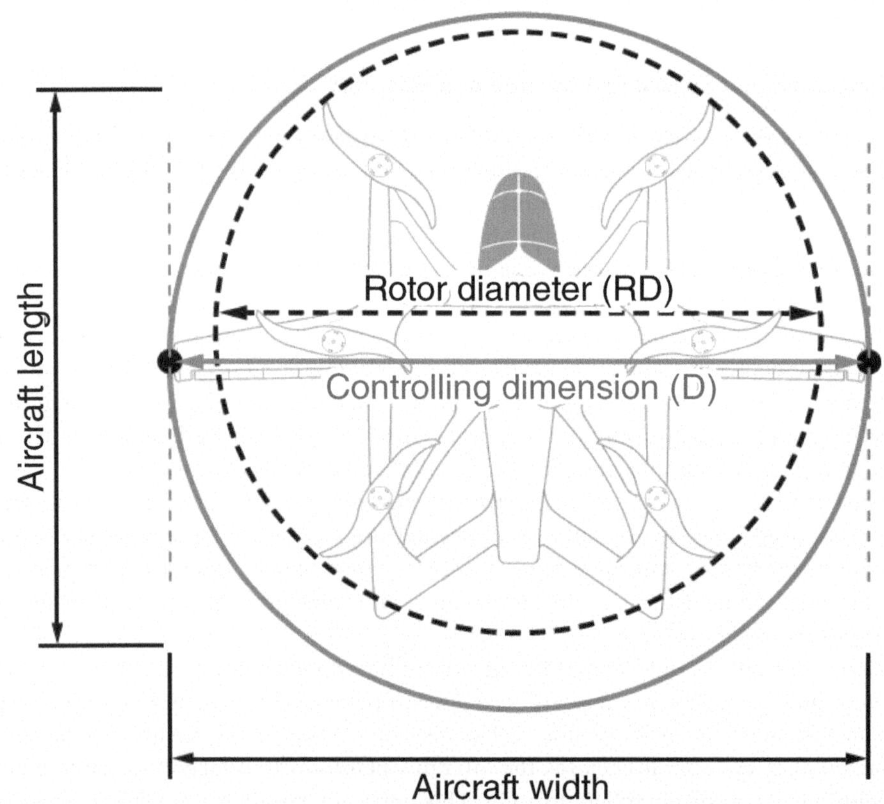

Figure 4.6 The D of the design aircraft. *Source*: Federal Aviation Administration/Public domain.

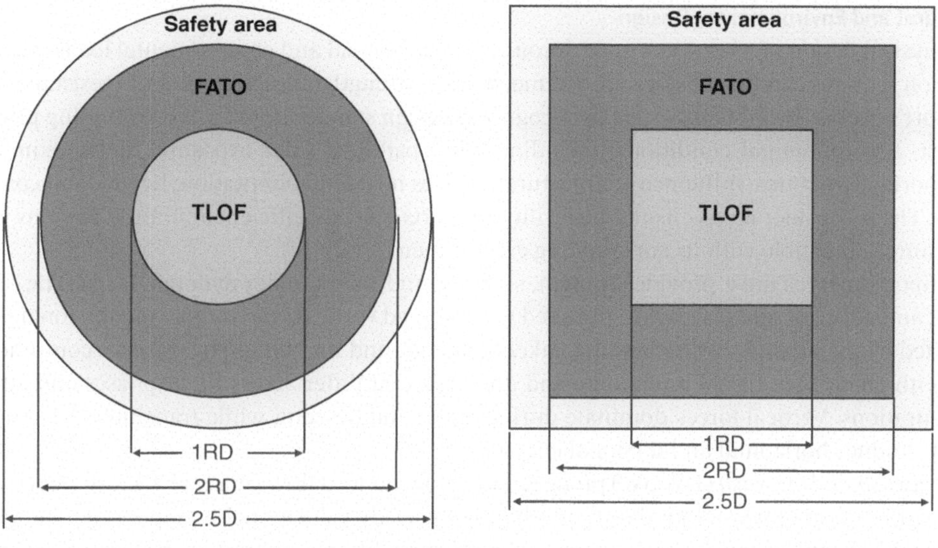

Figure 4.7 Nested relationship of the TLOF, FATO, and safety area. In this design, the TLOF and FATO scale with RD. The safety area scales with D.

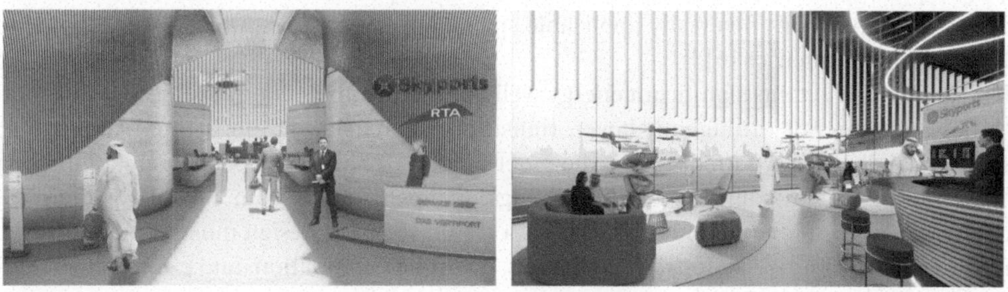

Figure 4.8 Artist's rendering of passenger reception (left) and boarding area (right). Courtesy of Joby Aviation.

These three zones define the airspace volume supporting safe maneuvering and flight operations. Limited line maintenance and pre-departure servicing may occur within adjacent designated airside bays. Full maintenance, diagnostics, and major repairs, however, are typically confined to larger vertiport facilities that incorporate dedicated hangars and specialized support equipment.

Next, the *interface* zone forms the transitional space between aircraft operations and passenger or cargo handling. It is a controlled-access zone with boarding gates, security screening points, ground-crew staging areas, and charging infrastructure. This domain's layout balances proximity for efficiency with separation for safety, maintaining controlled movement between the aircraft operating area and passenger and cargo processing zones.

Finally, the public-facing terminal and access realm make up the *landside* zone and connect the vertiport to the area's broader transportation network. Besides intermodal connectivity, the public-facing terminal contains passenger-handling functions such as check-in counters, waiting areas, and passenger amenities (Figure 4.8). All three domains operate as components of a continuous system. Inefficiency or congestion within any one domain limits throughput and time savings across the entire operation.

Structural and Environmental Design

The physical design of a vertiport must account for operational and environmental loads. Material selection requires careful consideration of mechanical strength, durability, and fire resistance. The vertiport's overall layout is governed by recognized design standards and the surrounding physical context. Environmental conditions, including wind patterns, solar exposure, and proximity to neighboring structures, influence architectural choices related to orientation, height, and configuration. These considerations ensure the facility supports safe and efficient aircraft operations while remaining compatible with its surrounding environment.

Vertiport surfaces must provide strength, stability, and safety under dynamic loads. The TLOF, FATO, and adjacent safety area are subjected to combined vertical, horizontal, and torsional forces imparted by the aircraft during landing, takeoff, and ground handling. These forces continuously vary with changes in thrust magnitude and direction that differ across flight phases and aircraft configurations. Vertical forces dominate during hover and descent, while transition or tilt operations introduce horizontal and torsional effects.

Downwash and outwash (DWOW) produced by rotor-induced flow generate 3D turbulence characterized by strong pressure gradients and highly unsteady velocity fields (Figures 4.9 and 4.10). The resulting aerodynamic effects, along with debris uplifted and entrained by the flow, create hazardous conditions that vertiport surfaces and surrounding structures must be designed to withstand, and that require the protection of people in adjacent areas. Available full-scale measurement campaigns provide useful insight into the magnitude and persistence of these effects but do not yet define a complete or universally applicable spatial envelope across aircraft classes, operating modes, or urban geometries [5].[1]

Observed DWOW retains significant energy well beyond the immediate landing and safety areas and follows the aircraft throughout hover, transition, approach, and departure. Although flow intensity generally decays with distance, the outer bounds of hazards to people remain uncertain, particularly in built environments with buildings, parapets, and terrain capable of redirecting, concentrating, or recirculating induced flow. As a result, vertiport design must adopt conservative assumptions that prioritize pedestrian protection, controlled circulation, and exclusion zones that extend beyond formally defined operational surfaces.

These high-energy flows interact with parapets, façades, and nearby structures. The distribution and intensity of these flows depend on vehicle weight, configuration, and height above the

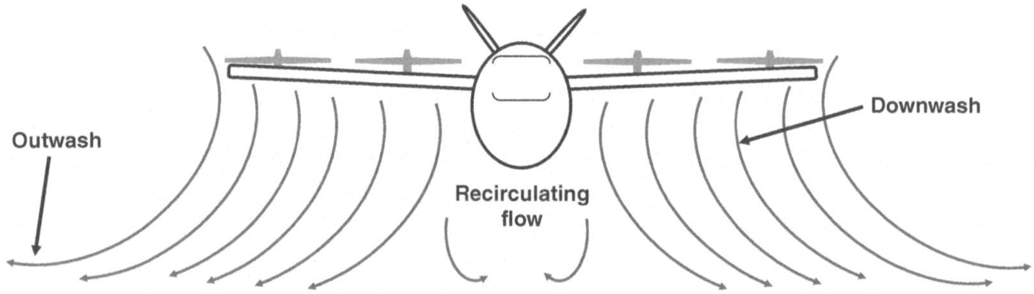

Figure 4.9 Regions of DWOW produced by an eVTOL.

1 Representative DWOW measurements for selected eVTOL configurations are summarized in DOT/FAA/TC-24/42, Electric Vertical Takeoff and Landing (eVTOL) Downwash and Outwash Surveys. The report emphasizes that available datasets are limited in scope and that the spatial extent and human-exposure implications of DWOW remain an active area of research, particularly in urban environments.

Figure 4.10 The effect of DWOW from a HH-60 Jayhawk helicopter's main rotor on the underlying water surface. *Source*: U.S.Coast Guard East / Public Domain.

landing surface, along with the geometry of the surrounding environment. Because the induced flow moves with the aircraft during approach, air taxiing, and departure, these effects must also be evaluated dynamically. Open-edge decks, ventilated parapets, and sufficient setback distances help dissipate induced airflow before it becomes trapped and recirculated. Vegetative barriers, when carefully selected and managed to avoid wildlife attraction, fencing, or graded berms may also be employed to shield pedestrians and adjacent properties while maintaining sight lines and emergency access. Simply put, the vertiport surface and supporting structure must be capable of sustaining both the maximum takeoff weight of the reference aircraft and the dynamic effects associated with uneven lift distribution, vibration, and downwash.

For all these reasons, vertiport decks and their supporting structures must be engineered to resist high load factors, exposure to rotor-induced flow, rapid thermal cycling, and various forms of precipitation. Cyclic thrust loading produces dynamic stresses that vary over short timescales. To prevent long-term material degradation, vibration isolation systems and fatigue-resistant joints are incorporated at key structural interfaces, particularly in elevated or rooftop installations with lower structural stiffness capable of amplifying the dynamic response (Figure 4.11). Rooftop and elevated facilities also introduce an additional requirement to ensure the host structure can sustain aircraft loads without compromising building integrity or creating adverse airflow disturbances that can affect aircraft stability during low-speed operations (Figure 4.12). In dense urban settings, surrounding buildings can channel, accelerate, or recirculate winds, producing localized flow fields that interact with rotor-induced effects. These urban-canyon interactions are addressed in greater detail in Chapter 7, where atmospheric flow modification and its operational implications are examined explicitly.

Material compatibility is important where composite panels interface with reinforced concrete or steel framing because differing thermal expansion rates can produce localized stress concentrations. Reinforced concrete and hybrid composite decking systems are commonly used for this because they combine rigidity with durability and allow the integration of internal drainage networks. These surface materials must maintain adequate friction in wet or icy conditions. Finally, the National Fire Protection Association (NFPA) standard 418 [6] establishes fire-resistance criteria

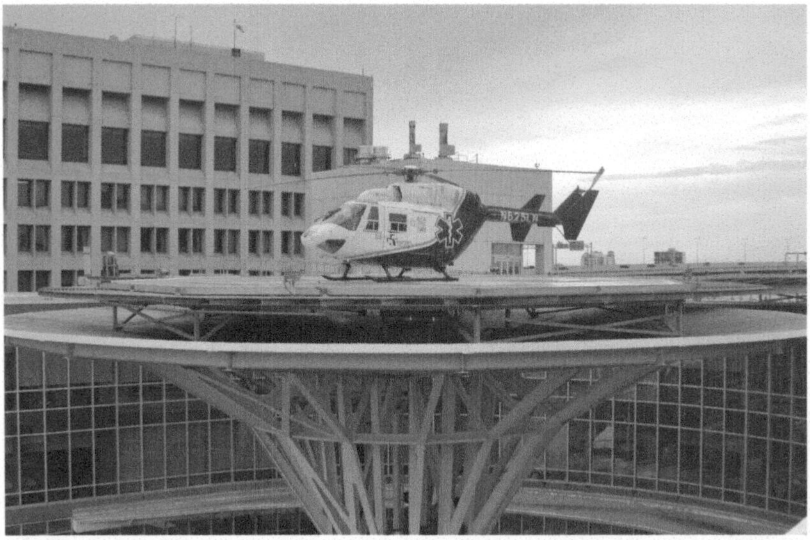

Figure 4.11 Supporting structure of a helipad at Baptist Medical Center in Jacksonville, Florida. *Source*: MikesMegapixels / Wikimedia Commons / CC BY 3.0.

Figure 4.12 Artist's rendering of a rooftop municipal vertiport. *Source*: NASA / Public Domain.

for primary structural elements exposed to potential fuel or electrical hazards. These criteria apply to both electric vertical takeoff and landing (eVTOL) and conventional rotorcraft operations at mixed-use pads. Such standards ensure the pad supports safe operations and remains stable during the critical minutes of emergency response.

Fire protection and emergency access are key to the physical design of any vertiport. In the event of an accident or incident, the facility must be designed so that first responders can reach the pad from multiple directions. This poses a particular challenge for rooftop vertiports on parking structures, where limited vehicle clearance can restrict the access of emergency-response equipment. Guidance generally emphasizes redundancy by recommending that primary access routes and alternate egress paths be separated as far apart as practicable, with many vertiport and helideck guidance documents specifying roughly perpendicular orientations to preserve access if one route becomes obstructed.

Fire protection strategies must also reflect the evolving risk profile of electrically powered aircraft. As discussed in Chapter 2, high-voltage systems and lithium-based energy storage introduce thermal hazards that differ fundamentally from those associated with conventional aviation fuels. Current industry practice does not assume that battery thermal events can be rapidly extinguished. Instead, design guidance increasingly emphasizes containment, cooling, and protection of surrounding assets while the event runs its course. This reality has direct implications for vertiport location, layout, and operation. Rooftop installations, in particular, must account for extended event duration, radiant heat exposure, runoff management, and limited access for heavy firefighting equipment. Foam-based systems, standpipes, and portable extinguishers remain important for controlling secondary fires and protecting egress routes, but they function as part of a broader hazard-management strategy instead of a definitive means of extinguishing battery fires. Annex D of NFPA 418 [6] and NFPA 855 [7] provide initial design guidance for mitigating these hazards through a layered defense in which suppression systems, drainage patterns, and structural fire resistance operate as an integrated safety network.

Because fire suppression systems, energy-storage hazards, and rainfall can introduce water, foam, or spilled fluids to the deck, managing surface runoff is yet another essential part of vertiport safety design. Operational surfaces should be slightly pitched to direct fluids away from passenger areas, egress points, and energy-storage locations. Proper grading prevents the accumulation of flammable liquids or standing water that could freeze or compromise surface friction. For rooftop sites, the deck drainage system must integrate with the host building's stormwater system and ensure that runoff containing foam or other contaminants can be safely contained and treated. In coastal or flood-prone regions, designers may employ elevated perimeter channels or detention basins to prevent cross-contamination with municipal drainage infrastructure.

4.2 Urban Integration and Land-use Planning

AAM alters a century-long spatial relationship between aviation and the city. For much of aviation's history, aircraft operations have been located on the outskirts of major urban areas, separated from population centers. This is by design to manage noise, risk, and the extensive approach and departure airspace required by conventional fixed-wing operations. Yet, the emergence of distributed, high-frequency vertical and short-field takeoff and landing AAM operations shifts portions of aerial activity from concentrated runway-based airports, many occupying large land parcels to accommodate runways and protected airspace, toward a more networked framework spanning urban, suburban, and regional contexts. Accordingly, moving aviation activity toward urban and suburban areas requires planners to carefully evaluate how the vertical dimension of mobility connects with the horizontal structure of streets, transit corridors, activity districts, and neighborhoods. Areas that have not previously supported aviation must now accommodate approach paths, safety zones, and multimodal access links. These conditions influence how skylines evolve and public spaces are valued, along with broader questions associated with access, equity, and the environment. The ultimate objective is to achieve functional compatibility where aerial operations enhance urban mobility without diminishing the safety, livability, or community well-being.

To successfully meet that objective, thoughtful planning is required. *Land-use planning* refers to the systematic allocation, regulation, and organization of land activities within a community to guide development patterns and manage competing spatial demands. Land-use planning can help communities achieve environmental, social, and safety goals. For AAM and vertiports, land-use planning begins with assessing how the facility connects to existing urban systems.

This is especially true for metropolitan areas where each vertiport must integrate smoothly with its immediate surroundings. Vertiports placed within mixed-use environments can draw upon established utility corridors, stormwater networks, and renewable-energy assets, allowing the facility to operate within the existing infrastructural framework. Early-stage thermal and acoustic modeling can inform how the facility should be oriented and configured to reduce environmental effects on nearby residences and public spaces. Modular and adaptive construction strategies further support this integration by allowing the facility to expand or reconfigure as vehicle geometries, charging requirements, or service patterns change. When these site-level considerations are coordinated with broader land-use objectives, the vertiport functions as a coherent component of the city's mobility ecosystem that is fully embedded within existing urban systems.

Integrating AAM infrastructure into the metropolitan fabric requires a structured approach for evaluating how each facility interacts with the surrounding environment. Because AAM facilities are introduced into urban contexts that already support dense transportation, utility, and social systems, their impacts cannot be assessed in isolation. As a result, **land-use planning for AAM must go beyond simply assessing whether a location can support safe operations but also how that location aligns with surface-transportation systems, utility capacity, and community expectations**. In this section, we consider how this evaluation process is completed via a concept called *compatibility*.

4.2.1 Compatibility

Compatibility provides a systematic framework for determining how effectively a vertiport can function within the physical, operational, and social conditions of its setting. It includes spatial and systems dimensions, meaning that a compatible site must support reliable aviation operations while aligning with the land-use, transportation, utility, and community expectations that define the surrounding area. Compatibility, therefore, functions as both a planning principle and a performance criterion. It establishes where a vertiport can be located and how that facility interacts with the broader urban and regional system. All this means, in simple terms, compatibility defines the relationship between a vertiport and its surrounding environment. Compatibility is evaluated across four interdependent dimensions, each described in turn (Figure 4.13):

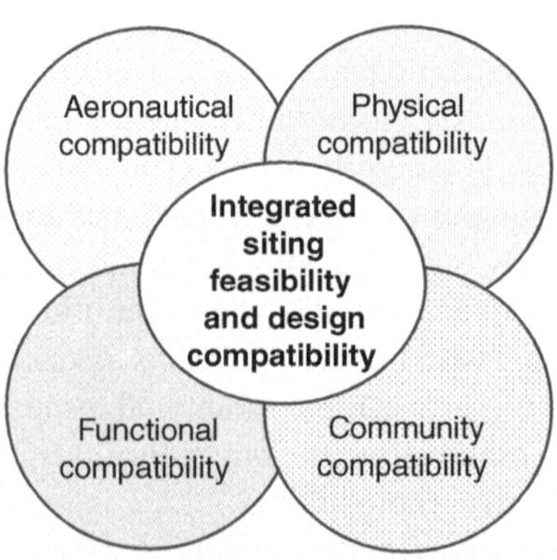

Figure 4.13 Conceptual framework illustrating the four dimensions of vertiport land-use compatibility that together support integrated siting feasibility and design decision-making.

1. *Aeronautical compatibility* ensures that protected operational airspace remains unobstructed by buildings, infrastructure, or terrain. Well-defined 3D volumes that safeguard approach and departure paths must be preserved for both existing and future development scenarios. This includes consideration of surrounding airspace classifications, proximity to nearby airports or heliports, and coordination requirements associated with existing air traffic control structures. Aeronautical compatibility also examines local wind patterns and turbulence, including rotor- and propeller-induced DWOW, along with other atmospheric factors that influence departure, approach, air taxiing, and hover performance.
2. *Physical compatibility* focuses on the relationship between the vertiport and the surrounding built environment. Vertiports are often most feasible in places where infrastructure and activity are already concentrated. Therefore, this dimension evaluates whether local infrastructure can support AAM operations without exceeding capacity or compromising reliability. The capacity of utility corridors, roadway networks, stormwater infrastructure, and emergency-response routes form the core criteria for assessing physical compatibility.
3. *Functional compatibility* examines how effectively the vertiport integrates with surrounding surface-transportation systems (buses, commuter rail, taxis, etc.). Efficient multimodal connectivity requires alignment with transit corridors, availability of proximate public transport, and coordinated arrangement of curbside, pedestrian, and micromobility (e.g., bicycles, e-bikes, and scooters) access points. Here, the objective is to minimize dwell time, eliminate redundant transfers, and produce continuity across air and ground travel.
4. *Community compatibility* evaluates how the facility interacts with the social and perceptual factors of the surrounding area. Community compatibility includes aspects such as noise exposure, visual prominence, safety perception, and the contextual fit of aviation activity within established neighborhood patterns. Early coordination with municipal planning authorities and direct engagement with residents help align operational intent with community values, improving acceptance and reducing the likelihood of subsequent conflict.

Together, these four dimensions form a comprehensive framework for assessing how vertiports integrate with the landscapes, infrastructure systems, and communities they serve. When used collectively, the four dimensions support vertiport site selection and design decisions that strengthen operational safety, transportation continuity, and urban quality.

4.2.2 Siting

Siting applies these compatibility principles to specific locational decisions through a staged evaluation that moves from broad regional screening to parcel-level analysis. The objective of siting is to identify locations that support safe operations, provide accessible mobility connections, and offer adaptability as AAM technologies mature. Generally, there are four major siting stages (Figure 4.14):

Figure 4.14 Vertiport siting process that progresses from region-scale screening to parcel-scale feasibility analysis and system-level decision-making.

Siting stage 1: *Strategic screening*: During this first stage, planners identify general areas (or districts) that satisfy operational and accessibility criteria. Protected operational airspace, clear approach and departure corridors, and proximity to established transportation routes define the initial boundaries of feasibility. At this scale, planners increasingly rely on data analytics that integrate mobile-device travel records, aggregate traffic flows, and demographic indicators to identify high-demand corridors. Additionally, using anonymized origin–destination, temporal congestion patterns, and travel-purpose datasets, mobility analytics clarify where a vertiport will most effectively contribute to system performance. These analytics distinguish between business and leisure flows, revealing differences in time sensitivity and income-linked accessibility. They also identify corridors where surface travel incurs significant delays. Basing land-use decisions on data-driven mobility behavior helps ensure that prospective locations align with established demand patterns and concentrate activity where high-frequency service can be supported.

Siting stage 2: *Area refinement*: This second stage establishes whether a district provides the structural and environmental conditions necessary to support reliable aircraft operations. It narrows the focus to districts that meet baseline airspace and terrestrial requirements. Analysts evaluate whether existing infrastructure can support AAM activity without imposing undue strain. Electrical-grid capacity, substation proximity, roadway access, and emergency-response reach determine if an identified district can accommodate sustained operations. Local topography, building geometry, and the microclimate are assessed to confirm that approach paths and hover envelopes will remain viable.

Siting stage 3: *Site evaluation*: The third stage examines individual sites within identified districts. Planners assess the structural capacity of the underlying soil or building, available footprint, slope, and drainage conditions to determine if a site can physically support a vertiport. Spatial modeling tools are used to verify that the TLOF, FATO, associated safety area, and approach and departure paths fit without obstruction, that protected operational airspace remains viable under current and forecasted development patterns, and that height penetrations relative to applicable obstruction-evaluation surfaces (e.g., Part 77 surfaces [8]) are appropriately managed. By this stage, planners have shifted their focus from network-scale planning to determining whether the physical layout of the site can realistically support safe and predictable passenger movement. Planners assess pedestrian circulation, curbside capacity, and connections to public transit, with particular attention to how DWOW influence allowable circulation paths, queuing locations, and exclusion zones, and use thermal and acoustic modeling to quantify how operations may influence nearby residences and public spaces.

After site-specific feasibility has been established, planners evaluate site-specific environmental conditions, such as stormwater routing, peak-flow behavior, and potential contamination risks. In coastal or flood-prone regions, siting requires additional assessments of elevation, storm-surge exposure, and the design of perimeter barriers. Further, local grid constraints are also examined as the need for high-power charging can exceed the capacity of existing feeders. If a site's electrical grid lacks multiple supply paths or has insufficient capacity to support high-power charging, planners may explore hybrid electrical configurations, including pairing the vertiport with adjacent renewable-energy assets, dedicated on-site energy storage, or modular and portable charging systems deployed to supplement fixed infrastructure during early operations or phased buildout.

Siting stage 4: *Multicriteria evaluation*: The final stage synthesizes aeronautical, infrastructure, environmental, and community-based criteria into a multicriteria evaluation that ranks sites according to operational feasibility, cost, perceptual effects, and alignment with municipal planning objectives. In practice, this synthesis is often implemented using weighted evaluation matrices developed in coordination with key stakeholders, allowing criteria importance to be explicitly represented and trade-offs to be compared transparently across candidate sites. This approach clarifies how trade-offs are resolved and ensures that selected sites support near-term service requirements and long-term expansions. Because AAM remains an evolving field, siting decisions must preserve flexibility so facilities can adapt to emerging aircraft geometries, increasing power demands, and changing service models.

4.2.3 Community and Environmental Considerations

AAM environmental considerations will be more thoroughly explored in Chapter 7, and community considerations will continue to be addressed as relevant throughout the book. However, a brief word about both is useful here because community and environmental considerations can be (and often are) addressed via thoughtful land-use planning efforts.

Community considerations stem from how residents perceive aircraft operations and ground activity. Predictable flight paths, consistent scheduling, and publicly communicated operational plans all strengthen public acceptance and can reduce concerns over noise or safety. The location of a vertiport relative to residential areas, schools, parks, and other sensitive land uses strongly influences community perception. Early coordination with municipal leadership and planning agencies, along with structured engagement with neighborhood organizations, allows planners to identify and resolve concerns before design decisions become firm.

Community engagement provides the framework that links technical analysis to local expectations. Clearly communicating safety procedures, noise-mitigation strategies, and operational parameters helps build trust and reduces apprehensiveness. Governments, AAM manufacturers, and operators, among others, can all actively participate in community engagement efforts. Engagement should occur early and be maintained throughout design, construction, and early operations so that community feedback can inform refinement of flight procedures, ground-access arrangements, and other operational elements.

Environmental considerations apply at both the network and site scales. At the network scale, each facility contributes to cumulative acoustic exposure along approach and departure paths, as well as to the visual presence of low-altitude aircraft within urban skylines and previously unobstructed airspace. Changes to skyline views, the frequency of overflight, and the proximity of aircraft to residential windows can influence perceptions of privacy and tranquility, even in the absence of elevated noise levels. Accordingly, acoustic models must incorporate the frequency content and tonal characteristics specific to eVTOL propulsion, along with operational variables such as climb profiles and approach angles. Although electric propulsion eliminates direct emissions at the point of use, life cycle effects related to power generation, battery production, and end-of-life handling all must be evaluated, as we will see in Chapter 7.

How, then, do planners address such concerns? The answer starts with proper zoning and regulation.

4.2.4 Zoning and Regulation

After technically feasible sites have been identified, and environmental and community considerations are understood, the planning process turns to the relevant regulatory framework that governs how a vertiport can be authorized and constructed. At this stage, the feasibility of a site depends as much on the structure of local regulation as on its geometry or access characteristics, and planners must evaluate how zoning ordinances, building codes, and airspace protections interact to support or constrain vertiport development.

Zoning requirements tell us where vertiports may be located and the dimensional conditions under which they must be constructed. Zoning establishes the precise land-use, height, and setback parameters that determine how a vertiport can occupy its parcel and maintain compatibility with adjacent development. Zoning's primary role is to support functional compatibility by aligning land-use permissions, height controls, and setbacks with protected airspace needs, surface-access requirements, and adjacent land uses.

In most jurisdictions, zoning ordinances and comprehensive plans already contain provisions for aviation uses (e.g., airports and heliports). These frameworks can be extended to AAM by defining vertiports as a distinct, but related, land-use classification. Some civil aviation authorities have chosen to treat vertiports as an evolution of existing heliport classifications, creating continuity with established facility types instead of establishing an entirely separate category. Both approaches allow new facilities to be governed within familiar regulatory structures while adapting performance criteria to the smaller scale and distributed nature of AAM operations.

Height and setback limitations are the most obvious zoning requirements that affect AAM facilities. Height restrictions can prevent surrounding developments from encroaching on the protected airspace that supports approach and departure operations, and lateral setbacks separate vertiports from land uses that are sensitive to noise, pedestrian flow, or ground-handling activity. The relative height of a vertiport with respect to surrounding buildings also shapes local wind fields, influencing turbulence, shear, and recirculation patterns within approach, departure, and hover envelopes. These parameters are often set using scaling rules that tie allowable heights and setback distances to the size and operational needs of the reference aircraft, allowing regulations to adapt as vehicle designs change. In dense urban settings, understanding these wind effects may require site-specific characterization supported by additional sensing or temporary measurement campaigns to inform safe and reliable operations.

Zoning provides a primary mechanism for integrating noise, visual, and environmental constraints into vertiport siting and operation. Noise and visual considerations define how vertiport operations interface with nearby land uses. Quantitative noise-exposure limits, discussed in Chapter 7, establish when residential or educational areas require additional buffers to maintain acceptable acoustic conditions. Visual impacts follow a similar logic since lighting systems must provide operational visibility while avoiding glare or light trespass onto adjacent properties. Architectural measures (e.g., building orientation, vegetative screening, and façade treatment) can serve as zoning-compatible mitigation strategies that moderate the visual presence of the facility and enhance community acceptance (provided, of course, these interventions do not interfere with protected airspace or operational performance).

Because AAM remains an emerging mode of transportation, some jurisdictions may apply interim regulatory mechanisms such as conditional-use permits or temporary authorizations. These instruments allow local governments to evaluate initial vertiport operations and collect data before establishing permanent zoning overlays.

Regulatory coordination is what connects aeronautical standards and urban development. When zoning ordinances, building codes, and airspace protection criteria are aligned, vertiports become an integrated component of the urban system. Such harmonization enables AAM surface infrastructure to evolve safely and predictably as cities grow and service models mature.

4.2.5 Connectivity and Accessibility

A vertiport achieves operational effectiveness when it functions as a well-connected component of the broader urban transportation network. *Connectivity* describes the structural relationship between a vertiport and surrounding transportation systems. *Accessibility* describes how efficiently people, equipment, and support vehicles can reach and move through the facility with minimal delay. Together, these parameters determine whether AAM provides meaningful reductions in total travel time.

Connectivity starts at the network scale. Vertiports should be located along established mobility corridors that either concentrate recurring travel demand or support economically viable, time-sensitive connections, such as regional rail alignments, arterial highways, or major bus interchanges. When positioned within these corridors, each vertiport functions as a node in a distributed system that balances aerial and surface capacity. Planners rely on analytic tools that integrate travel-demand models, traffic-flow data, and demographic information to forecast route utilization and align aircraft movements with patterns of urban activity.

Accessibility starts with the movement of people, goods, and support services into and through the vertiport. First- and last-mile access establishes the initial connection between the facility and the surrounding street network. This connection requires clear pedestrian pathways, proximate public-transit stops, and designated pickup and drop-off areas for TNCs, shuttles, and taxis. All this must be carefully organized to prevent traffic congestion and preserve routes for emergency response. Vertical-circulation components, including elevators, ramps, and stairs, should be integrated into the structural core so that passenger flow remains direct, predictable, and protected from weather.

Once external access is established, the internal circulation of passengers, cargo, and service functions must support efficient movement between ground modes and aircraft boarding. Circulation pathways require coordinated spatial planning so that boarding, security, passenger-processing, and service operations do not interfere with one another. This coordination maintains throughput and minimizes dwell time during peak operational periods.

Placing the vertiport within established public-transportation corridors that already organize movement through the city strengthens its accessibility. Co-located rail platforms, bus stops, or shared-mobility areas enable mode changes without unnecessary transfers or extended walking distances. Wayfinding, lighting, and climate protection function as components of the passenger-handling system that reinforce safety, operational predictability, and user comfort.

Operational accessibility also depends on the coordinated movement of power, materials, and information through the vertiport. Electrical infrastructure, data networks, and maintenance-vehicle access must be organized concurrently with the planning of passenger circulation to prevent future spatial competition over limited space. Charging equipment, storage areas, and service bays require placement strategies that minimize aircraft repositioning and maintain clear separation between public areas and operational zones. Whether these systems are aligned determines if aircraft can achieve rapid turnaround and if the facility can sustain high-frequency operations.

With connectivity and accessibility decisions defining how a vertiport operates within the transportation network, community and environmental considerations define how that same facility is experienced by the people and ecosystems surrounding it. The long-term viability of AAM depends on ensuring that operational benefits are achieved without imposing unacceptable acoustic, visual, or environmental burdens, and that the user experience is pleasant.

4.2.6 Planning Process

The AAM infrastructure planning process translates aeronautical, structural, community, environmental, and regulatory requirements into a structured sequence of analyses that guide site selection and facility design. This progression from conceptual objectives to technical evaluation forms the operational backbone of AAM infrastructure planning. All the principles we have addressed thus far define the performance expectations for a feasible and context-responsive vertiport. However, it is now helpful to take a step back and see how the planning process enables these requirements to be implemented.

This multistage framework progresses from regional considerations to parcel-level evaluations and applies progressively higher-resolution data at each stage. This final subsection illustrates how planners convert broad mobility objectives into specific site and facility decisions through a structured and technically grounded methodology (Figure 4.15).

Planning begins by clearly defining system-level objectives. These objectives specify how AAM services are expected to function within the regional mobility network, including intended trip purposes, operational frequency, accessibility goals, and expected fleet characteristics. These definitions establish the baseline against which all subsequent analytical findings are evaluated.

After system objectives are established, planners assemble the analytical tools required to evaluate potential locations. This includes travel-demand models; geographic information system (GIS)-based airspace assessments; land-use and zoning inventories; environmental screening layers such as flood-risk maps, wetlands designations, and ecological-constraint overlays; noise-exposure datasets; and utility-capacity data. Integrating these tools ensures that early decisions are grounded in regional movement patterns, physical constraints, and environmental considerations and prevents parcel availability from dictating the direction of early planning.

The next stage uses these tools to identify opportunity zones that meet baseline airspace, access, and land-use criteria. At this stage, planners compare airspace geometry with surface-access characteristics and evaluate land-use compatibility findings alongside environmental constraints such as flood exposure, stormwater pathways, and areas sensitive to noise. This comparison allows planners to eliminate zones where core requirements cannot be reconciled and to concentrate attention on areas with a realistic potential to support long-term operations.

With opportunity zones identified, planners evaluate how candidate facilities would perform within the emerging AAM network. This intermediate-scale analysis examines system efficiency, multimodal connectivity, and equity. The goal here is to determine if a proposed location improves overall network performance and whether its placement supports neighborhoods with unmet mobility needs. Through this evaluation, planners ensure that network development reflects collective mobility objectives versus isolated market opportunities.

The final stage converts these analytical findings into implementable recommendations. This synthesis identifies priority locations, outlines necessary regulatory or zoning actions, and organizes phasing for design and construction. Because AAM technologies continue to evolve, final recommendations should consider provisions for expansion envelopes that explicitly accommodate uncertainty in future aircraft performance characteristics, phased implementation that aligns infrastructure growth with

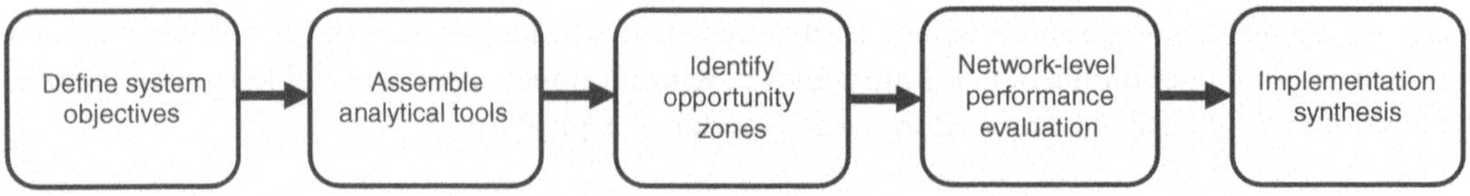

Figure 4.15 Sequential stages of the vertiport planning and evaluation process.

real operational demand, and regulatory requirements written in terms of the safety and performance outcomes that a vertiport must achieve without prescribing specific structural layouts. At present, the full range of potential AAM vehicle geometries, mass properties, downwash characteristics, power requirements, and operational envelopes remains only partially defined. Consequently, planning for expansion capacity provides resilience against future aircraft that may impose greater spatial, structural, or energy demands than early design assumptions anticipate. This integrated approach ensures that facilities remain adaptable as aircraft configurations and energy systems mature.

Executing this structured planning process supports vertiport infrastructure that emerges from a coherent integration of aeronautical logic, urban form, environmental stewardship, and community considerations. The resulting facilities are technically feasible, socially acceptable, and positioned to operate as stable components of the broader transportation ecosystem.

Simply put, the planning process establishes where vertiports can be located and how they integrate with the surrounding urban and regulatory environment. Once these spatial and operational foundations are in place, attention turns to the energy infrastructure that enables sustained AAM operations. Charging and refueling infrastructure determine how rapidly aircraft can be turned around, how reliably service can be scheduled, and how each facility supports the wider energy network.

4.3 Charging and Refueling Infrastructure

Beyond a place to land and take off, AAM aircraft also need a reliable means of replenishing energy between flights. Unlike conventional aviation, where fuel is delivered in bulk to concentrated airport hubs and aircraft can be refueled within minutes, AAM relies heavily on electrically powered vehicles that must recharge within a distributed network of facilities operating at a high cadence. Therefore, charging infrastructure constitutes a core operational component (Figure 4.16). The rate, capacity, and reliability of the charging system govern daily flight frequency, fleet size requirements, and the overall economics of service delivery.

Figure 4.16 Beta Technologies charging, thermal-management, and cabin conditioning cubes. Courtesy of Beta Technologies.

Electric propulsion introduces support requirements that differ fundamentally from those used in conventional aviation. In addition to restoring state of charge, energy replenishment also involves managing battery temperature, maintaining cell safety margins, balancing load on local electrical feeders, and coordinating with broader grid operations. These factors impose time, power, and thermal constraints that influence how each aircraft is scheduled and how each vertiport is designed.

Before examining specific charging configurations, it is useful to identify the requirements that govern AAM energy infrastructure. Charging systems must deliver sufficient power within the turnaround intervals dictated by high-frequency operations, maintain battery health over repeated charging cycles, and integrate cleanly with local and regional electrical networks. These requirements establish the technical foundation for evaluating charging architectures, grid-integration strategies, thermal and safety considerations, and the operational models developed in the following subsections.

4.3.1 Electrical Charging Fundamentals

AAM charging systems must deliver large quantities of energy within short intervals while maintaining the aircraft battery within safe thermal and electrical limits and drawing power within the capacity and power-quality constraints of the local electrical grid. These conditions establish high power, voltage selection, current limits, resistive heating, and thermal management, the core electrical principles that govern the design of AAM charging systems. Because AAM operations require rapid turnarounds, distributed vertiport locations, and repeated high-power cycles, charging systems must satisfy technical constraints that are more demanding than those found in most ground-vehicle charging applications. The following fundamentals provide the analytical basis for understanding how charger architecture, aircraft battery systems, and vertiport electrical networks are configured.

Voltage, Current, and Power Delivery

The rate at which energy is transferred during charging is described by the instantaneous electrical power, $P[\mathrm{W}]$,[2]

$$P = IV \tag{4.1}$$

where $V\left[\mathrm{V}\right]$ is the applied voltage and $I\left[\mathrm{A}\right]$ is the delivered current. High-power charging requires that the product of voltage and current be large. For a fixed power level, increasing the applied voltage reduces the required current. This reduction in current decreases resistive losses and the required cable mass and connector size. For this reason, AAM chargers operate at significantly higher voltages than typical ground-vehicle chargers, enabling compact interfaces compatible with the geometric constraints of AAM aircraft. The voltage levels available to support these architectures vary internationally as a function of regional electrical standards and distribution practices. These differences influence how vertiport charging systems interface with local electrical grids, even though the underlying power relationships remain unchanged.

Resistive Heating and Thermal Constraints

Whenever current flows through a conductor, electrical resistance produces heat. The amount of heat generated in a given time interval is approximately proportional to the square of the current multiplied by the resistance, $R\ \left[\Omega\right]$,

$$\frac{Q_{\mathrm{heat}}}{\Delta t} \sim I^2 R \tag{4.2}$$

2 Throughout the chapter, parameters are placed within parentheses (); units are placed within square brackets [].

Here, Q[3] [J] represents the total heat energy produced within the conductor or interface during the time interval, Δt. When considered alongside Eq. (4.1), this relationship illustrates why high-voltage charging architectures are advantageous for AAM operations: for a fixed power level, increasing the applied voltage reduces the required current, and lower current directly reduces resistive heat generation. The squaring of current in Eq. (4.2) shows that even a modest increase in current produces a disproportionately higher heating load. Consequently, high-power charging requires cables, connectors, and battery terminals that can withstand repeated thermal cycles without degradation. Thermal limits inside the battery also restrict allowable current, creating an upper bound on the charging rate regardless of the charger's rated power.

Charge Transfer and Time

During charging, the total electric charge delivered to the battery, Q_{charge} [C], is given by:

$$Q_{\text{charge}} = It \tag{4.3}$$

where $t\left[\text{s}\right]$ is the duration of charging. To replenish a significant portion of the aircraft's usable energy within short turnaround intervals, a large amount of charge must be delivered in a short period. This requires a high average charging current. However, as seen in Eq. (4.2), high current generates heat in both the battery cells and charging hardware because electrical resistance converts part of the electrical energy into thermal energy. The resulting progressive temperature rise is referred to as *thermal buildup*. If any component becomes too warm, charging current must be reduced or paused to protect the battery and maintain voltage, current, and temperature safety margins.

Further, lithium-based batteries used in AAM aircraft do not accept charge at a constant rate, even when thermal limits permit high current. Charging begins with a constant-current phase during which voltage rises, but once the cell voltage approaches its chemistry-defined upper limit, the charger must hold the voltage constant to prevent overvoltage. When voltage is subsequently fixed during the next phase of charging, internal battery resistance causes the current to taper, as illustrated in Figure 4.17, extending the total time required to reach a target state of charge. As discussed in Chapter 2, battery chemistry introduces an additional constraint on allowable charging current and thermal margins. Lithium-polymer chemistries, offering high specific power, generally exhibit narrower thermal and safety margins during charging than the lithium-ion chemistries commonly used in electric ground vehicles. As a result, chemistry selection directly limits acceptable heat generation and further constrains charging rates beyond purely electrical considerations. This tapering becomes more pronounced as cell temperature increases.

Turnaround time emerges from these coupled electrical and thermal processes. For high-frequency AAM operations, this interval influences the number of missions each aircraft can complete in a day, the size of the required fleet, and the number of chargers that a vertiport must provide to sustain scheduled service. A shorter turnaround interval improves pad utilization and network efficiency, while longer intervals increase aircraft dwell time and raise the number of aircraft needed to maintain a given level of service. Since vertiports often serve multiple aircraft on coordinated schedules, any increase in charging time extends the period each aircraft occupies a charger. These extensions can propagate through the network, potentially creating delays or necessitating additional chargers to stabilize operations.

---

3 The symbol Q appears twice in this subsection and represents different physical quantities in different contexts. In Eq. (4.2), Q denotes the thermal energy generated in a conductor in terms of joules, J. In Eq. (4.3), Q denotes electric charge delivered to the battery in terms of coulombs, C. The meaning follows directly from the governing equation in which the symbol is used.

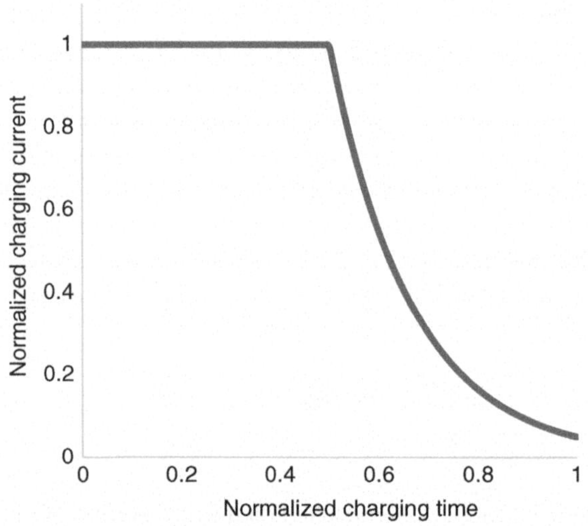

Figure 4.17 Conceptual battery charging profile illustrating the transition from constant-current charging to the constant-voltage phase. As the battery approaches its upper voltage limit, charging current tapers, extending total charge time despite continued energy delivery.

Why AAM Chargers Differ from Ground-vehicle Chargers

Several factors distinguish AAM chargers from those used in electric ground vehicles. AAM aircraft expend energy at a significantly higher rate, particularly during takeoff, climb, and landing, compared to urban driving cycles. Although cruise often accounts for a substantial portion of total mission energy, the vertical and transitional phases produce steep, concentrated energy draws that drive higher recharge requirements following each flight. High-frequency service further intensifies these demands because energy must be replenished quickly to maintain scheduled operations. However, aircraft mass and geometric constraints limit the size of onboard thermal-management systems, so chargers must contribute to maintaining safe cell temperatures during high-rate charging. Accordingly, AAM charging systems actively contributes to thermal management by keeping cell temperatures within allowable limits.

Many vertiports operate on distribution feeders that possess limited capacity. This requires careful integration with local electrical networks and constrains how much power can be delivered at any given time without affecting local power quality. These combined factors create requirements for high-voltage, high-power, and thermally controlled charging systems whose capabilities exceed those of most ground-vehicle chargers.

Safety, Control, and Communication Requirements

High-power charging relies on precise coordination between the aircraft and the charging system (Figure 4.18). AAM chargers use real-time communication to regulate voltage and current, verify connector engagement, monitor temperature, and detect anomalies. Electrical protection systems must prevent overcurrent events, ground faults, or voltage disturbances from reaching the aircraft battery. Safe disconnection requires that the current be reduced to zero before the connector is removed. These control and communication functions protect both the aircraft and the infrastructure and form the basis for interoperable charging across mixed aircraft fleets and charger architectures.

Figure 4.18 Charging the Beta ALIA CX300. Charging interfaces and thermal-management strategies vary across AAM aircraft, requiring vertiport infrastructure to accommodate multiple electrical and operational architectures. *Source*: BETA Technologies / Brian Jenkins / CC BY SA 3.0.

These electrical fundamentals explain operating constraints governing AAM charging systems. But how do these requirements shape the layout and electrical architecture of a vertiport's power distribution system? We answer that question in the next subsection.

4.3.2 Vertiport Electrical Architecture

Reliable AAM operations depend on the ability of a vertiport to receive electrical power from the local distribution feed, condition that power through on-site equipment, and deliver it to aircraft at the required voltage and current (Figure 4.19). This imposes different and often more demanding performance requirements on vertiport electrical architecture than at conventional airports. AAM charging requires much higher instantaneous power, more stringent voltage-regulation and power-quality control, along with equipment designed to withstand repeated high-power cycles. High charging power, repeated duty cycles, and sensitivity to local grid capacity make the configuration of service entrances, transformers, switchgear, distribution panels, and charging cabinets vital to facility design.

A vertiport's connection to the distribution grid begins at the *service entrance*, where three-phase power is received at the local distribution voltage. Urban feeders commonly deliver several kilovolts to several tens of kilovolts depending on regional standards. This incoming power typically passes through a *transformer* that adjusts the incoming medium voltage to the level required by the charging equipment. Transformer sizing depends on the expected charging load, the anticipated number of simultaneous charging events, and the duty cycle imposed by scheduled operations. Downstream of the transformer, the incoming power is then routed into *switchgear* that isolates circuits, provides overcurrent and short-circuit protection, and permits safe operation and maintenance of downstream equipment. Because high-power AAM vehicle charging introduces rapid and high-magnitude load changes, both the transformer and switchgear must maintain acceptable

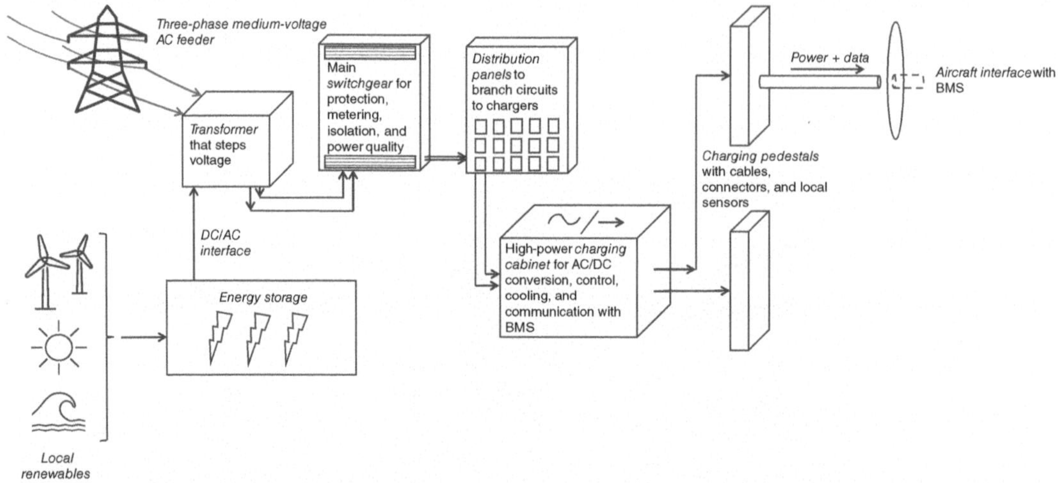

Figure 4.19 Conceptual layout of vertiport electrical infrastructure from the utility connection to the aircraft charging interface.

voltage levels on the serving feeder and limit the propagation of local disturbances into the utility's distribution system.

From the switchgear, power is routed to *distribution panels* that serve both the charging system and auxiliary loads that support ancillary vertiport operations. These panels divide the incoming supply into feeder circuits sized for expected charger loads, as well as circuits serving lighting, ventilation, security systems, passenger-handling equipment, and other building services. Feeder conductors must accommodate the continuous and peak currents associated with fast-charging events and must do so without causing unacceptable voltage drops. Panels and conductors must also satisfy thermal limits to avoid excessive heating during sustained high-demand periods across all connected loads, not only the chargers.

Charger cabinets serve as the power-conversion interface between the vertiport electrical system and the aircraft. These cabinets convert the alternating current (AC) supplied by the distribution panels into direct current (DC) suitable for charging the aircraft battery. During this conversion, the cabinet regulates voltage and current so that delivered power remains within the operational limits defined by the aircraft battery management system (BMS).

Most AAM aircraft rely on DC fast charging for routine operations, although some designs may also accept lower-power AC charging for maintenance or extended dwell periods. In addition to AC–DC conversion, the cabinet houses the sensing, communication, and control functions required to coordinate charging power with the BMS. Because AAM chargers must deliver high power repeatedly throughout the day, each cabinet includes cooling systems that maintain safe operating temperatures for the power-electronics hardware.

Power delivery at the *charging pedestal* completes the electrical pathway. Chargers must maintain stable voltage, regulate current within battery acceptance limits, and coordinate thermal management with the aircraft. They must also remain coordinated with the upstream distribution system so that switching transients, ground faults, or voltage fluctuations do not propagate into the aircraft or back into the grid. Communication links between the charger, the aircraft's BMS, and the vertiport's energy-management software coordinate charging so that voltage, current, and power levels remain within battery safety limits, grid capacity constraints, and the turnaround times set by the operating schedule.

Redundancy is an important consideration in vertiport electrical design because charging interruptions reduce operational capacity immediately. Redundancy can be provided by multiple utility feeders, parallel transformer configurations, or on-site stationary energy storage that allows partial operations to continue during upstream disturbances. The appropriate redundancy level depends on regional reliability standards, fleet size, charger count, and economic trade-offs between capital costs and desired operational continuity.

4.3.3 Charger Design and Interoperability

AAM charging systems must interface with multiple aircraft types varying in battery voltage, maximum allowable current, connector geometry, and BMS communication requirements. Therefore, charger design must satisfy two objectives. First, it must deliver electrical power within the thermal, voltage, and current limits imposed by the aircraft battery. Second, it must provide an interoperable physical and communication interface that allows a shared vertiport to serve mixed fleets without compromising throughput or safety.

AAM chargers combine high-power electronics, cooling hardware, and digital control systems. The charger converts the AC supplied by the vertiport distribution panels into DC suitable for fast charging. Conversion stages regulate output voltage and current with sufficient precision to remain within the limits communicated by the aircraft's BMS. Because AAM charging routinely requires power levels exceeding those of most ground-vehicle chargers, the power electronics must withstand repeated high-power charging cycles without excessive temperature rise that would reduce allowable charging current or compromise battery safety margins. Integrated cooling systems stabilize internal temperatures so that the charger can maintain continuous operation even during periods of high demand.

The charger must also detect the state of the aircraft battery, identify allowable voltage and current ranges, manage thermal constraints, and negotiate charging rates in real time. These functions rely on digital communication between the charger and the aircraft BMS. Communication protocols must provide reliable identification of the aircraft, safe initiation and termination of charging, fault detection, isolation of abnormal conditions, and protective shutdown if limits are exceeded.

The physical interface between the aircraft and charger must be compact, lightweight, and capable of carrying high current safely. With most AAM aircraft relying on DC fast charging, significant power transfer is concentrated through a relatively small connector (Figure 4.20). Therefore, connector geometry must minimize contact resistance, support secure mechanical locking, and include local sensing for temperature, voltage, and connector engagement. Because AAM aircraft vary in size and ground-handling configuration, the connector must also accommodate differences in mounting height, space clearance requirements, and ground-crew accessibility. As a result of these factors, connector design must consider both electrical performance and the operational layout of charging positions on the vertiport.

Interoperability Across Mixed Fleets

Shared vertiports will host aircraft from multiple manufacturers, so interoperability is essential to efficient infrastructure use. *Interoperability* refers to a charger's ability to serve any aircraft whose electrical and communication requirements fall within an agreed-upon standard. Industry working groups are developing common connector geometries, communication layers, and safety functions that allow chargers to identify aircraft, negotiate charging limits, and coordinate disconnection procedures without relying on proprietary protocols. Interoperable designs thus reduce the need for aircraft-specific charger equipment, improve vertiport throughput, and simplify maintenance.

Figure 4.20 Charging connector for the Beta Technologies charging cube. Courtesy of Beta Technologies.

Interoperability also supports scalability. When chargers follow a shared standard, manufacturers can introduce new aircraft into service without requiring upgrades to existing infrastructure, provided the power levels and communication requirements remain within the supported range. This reduces uncertainty for vertiport operators and enables long-term planning based on stable electrical and operational assumptions.

4.3.4 Fast Charging, Managed Charging, and Energy Scheduling

High-frequency AAM service depends on the ability to replenish aircraft energy at a rate that supports high-tempo operations without degrading battery health or exceeding local grid capacity. The interaction between fast charging, managed charging, and operational energy scheduling determines the achievable throughput of a vertiport and the long-term reliability of the energy infrastructure supporting it.

Fast charging shortens turnaround intervals by delivering energy at the highest rate the battery can safely accept. This strategy is essential for maintaining high-tempo operations, particularly when aircraft perform repeated short-range missions that return to the same vertiport frequently. However, fast charging raises battery temperature, accelerates cell aging when excessively repeated, and increases peak power demand on the serving feeder. These effects introduce constraints that must be balanced against the operational benefits of short turnaround times.

Managed charging moderates these constraints by shaping how charging power is delivered in real time. Instead of charging each aircraft at its maximum allowable rate, the vertiport's energy-management system modulates power levels across active charging sessions based on feeder conditions, aggregate demand, and equipment limits. By adjusting charging rates across multiple aircraft, the system prevents simultaneous power spikes that would stress the local grid or exceed transformer and switchgear ratings. Managed charging also enables operators to reduce charging power during periods of high feeder loading and increase power when capacity becomes available. This provides a predictable load profile for grid operators and reduces the risk that fast charging will cause voltage excursions or undesirable thermal cycles in the power electronics.

Energy scheduling integrates these strategies with aircraft mission planning by determining when and in what sequence aircraft access charging resources. The schedule considers which aircraft require immediate turnarounds, which can remain parked at a vertiport pad or designated parking position for longer periods, and which can defer charging until demand on the system subsides. Scheduling algorithms incorporate state-of-charge thresholds, expected trip durations, reserve requirements, and battery-temperature constraints to determine the most efficient sequencing of aircraft across available charging pedestals. When coupled with managed charging, these algorithms ensure that high-power events occur only when necessary and that total demand remains bounded by the battery's allowable operating limits and the feeder's deliverable capacity.

The combined use of fast charging for high-priority turnarounds and managed charging for fleet-level coordination creates a stable energy profile that aligns aircraft utilization with grid constraints. This coordination allows vertiports to support high flight frequency without excessive peak loading or accelerated battery degradation. Consequently, energy scheduling becomes a key determinant of vertiport throughput, grid stability, and long-term fleet availability, linking electrical-infrastructure design directly to operational performance.

4.3.5 Battery Swapping and Alternative Energy-delivery Architectures

Some AAM operational models may employ the use of removable battery modules in place of relying exclusively on direct high-power charging at the aircraft. *Battery swapping* replaces a depleted pack with a fully charged pack during the turnaround interval. In principle, this approach separates the time needed to charge a battery from the time needed to prepare an aircraft for its next flight. The concept can reduce pad occupancy, increase flight frequency, and relieve peak demand on the electrical feeder because charging can occur gradually in a dedicated storage area instead of at the aircraft on the pad. Similar decoupling strategies have been examined across multiple transportation domains over time, reflecting recurring interest in separating energy replenishment from vehicle utilization versus reliance on a single dominant architecture.

Swapping requires an aircraft architecture designed around modular energy packs. The packs must be structurally integrated into the airframe in a manner that maintains stiffness and crashworthiness, while also permitting rapid removal and replacement by ground crews or automated handling systems. Each module must include mechanical latching hardware, high-current electrical interfaces, and connectors that allow the BMS to authenticate, monitor, and control the module during flight. These interfaces must withstand repeated mating cycles without degradation and must maintain insulation, temperature limits, and vibration tolerance required for aviation use.

The vertiport infrastructure required for battery swapping differs from that used for fast charging. A dedicated battery-handling zone is required for the receipt, storage, charging, cooling, and inspection of the battery modules. Enclosures that house the modules during charging must manage heat rejection effectively, especially when multiple modules charge simultaneously and generate sustained thermal loads even at moderate power levels. The facility must also incorporate fire detection and suppression systems designed for lithium-based chemistries. Because the number of modules in circulation increases with flight frequency, the storage area must be sized so that a buffer of fully charged modules is always available without exceeding the capacity of the enclosures housing the batteries.

Energy-delivery scheduling remains essential within a swapping architecture. Modules can be charged during low-demand periods to reduce stress on the local electrical feeder, and the charging power can be modulated to maintain thermal stability and limit peak loads. The scheduling task then shifts from determining how fast a single battery must be charged at the aircraft to determining how a fleet of battery modules is charged, cooled, and rotated to match the pace of operations. Hence, this operational model provides operational flexibility but requires careful tracking of the state of charge and the health of each module to maintain reliability.

Swapping also introduces logistical considerations that influence vertiport design. Modules are heavy and require equipment such as lifts, robotic handlers, or guided carts to move safely between the charging area and the aircraft. The handling path must remain clear of passenger flows and integrate with safety zones around active landing pads. Procedures must ensure that each module is installed correctly, electrical connections are verified before departure, and the removed modules are returned to the storage area for inspection. Because every handling step introduces potential for error, swapping demands rigorous maintenance protocols and ground-crew training.

The feasibility of battery swapping depends on the design philosophy of the aircraft manufacturer, the expected mission frequency, and the economic trade-off between investing in additional battery modules and investing in higher-power charging infrastructure. Swapping can reduce the electrical capacity required at the aircraft interface, but it increases the number of modules that must be purchased, stored, and maintained. It also increases floor-area requirements at urban vertiports where space may be limited. For these reasons, many AAM concepts of operation evaluate battery swapping alongside fixed and mobile charging approaches as part of a broader energy-delivery portfolio, with each architecture aligned to specific operational needs and deployment contexts.

Beyond fixed charging and battery swapping, alternative energy-delivery architectures also include mobile or semimobile charging systems that supplement or temporarily replace fixed charging installations (Figure 4.21). These systems may be deployed during early operations, phased buildout, contingency scenarios, or at sites where permanent electrical upgrades lag operational demand. Compared to fixed installations, mobile charging units trade peak power capability and throughput for flexibility, relocatability, and reduced upfront infrastructure commitment. From a planning perspective, these architectures influence space allocation, traffic management, safety zoning, and energy scheduling strategies, and they are best evaluated as complementary elements within a broader vertiport energy ecosystem.

4.3.6 Site-level Integration and Grid Interaction

Site-level integration with the distribution grid determines the electrical capacity available to a vertiport, the limits on high-power charging, and the extent to which charging operations interact with the surrounding electrical network. These interactions govern whether high-power charging

Figure 4.21 Beta Technologies mini charging cube (left) supplying power to the CX300 aircraft (right). Courtesy of Beta Technologies.

maintains voltage stability, avoids overloading feeder infrastructure, and preserves the coordinated operation of protective systems.

The first determinant is *hosting capacity*, which refers to the amount of additional electrical demand a feeder can accommodate without producing excessive voltage drop, overloading of upstream transformers, or disrupting the coordinated behavior of protective relays, fuses, and breakers. Many urban feeders have limited spare capacity, especially in districts with increasing commercial or residential demand. High-power AAM charging can impose constraints that exceed the limits of the preexisting electrical infrastructure. Here, grid-impact assessments can help quantify available capacity, identify limitations in voltage regulation or thermal loading, and determine whether upgrades are required before vertiport construction commences.

Where hosting capacity is constrained, several strategies can help attenuate the impact of vertiport demand and local grid conditions. On-site energy storage can reduce peak loading by supplying part of the charging power directly and then replenishing its own charge during off-peak periods. Local renewable-energy assets, such as photovoltaic arrays, can offset a portion of daytime demand and reduce dependence on the grid during the most constrained hours. Managed-charging algorithms can schedule aircraft charging to avoid simultaneous high-power events and maintain compliance with feeder capacity limits. These strategies allow a vertiport to operate within the technical constraints of the serving grid while supporting high-frequency operations.

Power quality is another prominent component of site-level integration. Rapid changes in charging load can cause small, fast variations in voltage that may disrupt equipment on or near the vertiport. Accordingly, charging systems must include features that smooth the transition into and out of high-power operation, such as controlled ramp-up sequences and power-electronics designs that limit sudden electrical disturbances. These measures help maintain stable voltages throughout the facility and prevent electrical interference with other users on the same feeder. Coordination with the local utility ensures that charging operations remain within regional standards for acceptable voltage variations and electrical stability.

4.3.7 International Considerations

Although the physical requirements of aircraft batteries remain universal, the supporting infrastructure varies by region because power-system architecture, voltage standards, renewable-energy penetration, and permitting practices differ significantly around the world. AAM charging infrastructure must operate across a wide range of electrical, environmental, and institutional conditions. These regional differences in electrical-system architecture must be considered.

Some countries rely on three-phase distribution networks that deliver medium-voltage power at standardized levels, whereas other regions employ different voltage conventions and feeder configurations. Variations in nominal voltage, frequency, and transformer configuration influence how charging systems are connected to the grid and how power-conversion equipment is specified. These differences leave the fundamental electrical principles of charging unchanged; however, they directly shape equipment ratings, protective-device coordination, and the physical design of switchgear and transformers at the vertiport.

Renewable-energy penetration adds another dimension to international variability. Regions with high levels of solar or wind generation often experience larger fluctuations in power availability across the day. Vertiports located in these regions may rely more heavily on stationary storage to smooth local power delivery and maintain consistent charging performance. In contrast, regions with electrical systems that rely on continuously operating sources, such as large hydroelectric, geothermal, nuclear, or other plants that supply stable round-the-clock output (baseload generation), may support higher instantaneous charging loads but can face constraints on distribution feeders in dense urban settings.

Environmental conditions also affect energy-infrastructure design. Hot climates increase thermal loads on charging equipment and may require augmented cooling for charger cabinets and distribution hardware. Cold regions impose different constraints, including reduced battery acceptance rates at low temperatures and the need to heat certain components to maintain performance. Humid and corrosive environments introduce additional considerations related to moisture ingress, material degradation, and long-term reliability, particularly for outdoor electrical equipment and exposed structural components.

Institutional and permitting processes vary widely across nations. Some regions maintain highly centralized energy-planning structures that require detailed grid-impact studies before new high-power loads can be approved. Other regions permit more localized decision-making, which can accelerate deployment but demands close coordination between vertiport developers and local utilities. These institutional differences can influence project timelines and determine whether upgrades to feeders, transformers, or substations must precede vertiport construction.

Despite these variations, the fundamental engineering principles powering AAM charging remain constant: aircraft batteries must be supplied with stable voltage, controlled current, and safe thermal conditions; charger cabinets must convert power into the form and magnitude required by the aircraft; energy scheduling must preserve both battery health and grid reliability. International variation influences how these requirements are met without altering the requirements themselves.

4.4 Multimodal Transport Hubs: Connecting Ground and Air

As we have seen, AAM extends the vertical dimension of both urban and regional transportation networks, enabling faster point-to-point movement and improved accessibility. However, recall that the practical realization of these benefits depends on how efficiently travelers and goods transition between aerial operations and the ground-based mobility systems that complete most door-to-door journeys. In locations where AAM interfaces with dense surface-transportation activity, a vertiport achieves its full transportation value only if the facilities function as a multimodal hub in which aircraft operations, pedestrian circulation, public transit, micromobility systems, and private vehicles operate as coordinated elements of a continuous mobility chain. This concept builds directly on the land-use and compatibility principles introduced in Section 4.2.

Multimodal integration is essential for realizing the time-saving and reliability advantages that AAM may provide, particularly in metropolitan settings where transfer performance, congestion, and surface accessibility strongly influence total trip time. In regional or intercity applications, the same principles govern how aerial services link with regional rail terminals, intercity bus corridors, logistics nodes, and other surface-transportation gateways that distribute travelers and cargo across wider catchment areas. In all contexts, the degree of integration determines whether AAM operates as an isolated aviation service or a contributing part of a broader transportation ecosystem.

This section introduces the design and operational principles that allow multimodal hubs to function as effective interfaces between transportation modes. Emphasis is placed on spatial organization, operational coordination, and the relationship between multimodal facilities and their surrounding urban context.

4.4.1 Physical and Spatial Integration

Effective multimodal integration begins with the spatial relationships that connect a vertiport to the transportation networks that surround it. Internal and external *circulation*, understood here as the organized movement of people and goods through the facility and its surrounding pathways, determines how efficiently passengers and cargo transition between aerial and surface modes. When circulation paths, vertical movement systems, and ground interfaces are organized as a unified sequence, travelers experience a continuous transfer that preserves the time savings AAM affords. The same coherence must extend into the urban interface, where the connection between the vertiport, the surrounding street network, and adjacent buildings influences wayfinding and accessibility across the first and last segments of a journey. This subsection examines the spatial logic that enables such continuity. The discussion moves from the internal central circulation core that structures movement within the facility to the external placement of the vertiport relative to transit corridors that concentrate pedestrian and vehicular flows.

Continuous Passenger Circulation

Transfer distance remains a primary determinant of total trip time. Even modest increases in walking distance or unnecessary changes in elevation erode the time savings that motivate AAM use. Access paths between the vertiport and nearby transit platforms, sidewalks, and curbside zones must therefore minimize horizontal distance, avoid circuitous routing, and provide a direct and unambiguous sequence from ground level to the boarding area. The transfer between the terminal and the TLOF/FATO must respect required safety clearances yet remain compact enough to preserve efficient movement.

Passenger circulation must function as a continuous trajectory instead of a series of disconnected segments. Corridors must avoid abrupt directional changes, crossflows created by converging pedestrian streams, and spatial pinch points that interrupt movement patterns. When intermediate levels are used to accommodate ticketing, screening, or amenities, their placement must maintain the overarching alignment of the circulation spine so that passengers perceive a single, coherent path from entry to boarding. Predictability is reinforced through consistent wayfinding that uses uniform visual cues, protected interior walkways that guide movement, sightlines that provide clear views of the next route segment, and lighting that supports intuitive navigation across multiple levels (Figure 4.22).

Vertical circulation often forms a crucial component of this continuous movement sequence. Any urban vertiport that occupies an elevated structure requires passengers to move efficiently between street level and the boarding platform. Elevators, enclosed stairs, and escalators must be sized to support peak-period demand and permit safe, orderly evacuation under emergency conditions. Because demand in high-frequency AAM operations fluctuates over short intervals, the vertical spine must accommodate short-duration surges without degrading the user experience or slowing transfer performance.

Weather protection also influences circulation performance in elevated environments. Transitions from enclosed building cores to outdoor boarding platforms must shield passengers from wind, precipitation, solar exposure, and rotor-induced DWOW during aircraft operations. Canopies, vestibules, wind screens, and related architectural elements must be integrated into the circulation spine as integral components of the system and not appended secondarily, since consistent environmental shielding preserves steady movement and prevents delays caused by discomfort or shelter-seeking. Because these architectural elements can meaningfully influence local wind fields, vertiport design in dense or elevated settings requires site-specific wind-flow analysis

Figure 4.22 Airport wayfinding signage assisting in vertical circulation. *Source*: PaulSyms22 / Wikimedia Commons / CC BY-SA 4.0.

conducted by qualified specialists using computational or physical simulation methods that resolve structure-induced flow behavior in the vicinity of the TLOF and FATO. Enclosed bridges or vertically oriented access shafts can moderate passenger exposure to downwash and noise by providing protected transition spaces, but they must be positioned and shaped so they do not introduce localized turbulence, recirculation, or acceleration of flow that could degrade aircraft control margins during takeoff and landing or alter airflow patterns within the immediate operating envelope.

Safety Zoning and Operational Separation

Physical integration must be organized around a spatial framework that preserves operational safety. Pedestrian movement, ground-service functions, and aircraft operations will occur within close proximity at a vertiport, and safety zoning establishes the spatial logic that keeps these activities partitioned and organized.

The most critical boundary lies between aviation areas and public circulation. The TLOF, FATO, and the associated DCA require controlled access and obstacle-free surfaces implemented through physical barriers, controlled doors, and unmistakable visual cues that extend beyond signage alone. Because the DCA can extend beyond the immediate landing surface and vary with aircraft operating state, access control must account for both geometric boundaries and flow-induced hazard zones. This spatial clarity prevents inadvertent entry into active aircraft zones and ensures that ground-handling personnel access the apron only along designated, protected routes.

Ground-service functions require similar separation. Charging cabinets, maintenance shelters, and battery-swapping facilities must be positioned so that technicians can reach them without crossing public pathways. These support areas must reflect the operational tempo of the facility, particularly at high-utilization vertiports where equipment access must remain independent from the primary passenger circulation corridor. Cargo movement, likewise, requires its own circulation system. Carts, trolleys, or small autonomous ground vehicles require dedicated vertical access and loading paths that do not intersect with passenger areas and that accommodate turning radii, floor loads, and rapid access to delivery points.

Emergency-response access must remain fully independent of normal circulation. Fire lanes, staging zones, and evacuation routes must remain unobstructed, with vertical-access provisions that allow responders to reach aircraft and terminal areas without navigating through passenger queues. These pathways must remain aligned with the building's structural core to support direct and efficient evacuation during emergency conditions.

Safety zoning also informs the placement of multimodal interfaces. Curbside pickup zones, micromobility docks, and bicycle storage must not interfere with pedestrian routes or intrude into protected apron envelopes. Curbside areas used by private vehicles or shuttles must be located so that passengers do not cross operational roadways or move into zones influenced by rotor-induced flow. Similarly, lightweight micromobility devices must be sited outside regions affected by downwash to avoid hazards associated with displaced or airborne equipment.

Urban Interface and External Connectivity

The external interface between the vertiport and the urban environment determines how travelers access the facility and how its presence influences the surrounding public space. This interface must be designed purposefully as a functional component of multimodal performance and not treated as a residual boundary.

The curbside environment establishes the first movement vector in a passenger's journey. High-frequency AAM service produces concentrated arrival and departure pulses across private vehicles, TNCs, micromobility services, and transit shuttles. If these flows converge onto a single curb segment or driveway, delays propagate into the street network and degrade total trip performance. Distributed access points, dedicated pull-off lanes, and managed curb allocations preserve operational stability and must align with the circulation spine to ensure predictable movement from street level to boarding and back.

Where a vertiport is integrated into an existing building, the facility must connect to primary circulation paths instead of secondary corridors or isolated elevators. Direct alignment with the building's internal geometry reduces decision points, avoids crossflows with unrelated activities, such as retail or office functions, and maintains predictable routing. Structural considerations also govern this integration as the host building must support the additional loads associated with the vertiport deck, and this requirement dictates the placement of circulation cores, mechanical rooms, and staging areas.

The character of adjacent public space influences how individuals perceive and use the facility. Entrances and waiting areas must preserve visibility, safety, and comfort while accommodating the localized environmental and acoustic effects of nearby aircraft operations. Architectural and landscape treatments must mitigate downwash, noise, and solar exposure without intruding on protected airspace or altering approach geometry. In mixed-use districts, this interface determines whether the vertiport enhances or competes with the surrounding pedestrian realm.

Finally, accessibility within the immediate pedestrian interface between the vertiport and the surrounding transportation network must be treated as a primary design requirement. Pedestrian pathways must support direct, unobstructed movement between the vertiport entrance and adjacent modes without introducing conflicts with vehicular circulation. Transit stops must fall within an appropriate walking radius, and their spatial alignment with the vertiport must limit interaction with vehicular circulation. In regional or intercity implementations, these interfaces must support larger catchment areas and a wider range of passenger profiles, including those with luggage or mobility constraints.

4.4.2 Operational Coordination

Operational coordination describes the coordination of timing, ground processes, and user-facing systems that enable a multimodal vertiport to operate as a coherent, integrated system. While our discussion of physical integration has examined how people and aircraft must move, operational coordination addresses when these movements must occur and how they are synchronized. In a high-frequency AAM environment, even small mismatches in timing or ground handling can propagate quickly across the network and erode the travel-time advantages that AAM is intended to provide.

Temporal Alignment Across Air and Ground Modes

The temporal performance of a multimodal vertiport depends on the consistent alignment between aircraft cycles and surface-transportation rhythms. AAM operations typically occur in short, concentrated pulses aligned with arrival, turnaround, and departure at each stand. These pulses generate brief surges in passenger demand at boarding platforms, vertical movement elements, concourses, and curbside zones. If the arrival of transit vehicles, rideshare services, and private vehicles does not align with these passenger peaks, travelers encounter longer waiting periods or congestion in staging areas.

Effective temporal alignment begins with an integrated scheduling framework. Aircraft rotations, transit timetables, and expected passenger-arrival distributions must be planned as interdependent processes. Because early AAM operations are sensitive to localized weather conditions, including wind, visibility, and convective activity, scheduling frameworks must also account for increased likelihood of short-notice delays, ground holds, or service suspensions. In settings where rail or bus services operate at high frequency, the tolerance for timing misalignment is naturally larger because successive vehicles arrive often enough to absorb small deviations in aircraft timing. In locations with infrequent transit arrivals, this alignment requires careful planning of departure offsets, appropriately sized waiting areas, and managed curb allocations so that passengers arriving from one mode can reliably connect to another without excessive delay.

Internal operational activities must also follow predictable temporal patterns. Charging cycles, ground-handling tasks, and required safety checks determine the minimum achievable turnaround time. Variability in any of these tasks reduces schedule adherence. An effective operational plan incorporates buffers calibrated to observed variability in charging duration and ground operations, with margins derived from realistic operating conditions. When buffers are too small, cascading delays can emerge; when buffers are too large, infrastructure potential is underutilized. Data from early operations can provide the statistical basis for calibrating these margins so that the vertiport maintains both reliability and throughput.

Urban congestion introduces additional temporal uncertainty. Travel times for personal vehicles, rideshare services, and vertiport shuttles fluctuate throughout the day. This alters curbside arrival patterns. Traffic-signal coordination, designated pull-out lanes, and managed curb allocations can help dampen this variability and maintain a more regular flow that aligns with aircraft departure and arrival windows. In mature multimodal systems, predictive tools that combine traffic data, transit feeds, and aircraft status can refine recommended passenger arrival times and reassign gates or stands to preserve overall schedule integrity.

Passenger-flow Management and Staging

Passenger-flow management links operational timing to the physical circulation systems described in Section 4.4.1. Even when schedules across modes are well aligned, internal movement must remain continuous so that short-term demand peaks do not create chokepoints. The organization of staging and processing areas is central to this objective.

Processing steps such as ticket verification, security screening, and boarding-group organization must follow the dominant direction of passenger flow. When these functions are positioned out of sequence, travelers accumulate in locations that obstruct corridors or vertical movement nodes, increasing dwell times and reducing effective capacity. A clear, consistent order of operations reduces backtracking, prevents crossflows between opposing streams of passengers, and maintains the continuity required for rapid transfers.

Holding areas near boarding platforms regulate the release of passengers into the final pre-boarding sequence. The capacity and placement of these areas must reflect aircraft seat counts, turnaround cadence, and typical early-arrival behavior. Oversized holding areas may be acceptable when positioned so that waiting passengers are removed from through-movement paths. Undersized or poorly located holding areas, by contrast, allow queues to spill into circulation corridors, which degrades performance across the entire hub. When designed and located appropriately, holding areas buffer natural fluctuations in arrival without obstructing movement.

The interaction between passenger-flow management and vertical circulation is equally important. Elevators, escalators, and stairs must discharge passengers at rates consistent with downstream processing capacity. Over-delivery produces queuing and obstruction; under-delivery dilutes arrival pulses and weakens coordinated scheduling. These dynamics may be most pronounced in rooftop or elevated vertiports where vertical transitions form the primary conduit between ground modes and aircraft.

Ground-handling Coordination and Airside Interface

Ground-handling coordination determines how efficiently an arriving aircraft is prepared for its next departure and how seamlessly this preparation integrates with passenger movement. A clear, predictable sequence of airside tasks reduces variability and supports reliable scheduling. Once an aircraft reaches its designated stand, ground personnel must secure the vehicle, verify propulsion shutdown, establish safe rotor or propeller conditions, connect energy-delivery interfaces, including charging, and complete required inspections. These actions must be organized so that they do not conflict with deplaning or boarding and so that no single step becomes a recurring source of delay.

Energy replenishment is often the pacing element of ground handling, regardless of the underlying propulsion architecture. Energy-delivery systems, including charging, battery-swapping, or refueling architectures, must be positioned so that energy-delivery interfaces can reach the aircraft without repositioning the vehicle or crossing active passenger routes. For battery-electric systems, charging-curve tapering and battery thermal constraints limit how quickly usable energy can be added, and delays in initiating charging reduce scheduling flexibility on later flights. Maintaining a predictable energy-connection window becomes essential for preserving alignment across the multimodal network.

4.4.3 Urban and Regional Integration

Multimodal vertiports do not operate in isolation. Each facility occupies a specific position within an urban and regional network of streets, transit corridors, logistics chains, and patterns of land use. The spatial distribution of vertiports, the types of neighborhoods they serve, and the way surrounding development responds to their presence collectively determines whether AAM improves regional accessibility or reinforces existing imbalances in mobility. Therefore, urban and regional integration addresses both the spatial placement of multimodal hubs and the reciprocal relationship between these facilities and surrounding land-use patterns, travel behavior, and evolving community priorities.

Network Role and Catchment Structure

At the regional scale, multimodal vertiports can be evaluated as nodes within a hierarchy of transportation facilities, using analytical frameworks commonly applied in airport, rail, and intermodal planning. Some locations are planned or assessed as primary nodes intended to interface with major employment centers, regional rail terminals, or long-distance bus stations. Others are assessed as secondary facilities that serve more localized passenger or cargo demand within neighborhoods, suburban centers, or industrial districts. This hierarchical distinction provides a planning lens for examining how vertiport location influences flight patterns and the distribution of surface-access flows feeding each facility.

Each facility serves a catchment area, which is the geographic region from which travelers can reasonably access the vertiport within an acceptable time. Catchment size depends on surface-transportation speed, transit availability, and the quality of walking and cycling connections. In dense urban cores, catchments may be compact but highly populated because transit and micromobility provide rapid access. In suburban or exurban settings, catchments may extend farther, but access relies more heavily on private vehicles or specialized shuttles. Planners must evaluate these catchment characteristics explicitly, since they determine which populations can realistically use AAM services and how much surface travel precedes and follows each flight.

Network integration also requires that vertiports are evaluated in relation to existing transportation gateways to avoid unintended duplication and to clarify their intended role within the broader system. In many cases, effective planning positions vertiports to complement established hubs, extending the reach of surface networks and improving system efficiency. A facility placed adjacent to a regional rail station can act as an aerial extension of the rail network, enabling passengers to shift from long-distance travel to localized aerial distribution. A vertiport anchored to a bus rapid transit corridor can expand that corridor's reach into areas lacking direct rail service. In other contexts, vertiports may be deliberately sited to serve specialized markets or mission-specific demands, such as premium services or time-critical logistics, where overlap with existing gateways reflects economic or operational priorities. When these relationships are deliberately planned, the result is a cohesive multimodal network. If these relationships are ignored, they risk diverting riders from essential surface modes without yielding net regional mobility gains.

Redundancy and resiliency at the network scale also influence integration quality. A system that depends on a single hub becomes vulnerable to local disruptions, such as severe weather, grid faults, or facility-specific incidents. Distributing vertiports across multiple corridors, while maintaining strong multimodal connections at each location, creates alternative routing options during temporary interruptions. In this way, AAM contributes to regional transportation resilience by avoiding the creation of new systemic vulnerabilities.

Land Use, Station-area Form, and Development Patterns

The introduction of a multimodal vertiport modifies the development potential of the surrounding area. Just as heavy rail or bus rapid transit stations often stimulate transit-oriented development, high-quality AAM access can increase the attractiveness of adjacent parcels for office, hospitality, logistics, or mixed-use projects. Consequently, the resulting development pattern influences both economic outcomes and the facility's ability to support long-term mobility and community objectives.

Planning for the area surrounding a vertiport must consider land-use compatibility across both operational and community-facing dimensions. Vertiport activity introduces requirements related to acoustic conditions, safety buffers, protected airspace, and access needs, and these requirements limit what types of development can occupy nearby parcels. At the same time, the existing pattern of land uses around the site influences who will use the facility, at what times of day, and for what travel purposes. Concentrations of employment and institutional uses may generate strong peak-period demand, while mixed residential and commercial districts may produce more distributed, off-peak usage. Aligning land-use patterns with anticipated service rhythms increases stability in demand and supports efficient scheduling.

Building form near multimodal hubs also matters. Mid-rise and high-rise structures adjacent to a vertiport can generate obstacles and flow disturbances that influence approach and departure paths. Height and massing must be coordinated to protect airspace and preserve safety zones. Conversely, building design can strengthen multimodal performance when podium levels incorporate vertiport access, when vertical-circulation cores align with structural systems, and when ground-level uses sustain active pedestrian presence without obstructing movement routes.

Parking and curb space demand constitute additional land-use concerns. Multimodal hubs attract vehicles that drop off or pick up passengers and cargo. If surrounding development increases without coordinated curb space or parking, localized congestion can erode the travel-time advantages of AAM. Consequently, land-use planning must reserve space for staging, micromobility storage, and loading functions that match the operational cadence of the facility.

Equity, Community Outcomes, and Long-term Adaptation

Urban and regional integration also involves equity considerations. Vertiport location, AAM service pricing, and the quality of connecting surface transportation collectively determine which communities experience improved mobility and which do not. Without intentional planning, new AAM infrastructure may aggregate in already advantaged districts, exacerbating existing disparities in travel time, access to employment, and shouldering of environmental burdens.

Evaluating equity at the network scale requires mapping proposed vertiport locations against demographic indicators such as income, car ownership, transit dependency, and current travel times to major employment centers or essential services. Locating facilities along corridors with limited existing high-capacity transit can reduce structural disadvantages in access. Conversely, a network that concentrates vertiports in already well-served central business districts may offer marginal regional benefit.

Long-term adaptation signifies the final dimension of integration. The spatial distribution of vertiports, the design of surrounding districts, and the communities served will not remain fixed. Aircraft performance will evolve, routes will be adjusted, and urban development pressures will shift. A network that is tightly coupled to a single vehicle type or a narrow service model becomes difficult to adapt when technology or demand changes. By contrast, vertiports that preserve expansion space, flexible layouts, and scalable electrical infrastructure can accommodate new aircraft geometries, higher service frequencies, or new operating concepts without burdensome reconstruction.

At the regional level, planning frameworks can support adaptation through corridor-based strategies that coordinate vertiport development across multiple sites. Identifying priority aerial corridors that link major activity centers, and reserving a range of candidate nodes along those corridors, allows future vertiports to be added or reconfigured as operations mature. This corridor-based approach aligns AAM with long-term regional plans for transit, land use, and regional resilience and reduces the risk that early siting decisions constrain future options.

4.5 Summary

AAM depends on a robust, coherent, and rigorous infrastructure system. Vertiports form the core physical interface between aerial operations and the surface environment. Their spatial configuration, structural design, and internal organization determine how safely and efficiently aircraft can land, recharge, and reenter service. These facilities must accommodate demanding operational loads, integrate charging or refueling systems capable of supporting high-frequency missions, and provide clear, predictable movement pathways for passengers and cargo.

Land-use planning, zoning frameworks, and multimodal connectivity define how each facility contributes to the broader mobility ecosystem. Vertiports must align with protected airspace, established transportation corridors, and local utility capacity while supporting equitable access and minimizing environmental and community impacts. Charging and refueling infrastructure must integrate with local electrical grids through managed charging strategies, and, in some cases, alternative architectures such as battery swapping. When these components operate seamlessly together, multimodal hubs link aerial service with rail, bus, ridesharing, micromobility, and pedestrian systems in a continuous sequence. The effectiveness of this integration establishes whether AAM functions as an isolated aviation option or as a coherent element of a broader metropolitan and regional transportation network.

References

1 Federal Aviation Administration (FAA). (2024). Engineering Brief No. 105A: Vertiport Design. Washington, DC: U.S. Department of Transportation. https://www.faa.gov/airports/engineering/engineering_briefs

2 European Union Aviation Safety Agency (EASA). (2023). Prototype Technical Specifications for Vertiports. Cologne, Germany: EASA. https://www.easa.europa.eu/en/document-library/general-publications/prototype-technical-design-specifications-vertiports

3 Civil Aviation Safety Authority. (2022). Advisory Circular AC 139.V-01: Guidance for Vertiport Design. Canberra, Australia: Civil Aviation Safety Authority (CASA). https://www.casa.gov.au/

4 General Civil Aviation Authority. (2025). Civil Aviation Regulation (CAR) AGA Part VFI (Vertical Flight Infrastructures). Abu Dhabi, UAE: General Civil Aviation Authority (GCAA). https://www.gcaa.gov.ae/en/epublication/civil-aviation-regulations

5 Federal Aviation Administration. (2024). Electric Vertical Takeoff and Landing (eVTOL) Downwash and Outwash Surveys (DOT/FAA/TC-24/42). Washington, DC: U.S. Department of Transportation.

6 National Fire Protection Association. (2024). NFPA 418: Standard for Heliports and Vertiports. https://www.nfpa.org/

7 National Fire Protection Association. (2024). NFPA 855: Standard for the Installation of Stationary Energy Storage Systems. https://www.nfpa.org/

8 Federal Aviation Administration (FAA). (2024). 14 CFR Part 77: Safe, Efficient Use, and Preservation of the Navigable Airspace. Washington, DC: U.S. Department of Transportation. https://www.ecfr.gov/current/title-14/chapter-I/subchapter-E/part-77

5

Regulatory and Policy Considerations

Lauren Lacey Haertlein[1]

> *Planes do not wander about in the sky like vagrant clouds. They move only by federal permission, subject to federal inspection, in the hands of federally certified personnel and under an intricate system of federal commands. Northwest Airlines, Inc. v. Minnesota, 322 U.S. 292, 303 (1944) (Justice Jackson, concurring)*

Regulations are rules issued by an authoritative body, often a government agency, to govern a particular subject or type of conduct. For example, regulations may address how a specific activity or industry sector may operate. Aviation is widely recognized as one of the most heavily and comprehensively regulated industries in the world.[2]

And around the world, these aviation regulatory landscapes are evolving to support the introduction and integration of advanced air mobility (AAM) aircraft and operations. Many of the current aviation regulations were not originally written with AAM in mind. Regulatory bodies are working to address the necessary updates and, in some cases, creating new regulations to support AAM. But regulatory processes are often slow and notoriously struggle to keep up with rapid technological advancements.

This chapter focuses on the regulatory and policy considerations surrounding AAM. It uses the United States as a lens through which to examine the challenges and opportunities for AAM regulation. As the largest aviation market in the world, with the busiest and most complex airspace, developments in the US regulatory framework for AAM can be influential for both the global industry and other authorities.

Before we start, it is important to note that AAM regulation is a rapidly evolving area of law. By the time this chapter is published, some of the information in it may already be outdated!

---

1 All information relied on in this chapter was drawn from publicly available sources. Any opinions expressed in this chapter are solely the author's own and do not necessarily reflect the views of the author's employer and affiliations.

2 H.R. REP. NO. 103-525, pt. 2, at 6 (1994), reprinted in 1994 U.S.C.C.A.N. 1644, 1647 (recognizing that aviation products are regulated "to a degree not comparable to any other" industry).

5.1 The US Regulatory Framework for AAM

5.1.1 Federal AAM Legislation

In the United States, aviation is principally regulated at the national level by the federal government. State and local government regulation of aviation generally is limited to complementary roles and areas traditionally reserved for state and local authorities—such as police powers to promote and protect health, safety, general welfare, and land use (e.g., planning and zoning). Many of these local activities, such as land use and zoning, were already touched upon in Chapter 4.

Although several federal agencies regulate aspects of aviation, the **US Department of Transportation** (DOT)—and the **US Federal Aviation Administration (FAA)**, an agency within the DOT—have the sole and exclusive authority to regulate civil aviation safety.[3] This includes, but is not limited to, the regulation of aircraft design and manufacturing, operations, and maintenance; the regulation of airspace; and the qualification and certification of pilots, dispatchers, and maintenance technicians. The FAA also regulates and operates the US air traffic control (ATC) system.

Although AAM implicates novel aircraft and operations, it generally falls within the FAA's *existing* authority to regulate civil aviation. Nonetheless, existing federal aviation regulations were not created with AAM in mind. Successfully integrating AAM into the US national airspace system (NAS) requires updating federal laws to fill gaps and resolve conflicts in current legislation and regulations. In some cases, it requires creating completely new rules specifically for AAM. Both the legislative and rulemaking processes, however, are complex and can take years to complete—making it difficult for the law to keep pace with rapid technological advancements in AAM. As of March 2026, AAM-specific federal legislation remains relatively limited.

In 2022, the US Congress passed the first federal legislation defining the concept of AAM, **the Advanced Air Mobility Coordination and Leadership Act**.[4] Specifically, this law defined AAM as "a transportation system that moves people and property by air between two points in the United States (U.S.) using aircraft with advanced technologies, including electric aircraft, or electric vertical takeoff and landing (eVTOL) aircraft, in both controlled and uncontrolled airspace." The Act directed the Secretary of Transportation to establish an AAM Interagency Working Group (IWG), and tasked it with facilitating cross-agency collaboration and coordination to foster innovation, ensure safety, and promote a cohesive regulatory framework for the adoption and implementation of AAM in the United States. The AAM IWG was specifically directed to develop an AAM national strategy.

Two years later, Congress passed an even more significant and directive piece of federal AAM legislation when it included the first-ever **Advanced Air Mobility subtitle** (Subtitle B) in the **FAA Reauthorization Act of 2024**.[5] With the AAM subtitle, Congress made an express commitment to US leadership in AAM, stating "[t]he United States should take actions to become the global leader in advanced air mobility." This landmark, bipartisan legislation detailed funding priorities and FAA directives specifically designed to enable new entrants and foster innovation

---

3 49 U.S.C. § 44701; see also *Montalvo v. Spirit Airlines*, 508 F.3d 464, 472 (9th Cir. 2007) ("[W]hen we look to the historical impetus for the FAA, its legislative history, and the language of the Act, it is clear that Congress intended to invest the Administrator of the Federal Aviation Administration with the authority to enact exclusive air safety standards.").

4 Public Law No. 117-203, 136 Stat. 2227.

5 **FAA Reauthorization:** The FAA is subject to intermittent reauthorization by the US Congress, typically every five years. With reauthorization, Congress considers not only the FAA's authority and funding to operate, but also the agency's priorities. The 2024 Reauthorization extends through Fiscal Year 2028.

with streamlined processes and rules. It also included several provisions on airports, vertiports, and infrastructure, including provisions related to charging equipment and meeting future energy demands for electric aircraft. It further directed the FAA to initiate rulemaking and issue policies necessary for future commercial AAM operations—prioritizing aircraft design approvals and rules for operations and pilot requirements.

In addition to legislative efforts, the executive branch has also directed specific actions related to AAM. In 2025, the US President issued an executive order (EO)[6] titled "Unleashing American Drone Dominance," aimed at promoting the commercial development and integration of uncrewed aircraft systems (UAS) in the United States by redressing certain regulatory burdens and creating new incentives. Although its title emphasizes drone technology, the EO's directives extend beyond UAS. The EO specifically instructs the FAA to establish an eVTOL Integration Pilot Program (eIPP), a program for eVTOL operational integration projects, modeled after the FAA's UAS integration program ("FAA's BEYOND Program").[7]

The eIPP aims to accelerate the deployment of eVTOL operations in select, strategic US markets. To achieve this, the program calls on state and local government entities to form public–private partnerships with AAM companies and propose pilot projects to test and validate operational concepts and demonstrate the viability of eVTOL operations. These projects are intended to provide the FAA with operational data to help inform future regulatory efforts and accelerate the scaled commercialization of eVTOLs. In March 2026, DOT announced that it had selected eight projects from more than 30 applicants for the eIPP. The program is expected to run for three years after the first project becomes operational.

The FAA also has been leveraging lessons from other federal bodies' work on AAM integration, including the National Aeronautics and Space Administration's (NASA's) AAM research and the US Air Force's "innovation arm," AFWERX. Both NASA and defense organizations have played key roles in researching AAM integration and supporting technologies.

Additionally, at the conclusion of 2025, the US DOT released the National Strategy and Comprehensive Plan. Written by the AAM IWG, the National Strategy and Plan was the product of almost three years of work by over 100 subject matter experts from more than 25 federal agencies. The National Strategy outlines findings and recommendations organized into six "pillars of a strong AAM system": (1) airspace, (2) infrastructure, (3) security, (4) community planning and engagement, (5) workforce development, and (6) automation:

> "The future of aviation is taking flight. In the years ahead, AAM may redefine how people fly and how communities connect, ushering in a new era of possibilities. Guided by key recommendations, the U.S. Government will advance initiatives that promote accessibility, affordability, and aviation autonomy. This next chapter in aviation capability is coming to the skies—and the AAM National Strategy and Comprehensive Plan offer a closer look at how it will unfold."
>
> — *US DOT*

6 **EO:** An EO is a directive from the US President on the operations of the federal government. EOs are not legislation and do not require congressional approval, but they do have the force of law for the executive branch, which includes federal agencies.

7 The UAS Integration Pilot Program (IPP) was established by Presidential Memorandum in 2017 to bring together state, local, and tribal entities and industry to test and evaluate the integration of UAS operations. The program was designed to assist the FAA in developing new rules to support complex, low-altitude UAS operations. When the initial program concluded in 2020, the FAA established a new program, called BEYOND, to continue the partnerships and progress made under the IPP.

The National Strategy acknowledges that most of its recommendations will require many years of work to be successfully implemented. It emphasizes the importance of interagency and stakeholder coordination. It also discusses opportunities to facilitate more immediate objectives, such as using existing or repurposed infrastructure for near- and medium-term AAM operations, and leveraging existing air traffic modernization efforts to enable greater automation and advance airspace management.

5.1.2 Federal AAM Regulations

The FAA's rules for civil aviation, called the Federal Aviation Regulations, are found in Title 14 of the US Code of Federal Regulations (14 C.F.R., *Aeronautics and Space*).[8]

AAM, eVTOL, or Powered-lift?

The FAA's *Aeronautical Information Manual* (*AIM*) is the agency's official guidance on flight information and ATC procedures—containing the aeronautical knowledge necessary to fly in the US NAS. The *AIM* defines AAM as "a rapidly-emerging, new sector of the aerospace industry which aims to safely and efficiently integrate highly automated aircraft into the NAS. AAM is not a single technology, but rather a collection of new and emerging technologies being applied to the aviation transportation system, particularly in new aircraft types. Notional AAM use-cases include Urban Air Mobility (UAM), Regional Air Mobility (RAM), public services, large cargo delivery, and private or recreational vehicles." AIM 11-6-1. Many of these use cases will be discussed further in Chapter 8.

As the AIM notes, AAM is an umbrella term. AAM is not defined in FAA regulations; in fact, the term does not even appear in FAA regulations. FAA regulations also do not contain the term **eVTOL**. eVTOLs come within the FAA's regulatory definition of "aircraft": "a device that is used or intended to be used for flight in the air." 14 C.F.R. § 1.1. eVTOLs also meet the FAA's regulatory definition of **powered-lift**: "a heavier-than-air aircraft capable of vertical takeoff, vertical landing, and low speed flight that depends principally on engine-driven lift devices or engine thrust for lift during these flight regimes and on nonrotating airfoil(s) for lift during horizontal flight." 14 C.F.R. § 1.1.

5.1.3 FAA Regulations: Certification of AAM Vehicles

Because AAM vehicles are aircraft,[9] the approval or "certification" of AAM vehicles falls under the FAA's existing authority to regulate aircraft safety. Congress empowered the FAA to comprehensively regulate the safety of the designs and manufacturing of aircraft and aviation products. FAA regulations also govern post-certification maintenance, modifications to approved designs, and the safety of products in service throughout their lifespan.[10]

To meet its congressional mandate, the FAA issues several types of vehicle-related certifications, including:

---

8 Historically, the Federal Aviation Regulations have sometimes been referred to as "the FARs" and some sources may still use this nomenclature. The US DOT and the Office of the Federal Register, however, do not allow the use of "FAR" to refer to FAA regulations. Within the US government, FAR refers to the Federal Acquisition Regulations, which are the rules for government procurement. FAA Order 1320.46D, *FAA Advisory Circular System*.

9 14 C.F.R. § 1.1.

10 49 U.S.C. § 44704.

- an **airworthiness certificate**, which approves a specific aircraft for operation in flight;
- a **type certificate** (TC), which approves an aircraft (or engine or propeller) design; and
- a **production certificate** (PC), which approves the manufacturing (duplication) of an approved design.

Airworthiness Certification: Certification of Individual Aircraft

With limited exceptions, operating a civil aircraft in the United States requires an FAA airworthiness certificate.[11] An airworthiness certificate authorizes the operation of *a specific aircraft* in flight. The FAA issues an airworthiness certificate only if the aircraft meets its design requirements and is in a condition for safe operation.[12]

There are two different types of FAA airworthiness certificates: **standard** and **special**.

- A **standard airworthiness certificate** allows for aircraft operations with the fewest restrictions—including, for example, the transportation of people or property for compensation or hire. It also generally requires the aircraft to have an FAA-approved design or "type certificate." As discussed in further detail below, several AAM aircraft manufacturers are seeking type certification for their aircraft, but as of March 2026, the FAA had not issued any type certificates for civil AAM aircraft—accordingly, no AAM aircraft had received a standard airworthiness certificate.
- A **special airworthiness certificate** is issued in specific categories, the most common of which is "experimental." The FAA issues an experimental airworthiness certificate for a specific purpose or purposes, including research and development and exhibition activities. The FAA has issued special airworthiness certificates to AAM aircraft to allow them to operate for development and flight testing—including in pursuit of a type certificate—as well as for demonstrations—such as at air shows. Special airworthiness certificates generally come with operating limitations, such as a restriction on use for commercial operations.

Special Flight Permits

The FAA may issue a **special flight permit** (also known as a "ferry permit") for an aircraft that is *unable to meet applicable airworthiness requirements* but is capable of safe flight for specific operations. Special flight permits are temporary authorizations often issued for activities such as production flight testing, demonstration flights for potential customers, and transporting an aircraft to a base where inspections or repairs will be performed.

TC: FAA Design Approval

TC is the FAA's approval of an aircraft's design and all its component parts.[13] How does TC work? The process generally starts with the FAA conducting orientation or familiarization briefings with the type certificate applicant to understand the proposed product. The FAA then establishes the certification basis, identifying all applicable regulations and conditions that must be met for FAA approval. The applicant then submits a detailed certification plan, explaining how it will demonstrate compliance with the requirements outlined in the certification basis. Once the FAA approves that plan, the applicant implements it, conducting extensive tests and analyses. The applicant

11 14 C.F.R Subpart H, *Airworthiness Certificates.*
12 *Ibid.*
13 The FAA issues TCs for airframes, engines, and propellers.

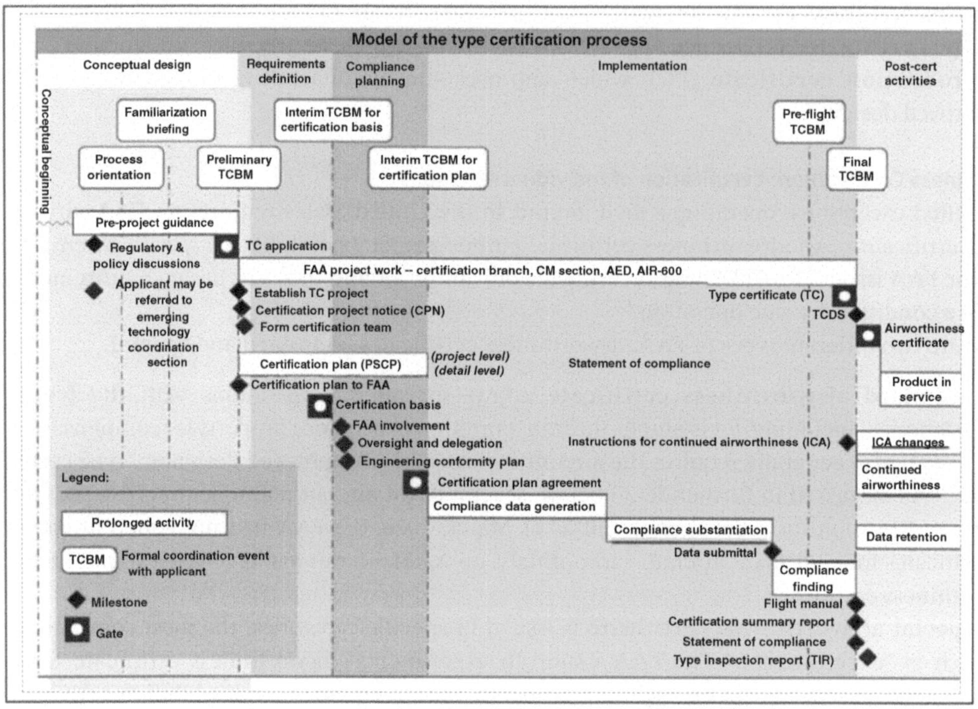

Figure 5.1 Typical TC process. *Source*: FAA Order 8110.4C Chg 7/Public domain.

generates, substantiates, and documents compliance data, and the FAA reviews it and makes findings for each certification requirement. See Figure 5.1.

The FAA has primarily focused its initial efforts on type certificating one specific kind of AAM aircraft: powered-lift aircraft, including "eVTOLs," with a pilot onboard. Several AAM manufacturers are seeking TC, but these aircraft include one or more novel design features not fully addressed by the FAA's existing TC regulations and processes. This has presented challenges for both industry and regulators.

In particular, the specific certification requirements or **airworthiness standards** that apply to a TC project depend on the kind of aircraft being certified—e.g., airplane, rotorcraft, glider, etc. For example, the airworthiness standards for airplanes are found in 14 C.F.R. Parts 23 and 25, for rotorcraft in Parts 27 and 29, and balloons in Part 31. Neither "AAM" nor "eVTOL" is an aircraft category, and at the time winged eVTOLs initially came on the stage, they did not fit neatly into the FAA's existing aircraft categories. Like rotorcraft, winged eVTOLs are capable of vertical takeoff and landing (VTOL) and low-speed flight, but like airplanes, they also are capable of wingborne flight. Consequently, the FAA established a *new* category for these aircraft, "powered-lift"—the first new category of aircraft created since modern helicopters in the 1940s.[14]

---

14 The powered-lift aircraft category existed previously, but was largely unused in the civil context. 14 C.F.R. § 1.1, *General definitions*. FAA regulations define "powered-lift" as a "heavier-than-air aircraft capable of vertical takeoff, vertical landing, and low speed flight that depends principally on engine-driven lift devices or engine thrust for lift during these flight regimes and on nonrotating airfoil(s) for lift during horizontal flight."

When the first civil powered-lift AAM aircraft applied for FAA TC, there were no airworthiness standards specifically for powered-lift aircraft. Initially, the FAA planned to use the existing requirements for small airplanes ("Part 23"[15]), along with **special conditions** to address novel features—under the FAA's **21.17(a)** certification rule.[16] Notably, the Part 23 small airplane airworthiness standards were recently updated in 2016 (the "Part 23 Rewrite") to replace strict, rule-based requirements with more flexible, performance-based standards.[17] Performance-based standards set the level of performance that a design must achieve without dictating *how* it should achieve it. The update to the small airplane rules was specifically aimed at fostering innovation by enabling the certification of new technologies and designs, and several powered-lift aircraft certification projects were initiated under this framework.

Special Conditions

A special condition is a regulation that applies to a unique aircraft design. The FAA issues special conditions when it determines that existing regulations do not contain adequate safety standards to address a new or unusual aircraft design feature. 14 C.F.R. § 11.19. Special conditions contain requirements to address the new or unusual feature and establish a level of safety equivalent to that of existing airworthiness standards. For example, the FAA issued special conditions for Cirrus's whole-aircraft ballistic parachute recovery system—the first general aviation parachute system ever produced in an FAA-certified aircraft. *Special Condition (23-ACE-88) for ballistic parachute; FAA letter November 25, 1997 for models SR20, SR22.* The FAA used special conditions to set requirements for the parachute system, including its performance, inspection, and operating limitations.

The FAA generally develops special conditions through its **Issue Paper** process. The FAA uses Issue Papers to resolve various technical, regulatory, or administrative issues in the certification process. For example, the certification basis in a TC project is a G-1 Issue Paper, which designates the applicable airworthiness and environmental regulations for a project. See AC 20-166A, *Issue Paper Process* (2014).

In 2022, however, the FAA announced a significant change in direction: instead of using the Part 23 small airplane rules and special conditions under 21.17(a), the agency decided that it will type certificate the first set of powered-lift aircraft under its **special class** process instead (14 C.F.R. § **21.17(b)**[18]). Using the "special class" process, the FAA issues airworthiness requirements for each individual certification project, compiled from applicable requirements and any other criteria the FAA finds necessary to provide an equivalent level of safety (ELOS) to existing standards.[19]

---

15 14 C.F.R. Part 23, *Airworthiness standards: Normal category airplanes.* Applicable to airplanes with a seating configuration of 19 passengers or less, and a maximum certificated takeoff weight of 19,000 lb or less.

16 14 C.F.R. § 21.17(a), *Designation of applicable regulations [for certification].*

17 See, e.g., *FAA Part 23 Amendment 23-64 Implementation Procedures Guide.*

18 Section 21.17(b) provides: "For special classes of aircraft, including the engines and propellers installed thereon (e.g., gliders, airships, and other nonconventional aircraft), for which airworthiness standards have not been issued under this subchapter, the applicable requirements will be the portions of those other airworthiness requirements contained in Parts 23, 25, 27, 29, 31, 33, and 35 found by the FAA to be appropriate for the aircraft and applicable to a specific type design, or such airworthiness criteria as the FAA may find provide an equivalent level of safety to those parts."

For AAM aircraft using more traditional airframe designs, the FAA may still use existing class regulations and special conditions (§ 21.17[a]).

19 ELOS means an alternative approach that achieves an equal or greater level of safety compared to what is achieved by compliance to the applicable regulations. Applicants may seek an ELOS finding from the FAA when they cannot strictly meet regulations as written.

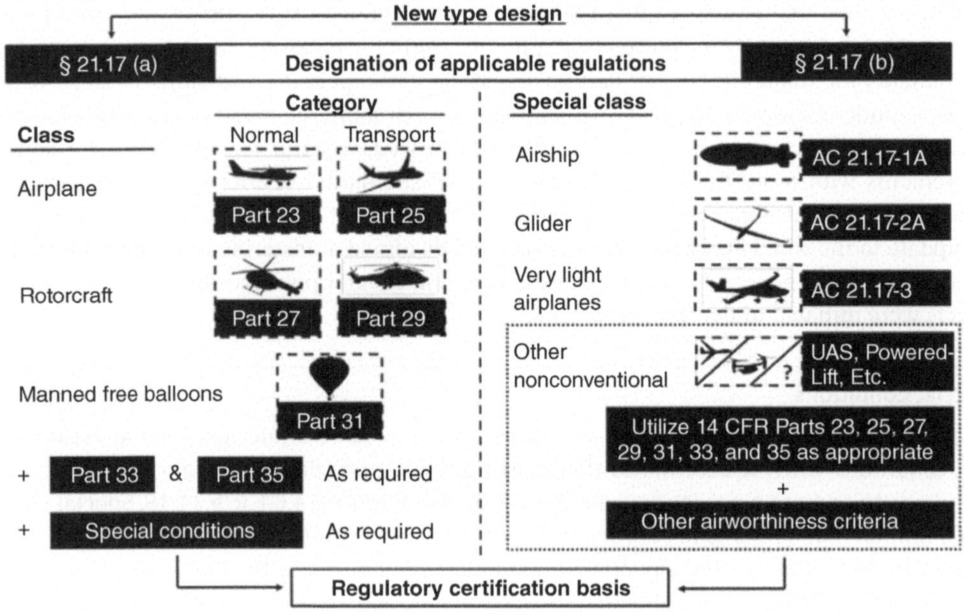

Figure 5.2 FAA TC paths. *Source*: Regulatory Gaps and Lack of Consensus Hindered FAA's Progress in Certifying Advanced Air Mobility Aircraft, and Challenges Remain, OIG Report AV2023037 (June 21, 2023)/ FAA/Public domain.

This policy change required the FAA to revisit the certification requirements for powered-lift projects already underway.[20] See Figure 5.2.

The FAA issued the first proposed airworthiness criteria under the special class process on November 8, 2022; the criteria was finalized in March 2024.[21] In July 2025, the FAA issued new guidance on powered-lift TC, based on its experiences with early applicants.[22] The guidance sets acceptable certification criteria for future projects under the special class process, aiming to provide greater clarity, predictability, and efficiency for applicants.

TC is typically an intensive, multi-year process, even without the regulatory complexities of new and novel designs.[23] The industry and FAA are continuing to learn from key experiences with early entrants, which are helping to shape the regulatory framework for this sector as it progresses toward the first TCs for powered-lift aircraft.

20 In 2022, the FAA was reportedly engaged in the certification process.

21 Airworthiness Criteria: Special Class Airworthiness Criteria for the Joby Aero, Inc. Model JAS4-1 Powered-Lift, 89 89 FR 17230 (March 8, 2024).

22 AC No: 21.17-4, *Type Certification—Powered-lift*. In 2025, the FAA issued guidance on TC of powered-lift, stating "This AC establishes a more efficient path in designating the type certification basis for certain powered-lift projects, as the FAA will not need to announce the criteria for each project in the Federal Register for notice and comment for those designs for which the FAA has deemed it appropriate to use the criteria in this AC with no additions or changes." AC 21.17-4.

23 FAA, *How Does the FAA Certify Aircraft?* https://www.faa.gov/aircraft/air_cert/airworthiness_certification#:~:text=The%20Boeing%20737%2D8/9,between%205%20and%209%20years.

FAA Designees

To assist the FAA in meeting its comprehensive regulatory responsibilities, Congress empowered the FAA to delegate to qualified persons or organizations the legal authority to act on the agency's behalf. In the case of individuals, these people are called "designees." Designees act as representatives of the FAA and are governed by the same standards, procedures, and guidance applicable to FAA employees.

Using designees for certain tasks allows the FAA to focus its limited resources on critical safety issues and challenges. The FAA uses designees in the certification of both aircraft and people. For example, engineering and flight test designees make findings about whether engineering data for proposed aircraft designs complies with airworthiness requirements. Medical designees or "AMEs" are authorized to issue medical certificates. An organization designation authorization (ODA) delegates certain certification authority to an organization.

Noise Standards

The FAA's regulatory responsibilities also include protecting "the public health and welfare from aircraft noise."[24] The FAA regulates aircraft noise as part of its type and airworthiness certification processes. Specifically, the FAA regulates the maximum noise level that an aircraft can produce through **noise certification standards**.[25] Aircraft are certified to the noise standards in effect at the time an application for TC is made. Maximum noise levels are classified according to "stage" designations: for example, for a jet, Stage 1 is the oldest and loudest stage, whereas Stage 5 is the newest and quietest. With few exceptions, all certified aircraft flying in the United States must be Stage 3 or quieter. For an airplane, the Stage 3 standards for takeoff, landing, and lateral ("sideline") noise measurements range from 89 to 106 dB, depending on the airplane's weight and the number of engines that it has.

The ban on older, louder aircraft has greatly reduced noise from aircraft flying in the United States today; however, aviation noise concerns persist. One of the questions around integrating AAM aircraft is what noise impact operations will have—in terms of both level and frequency. Actual and perceived noise impacts on the communities may be a crucial factor in achieving public acceptance of AAM operations, especially because some AAM aircraft are intended to operate closer to local populations than prior aviation operations.

Some AAM eVTOLs are being designed with specific features intended to address both the amount and the character of the sound they produce—including electric motors and unique propellers or blade shapes and sizes. With limited operations to date, however, there is limited real-world data on community perceptions of both hearing and seeing eVTOLs. The FAA has been collaborating with NASA on AAM noise research, including collecting data and developing noise modeling tools. In 2021, NASA conducted its first acoustic flight test on an eVTOL aircraft in partnership with Joby Aviation, capturing noise data from Joby's pre-production prototype aircraft.

As of December 2025, the FAA has not established any new noise requirements for certification of powered-lift aircraft. Currently, the FAA is applying existing noise regulations—which include standards for "tiltrotors," a class of powered-lift—to the new powered-lift aircraft applicants. If the FAA determines that existing noise standards are insufficient to certify a specific powered-lift

---

24 49 U.S.C. § 44715.

25 14 C.F.R. Part 36, *Noise Standards: Aircraft Type and Airworthiness Certification*. The FAA first implemented aircraft noise standards in 1969, and since then, has periodically adopted increasingly strict standards, leading to increasingly quiet aircraft.

project, the agency will publish specific requirements for that aircraft. Noise is more thoroughly discussed in Chapter 7.

Safety of Approved Products in Service

The primary purpose of the FAA's extensive aircraft design regulations is to ensure that aircraft are safe for flight. In service, however, previously undetected design and manufacturing defects are sometimes revealed. To maintain the safety of approved designs, the FAA monitors products throughout their lifespans. The FAA's **continued operational safety** (COS) requirements aim to ensure the integrity of an FAA-approved design, with regulations governing the processes for design changes, maintenance, and inspections.

If—through inspections, reports, accident data, or any other manner—the FAA becomes aware of a safety issue in an approved design and determines that the unsafe condition "is likely to exist or develop" in other products,[26] the FAA issues an **airworthiness directive** (AD). ADs are legally enforceable rules; owners and operators must comply with ADs for their aircraft to be considered airworthy and operate legally in the United States.[27]

Sometimes, the FAA finds that correcting the unsafe condition requires a design change. After the FAA issues a type certificate, a manufacturer cannot make changes to an approved design without further FAA approval—regardless of whether the change is to correct a safety issue, or to simply improve or update an approved design.[28] The process for approving changes to approved designs that would appreciably affect weight, balance, structural strength, reliability, operational characteristics, or other characteristics affecting airworthiness generally mirrors the original design approval process.[29]

Given the lengthy timelines for design certification and the rapid pace of innovation in AAM technologies, it is reasonable to anticipate that after initial AAM aircraft type certificates are approved, some AAM manufacturers may soon seek additional approvals for design changes to incorporate design refinements or updates to improve performance and expand operational use cases.

Production Certification: FAA Manufacturing Approval

An FAA **PC** authorizes a manufacturer to duplicate products under an FAA-approved design (e.g., a type certificate). Only an FAA design approval holder (e.g., type certificate holder)[30] can apply for a PC. Accordingly, many AAM manufacturers are expected to pursue production certification after achieving TC. A PC will allow AAM manufacturers to move from development and prototyping to scaling aircraft production and, ultimately, commercialization.

FAA production certification ensures a manufacturer can consistently produce airworthy products. To receive production approval, the FAA examines an applicant manufacturer's quality systems, including the organization, facilities, documentation practices, personnel, and supplier controls. Accordingly, a PC is organization specific; unlike a TC, it cannot be bought or directly transferred.[31]

The FAA's production certification regulations are largely agnostic to new and novel aircraft design features. They are intended to be scalable to the size and complexity of the product being produced. The stringent organizational processes involved, however, can be unfamiliar and

---

26 14 C.F.R. § 39.5, *When does FAA issue airworthiness directives?*

27 14 C.F.R. § 39.7, *What is the legal effect of failing to comply with an airworthiness directive?* § 39.9, *What if I operate an aircraft or use a product that does not meet the requirements of an airworthiness directive?*

28 14 C.F.R. § 21.91–21.101.

29 14 C.F.R. §§ 21.111–21.120.

30 Or an entity that has the rights to a design approval under a licensing agreement.

31 14 C.F.R. § 21.144.

especially costly and burdensome for manufacturers that are new to the aviation industry, like some AAM companies. Accordingly, achieving and maintaining production certification could present challenges to some AAM companies that do not have prior aviation experience.

Light Sport Aircraft

> "Removing these restrictions is necessary for the introduction of powered-lift and certain rotorcraft, e.g., electric vertical takeoff and landing (eVTOL), into the light-sport category.... These changes will also enable the development of new technologies, including electric, hydrogen, and hybrid engines and motors." MOSAIC, 90 Fed. Reg. 35034, 35069 (July 24, 2025).

In 2025, the FAA issued a final rule, "the Modernization of Special Airworthiness Certification (MOSAIC)," expanding the size and scope of **light sport aircraft** (LSAs)—with potential implications for AAM manufacturers.

What are LSAs? The FAA established the LSA category in 2004, with the goal of increasing access to flying. LSAs are subject to limitations intended to make them easier to fly and simpler to operate, such as maximum occupancy and weight restrictions. The FAA also created a new pilot certification for flying LSAs, "sport pilot," with reduced training requirements compared to other pilot certificates.

Some type certificated, standard airworthiness aircraft meet the definition of light sport (e.g., the Piper J-3 Cub). Other LSAs are *not* type-certified[32]—that is, they do not meet FAA airworthiness requirements, but rather are designed and manufactured to meet FAA-accepted *industry-developed consensus standards*. In contrast to FAA regulations, industry consensus standards tend to provide both greater flexibility and lower costs in aircraft design and manufacturing. Unlike the TC process described above, the FAA does not review, test, or approve the design of LSAs; instead, the manufacturer (design holder) is responsible for reviewing, testing, and approving the aircraft design under industry consensus standards, and attesting to compliance with the FAA.

Originally, LSA regulations excluded many AAM aircraft—for example, the original rule limited powered LSAs to having a single, reciprocating (piston) engine. With MOSAIC, the FAA significantly expressly enabled eVTOL manufacturers to utilize the category. MOSAIC removes regulatory restrictions on the number and type of propulsion. In MOSAIC, the FAA also specifically encourages industry to create new consensus standards for the design, production, and airworthiness of eVTOLs.

Under MOSAIC, LSA eVTOLs are restricted to two seats and generally prohibited from use in commercial transportation services (although MOSAIC does allow new types of aerial work, such as infrastructure inspections, to be conducted in LSAs). Accordingly, MOSAIC is not intended to provide a pathway for AAM commercial air taxi or cargo operations, but it nonetheless may lower barriers for certain AAM aircraft designed for private and flight school use.

In addition to expanding the light-sport market, MOSAIC is also expected to help increase both the pilot and maintenance technician pools. This is of particular interest to the AAM industry, as a significant number of pilots and maintenance professionals are expected to be needed in the coming decades to support the industry's projected growth. Indeed, Joby Aviation, for example, has already been leveraging the light-sport framework to train both pilots and maintenance technicians.

---

32 These LSAs may be either "special" (S-LSA), which are factory-built to a "ready-to-fly" condition, or "experimental" (E-LSA), which are built from kits or plans based on approved models.

5.1.4 FAA Regulations: Training and Certification of Pilots

The FAA's decision to create the new "powered-lift" aircraft category also triggered the need for new rules to address instructor and pilot training and certifications. How should pilots be trained and certified to fly aircraft that have characteristics of both helicopters and airplanes? In October 2024, the FAA answered the question, publishing a long-anticipated Special Federal Aviation Regulation (SFAR) on powered-lift pilot certification and training and operational rules, referred to as the **Powered-Lift Part 194 SFAR**.[33] A significant milestone in AAM regulation, the Powered-Lift Part 194 SFAR provides an essential foundation for initial commercial AAM operations, once the first aircraft achieves TC.

> **What Is an SFAR?**
>
> An SFAR is an interim rule that the FAA issues to address a unique situation. For example, the FAA has issued SFARs to prohibit flight in specific areas and address aircraft-specific safety considerations.
>
> In the case of the Powered-Lift Part 194 SFAR, the FAA anticipates that this rule will be in place for 10 years.

Under the Powered-Lift Part 194 SFAR, a pilot flying a powered-lift aircraft must hold one of the following FAA certificates: an airline transport pilot (ATP) and type rating for powered-lift category; a commercial pilot for powered-lift category; or a private pilot for powered-lift category. Because of the unique characteristics of individual powered-lift aircraft, such as different flight controls, the FAA will require individual powered-lift *type ratings*—or aircraft-specific training and testing—for pilots. The FAA expects the initial cadre of powered-lift flight instructors to come from AAM manufacturers, who will receive training and experience through test flight and crew training activities necessary for initial aircraft certification (i.e., TC).

A notable provision in the Powered-Lift Part 194 SFAR is the allowance for powered-lift pilots to train in aircraft with a *single pilot station and* a *single set of flight controls*. Prior to the rule's issuance, several eVTOLs were already being developed and seeking certification with only one pilot station, (such as Joby's S4 and Archer's Midnight). Legacy aeronautical experience rules have traditionally required training aircraft to have *two* sets of flight controls—one for the individual receiving instruction, and one for the instructor. Under the Powered-Lift Part 194 SFAR, however, pilots will be able to train in aircraft with a single set of controls, leveraging a combination of simulator experience, solo flights, and supervised operating experience.

What about remotely piloted or autonomous aircraft, such as Wisk's eVTOL? The Powered-Lift Part 194 SFAR does not apply to an aircraft without a pilot onboard; accordingly, it does *not* address pilot qualification or operational requirements for remotely piloted or autonomous AAM aircraft. Additional rules likely will be needed as aircraft with these characteristics seek certification and approach operations.

5.1.5 FAA Regulations: Certification of Operators

In the United States, most commercial passenger operations require the operator to have an FAA **air carrier certificate**. Like aircraft and pilot certification, previously existing operator

33 This final rule included new "Part 194": SFAR No. 120: Powered-Lift: Pilot Certification and Training; Operations Requirements.

certification rules did not account for some of the new and novel aspects of AAM aircraft and operations. Indeed, until the FAA updated its definitions in 2023 to include powered-lift, regulations only allowed for airplanes and rotorcraft to be used in air carrier operations.

Commercial AAM operations, including "air taxi" services, are expected to be conducted under **Part 135**—the FAA's existing operating rules for "commuter and on demand air carriers." Part 135 includes aircraft charter services and some scheduled services, with restrictions (e.g., NetJets). The Powered-Lift Part 194 SFAR does now allow for the use of powered-lift aircraft in scheduled airline service (e.g., Delta or FedEx), which are generally conducted under Part 121. Like Part 121, Part 135 rules include rigorous safety requirements, but it is less restrictive overall. For example, Part 121 requires pilots to have an ATP certificate (and, generally, 1500 hours of flight time), whereas certain Part 135 operations can be conducted with a commercial pilot certificate (minimum of 250 hours of flight time for a first officer and 500 hours for pilot in command for operations under visual flight rules).

To conduct interstate flights (that is, fly between states), Part 135 operators also must obtain economic authority from US DOT. There are two kinds of Part 135 operations: on-demand and scheduled or commuter. On-demand operators can conduct unscheduled (not on a fixed timetable or published route) and *limited* scheduled operations. Commuter authority allows for unlimited scheduled operations. Both on-demand and commuter have specific requirements and limitations associated with them. For example, airplanes used in on-demand operations can only have seats for 30 passengers or fewer.

In general, aircraft must have a type certificate to be used in Part 135 operations. Aircraft being used in Part 135 operations are required to undergo a conformity inspection by the FAA. Several AAM companies have chosen to pursue operator certification in advance of receiving aircraft TC—using traditional aircraft[34] to gain experience and lay the foundation for future commercial operations with AAM aircraft. For example, Joby received Part 135 air carrier certification in 2022, and Archer received Part 135 certification in 2024. These operators are expected to add their eVTOLs to their existing air carrier certificates through the FAA's review process after they each receive TC.

Aerial Rideshare and Part 380

AAM air taxi operations may include rideshare aggregation services: the sale of individual seats on an aircraft to the public, similar to an Uber pool. The DOT regulates the resale of individual seats on individual charter flights under "Part 380," as a "public charter operator." 14 C.F.R. Part 380. A public charter operator is considered an *indirect* air carrier, meaning they act as an intermediary between the customer seeking air transportation and the direct aircraft carrier (e.g., Part 135 operator) operating the flight. A public charter operator may also be a direct air carrier itself, or it may partner with a different entity to operate the flight.

By regulation, public charter operations under Part 380 are *not* scheduled operations, despite sometimes being marketed and sold with set times and locations. The regulatory requirements in Part 380 primarily focus on ensuring that customers are informed and protected. AAM operators intending to function as public charter operators will be required to operate in accordance with a DOT-approved charter prospectus, which includes documents that demonstrate compliance with regulatory requirements. A charter prospectus also includes a flight schedule. 14 C.F.R. § 380.28.

(Continued)

34 Part 135 regulations require that applicants have exclusive use of at least one aircraft that meets the requirements of at least one kind of operation.

(Continued)

Some entities arrange aircraft charters outside the Part 380 framework. For example, typical air charter brokers (regulated under Part 295) are first contacted by customers seeking charter flights, before seeking an air carrier to operate the flight. In contrast, a Part 380 operator has already arranged the aircraft and is seeking passengers.

Airspace Management

The FAA's regulatory authority includes managing the US NAS. This includes organizing aircraft operations in the airspace for safety and efficiency. Even without the introduction of new types of aircraft, the United States is already the busiest and most complex airspace system in the world: the FAA handles an average of 44,360 flights daily.[35]

How will AAM aircraft integrate into an already busy NAS? Like helicopters and airplanes, powered-lift aircraft can operate wherever the aircraft and the pilot meet the applicable operational requirements. In addition to pilot training and certification, the Powered-Lift Part 194 SFAR also addresses critical operating rules, such as minimum safe altitudes, visibility, and fuel (or for electric aircraft, energy) reserves for powered-lift aircraft. In general, the Powered-Lift Part 194 SFAR aims to provide a flexible, performance-based approach: For example, the rule applies helicopter fuel reserves requirements, minimum safe altitudes, and visibility and cloud clearance minimums—rather than stricter airplane requirements—to powered-lift aircraft when they are operating in vertical-lift modes and can demonstrate specific performance capabilities. In terms of ATC procedures, the FAA anticipates that early AAM operations will be able to be managed using existing rules and tools. For example, to operate in Class B airspace—the airspace around major airports—pilots generally must receive ATC clearance and aircraft must be equipped with an operating two-way radio, automatic dependent surveillance-broadcast (ADS-B) out, and an operable transponder with altitude-reporting capability. Initial AAM aircraft are generally expected to be piloted, operate under visual flight rules (VFR) conditions,[36] and comply with the existing communication, navigation, and surveillance requirements for the airspace in which they operate. Some AAM operations, such as in low-altitude and rural areas, are anticipated to occur outside of the traditional air traffic service areas and may require expanded surveillance capabilities.

Significantly, AAM is not the only integration challenge facing the FAA: the agency is also managing a rapidly growing commercial space industry, as well as an exponential increase in UAS (drone) operations—all of which contribute to the complexity of modernizing the NAS. The US has already begun a modernization effort to improve the safety, security, and efficiency of the ATC system to support expanded and more dynamic capacity in the future. As AAM operations mature and scale, particularly with the potential for increased autonomy, new regulations and/or procedures, as well as technology upgrades, may become necessary. The FAA anticipates that data gathered from early AAM operations will help to inform and shape these changes.

5.1.6 FAA Regulations: Maintenance

The FAA extensively regulates aircraft maintenance—everything from preventive maintenance to the rebuilding of aircraft to maintenance record-keeping and the training and certification of maintenance technicians. In 2022, the FAA released the first significant change to its rules for training

35 FAA, *Air Traffic By the Numbers*, https://www.faa.gov/air_traffic/by_the_numbers.

36 VFR are the rules for conducting a flight under visual conditions. VRF conditions means weather conditions equal to or better than the minimum for flight under VRF, expressed in terms of visibility, distances from clouds, and (cloud) ceiling.

aviation maintenance technicians (Part 147, Aviation Maintenance Technician Schools) since the 1940s to help maintenance training providers better align curricula with future workforce needs. Like pilots, training maintenance technicians for AAM aircraft will require the development of new standards to address unique characteristics, such as electric motors, high-voltage systems, large lithium-ion battery arrays, advanced avionics, and advanced composite materials. As AAM aircraft transition from development, testing, and TC to entry into service, industry, training providers, and regulators will need to work together to further enable and refine maintenance standards and advance skills development to support the AAM sector.

Some AAM aircraft may require fewer maintenance hours compared to certain traditional aircraft. For example, the electric motors that some AAM aircraft use generally contain significantly fewer moving parts than internal combustion engines, which can contribute to reduced maintenance needs. Additionally, some AAM maintenance activities may focus more on the replacement of modular, unserviceable components, such as batteries, rather than performing component repairs. AAM aircraft are also expected to incorporate increased levels of autonomy, which could enable increased use of predictive maintenance technologies. These advancements could offer greater insight into aircraft health, enhancing maintenance planning and efficiency. At the same time, the operational expectations for some AAM may demand faster turnaround times compared to traditional aviation operations and necessitate new approaches to maintenance.

5.1.7 FAA Regulations: AAM Infrastructure

As with traditional aircraft, AAM aircraft vary across different aircraft and manufacturers, implicating different infrastructure needs. Some AAM aircraft will use existing aviation infrastructure for takeoff and landing, including airports and heliports—with and without modifications. Additionally, specialized new sites are being constructed, purpose-built to accommodate powered-lift AAM aircraft: called "vertiports" (Figure 5.3).

Figure 5.3 Sample vertiport caution sign. *Source*: EB 105A, Vertiport Design, Supplemental Guidance to Advisory Circular 150/5390-2D, Heliport Design/FAA/Public domain.

The FAA classifies a vertiport as a type of heliport with distinct requirements. The FAA's Office of Airports is developing vertiport design and safety standards, beginning with interim guidance that the agency anticipates updating once it has more data from VTOL aircraft and operations.[37] The FAA's initial guidance covers new vertiports, as well as modifications to existing heliports and airports, and addresses topics including landing area geometry, lighting and markings, downwash and outwash protection, aircraft parking, charging, and electric infrastructure. Additionally, existing FAA regulations for aircraft takeoff and landing facilities and surrounding areas also generally apply to AAM infrastructure[38]—such as the regulations requiring FAA notification before certain airport-related construction, and detailing the process for studying obstructions to ensure safe and efficient use of navigable airspace.

New facilities requiring FAA approval and/or funding typically undergo FAA environmental review to assess the project's impacts. According to the National Environmental Policy Act of 1969 (NEPA), federal agencies, including the FAA, must conduct an environmental impact analysis for certain federal actions—that is, consider the environmental consequences and disclose significant impacts publicly. These actions can include airport development and aircraft route decisions by the FAA. As of 2025, the FAA is actively looking at opportunities to streamline its review processes, such as using Categorical Exclusions for applicable vertiport projects. Categorical Exclusions are for projects that normally do not have a significant effect on the human environment—for example, a project on an existing airport—and therefore do not require an environmental assessment or impact statement. In addition to FAA regulations, vertiports also may be subject to existing and developing state and local rules for safety and planning. This includes permitting laws, zoning laws, building and fire codes, and rules regarding noise and air pollution. State and local considerations will be discussed in the next subsection of this chapter.

A key consideration for AAM infrastructure is what is needed to support new propulsion systems and non-traditional aviation fuel sources. Many AAM manufacturers are exploring alternative fuels, including electric batteries and hydrogen. Standards for electric aircraft charging systems and hydrogen fuel storage at airports are still evolving nationally, as well as at the state and local levels. In addition to establishing infrastructure at vertiports, there is also the need to consider how electrical charging demands could impact the broader grid infrastructure and power supply. Other unique infrastructure considerations may be driven by the expectation that some AAM operations will occur at lower altitudes. For example, expanded infrastructure to support improved low altitude weather reporting and prediction may be needed.

5.1.8 The Role of State, Local, and Tribal Governments

Although vehicle, pilot, maintainer, and operator certification is the purview of the federal government,[39] state, local, and tribal governments have significant roles to play in AAM regulation. As discussed in the previous subsection, AAM will require infrastructure for taking off and landing, charging, and loading and unloading passengers and cargo. State and local authorities

37 The FAA had issued guidance on vertiport design in 1991, "based on civil tiltrotors modeled after military tiltrotor technology," but those aircraft were never used commercially and the FAA canceled that guidance in 2010. EB 105A.

38 14 C.F.R. Part 157, *Notice of Construction, Alteration, Activation, and Deactivation of Airports*; 14 C.F.R. Part 77, *Safe, Efficient Use and Preservation of the Navigable Airspace*.

39 State and local laws attempting to regulate aviation safety are *preempted*, meaning federal law, as the supreme law of the land, displaces state and local laws on this issue.

are responsible for certain regulations, licensing, and policies related to ground infrastructure and equipment, and they frequently operate or fund aviation facilities. Moreover, compared to existing aviation, AAM is expected to "localize" air travel—only increasing the importance of state, local, and tribal leadership—as it brings connectivity within and between communities, including some that previously lacked access to flight. In most states, land-use permitting is managed by local governments. Historically, when considering land use near airports, key factors include environmental impacts on nearby communities, such as noise and air pollution. There are also often state-level environmental regulations and natural resource departments with oversight and enforcement authority for environmental-related rules. Some states also require registration or licensing of aircraft and operators. These state and local powers can inherently influence where AAM aircraft operate. Additionally, although states generally cannot regulate aircraft noise directly, state and local authorities possess powers to mitigate aircraft noise impacts. For example, state and local governments can use zoning and permitting authorities to prohibit takeoff and landing sites in noise-sensitive areas.

State, local, and tribal powers can also significantly influence community acceptance, integration, and funding. As AAM is anticipated to bring aviation services much "closer" to local communities, state and local authorities play key roles in community education, outreach, and decision-making processes. US DOT reports, however, that there are "major challenges for community planning and engagement," including confusion regarding the roles, responsibilities, and authorities of state, local, and tribal governments in comparison to federal entities.[40] Several US states have proactively developed, or are in the process of developing, AAM rules and policies. Some of these efforts are coordinated across states, while others pursued independently.

Although individual states and localities may have different considerations and priorities, certain variability across state and local rules could present challenges for the development and commercialization of AAM services. One of the primary reasons for regulating aviation at the federal level is the need for uniformity. As Congress recognized, the interstate nature of aviation necessitates uniform and predictable regulation across states. Consequently, achieving an appropriate level of harmonization, or at least minimally viable interoperability, among state, local, tribal, and federal frameworks is crucial for successfully integrating AAM into the existing aviation ecosystem and enabling its growth at scale.

5.1.9 Accident and Incident Investigation

In the United States, the US National Transportation Safety Board (NTSB) is the lead organization primarily responsible for investigating and determining the cause of civil aviation accidents, including events involving AAM vehicles. An independent agency outside of US DOT, the NTSB is tasked with conducting independent investigations of transportation accidents; determining the facts, circumstances, and probable causes; and making transportation safety recommendations. The NTSB also conducts safety studies and special investigations on matters relating to transportation safety, and reviews appeals for certificate issues for pilots, maintainers, and mariners. Although empowered with broad investigative powers, the NTSB lacks regulatory and enforcement authority. Nonetheless, NTSB findings and recommendations can, and do, influence both industry and government actions.

40 DOT, *The Advanced Air Mobility National Strategy: A Bold Policy Vision for 2026–2036* (December 17, 2025).

Federal law requires the NTSB to investigate *every* civil aviation accident.[41] Unless a criminal investigation takes precedence, an NTSB investigation generally has priority over any investigation by another government department, agency, or instrumentality.

The NTSB has already updated accident and incident investigation regulations and procedures to accommodate UAS. In 2010, the NTSB revised its notification requirements to clarify their application to uncrewed aircraft accidents.[42] Initially, the NTSB applied a 300-lb weight threshold for uncrewed aircraft accident notifications, other than those involving serious injury or death.[43] In 2022, however, the NTSB replaced the weight threshold with a requirement that the aircraft hold an airworthiness certificate and sustain substantial damage.[44] The definition for uncrewed aircraft accidents could encompass certain AAM aircraft operating without a pilot onboard. The NTSB also requires uncrewed aircraft operators to notify of certain serious incidents, regardless of whether the aircraft has an airworthiness certification.

5.2 The International Regulatory Framework

The International Civil Aviation Organization (ICAO) is a specialized, intergovernmental agency of the United Nations. ICAO is responsible for developing and supporting global standards and recommendations for safe and efficient civil aviation among members.[45] ICAO's work includes establishing international **Standards and Recommended Practices (SARPs)** for civil aviation, including standards for aircraft design and operation, personnel, and equipment.[46]

ICAO's members are individual sovereign powers (referred to as "States"). Each State has exclusive sovereignty, or power, to make and enforce laws for the airspace above it. This authority extends to the regulation of foreign aircraft entering a State's territory. Generally, aircraft are registered in specific States, and pilots are required to hold a license issued by that State's civil aviation authority. International coordination and cooperation under ICAO and through agreements among States facilitate the ability of aircraft to fly across borders. For example, bilateral agreements establish reciprocal recognition of aircraft airworthiness certification between countries. Under such an agreement, once an aircraft is certified by the civil aviation authority of one country, the other country can validate that certification, allowing the aircraft to be operated within its jurisdiction.

---

41 The NTSB also investigates major accidents in other modes of transportation, conducts studies and special investigations on matters relating to transportation safety, issues transportation safety recommendations, and serves as the "court of appeal" for FAA or US Coast Guard certificate actions against aviators and mechanics, or mariners, respectively.

42 NTSB regulations require that aircraft operators notify the NTSB of aircraft accidents and certain serious incidents.

43 75 FR 51953 (citing similarity to the maximum weight for piloted, powered ultralight vehicles).

44 87 FR 42100. For reference, "*Substantial damage* means damage or failure which adversely affects the structural strength, performance, or flight characteristics of the aircraft, and which would normally require major repair or replacement of the affected component. Engine failure or damage limited to an engine if only one engine fails or is damaged, bent fairings or cowling, dented skin, small punctured holes in the skin or fabric, ground damage to rotor or propeller blades, and damage to landing gear, wheels, tires, flaps, engine accessories, brakes, or wingtips are not considered "substantial damage" for the purpose of this part." Part 830.2.

45 The 1944 Convention on International Civil Aviation, known as the Chicago Convention, established ICAO.

46 Only when a state incorporates a SARP into its own domestic laws and regulations does it become legally enforceable within that state. ICAO member states are obligated, however, to collaborate in achieving the highest practical degree of uniformity by incorporating SARPs into their national regulations.

Significantly, differences have emerged in how individual states' **civil aviation authorities**[47] are developing regulations for AAM. For example, Canada and Australia intend to use processes equivalent to the FAA's 14 C.F.R. § 21.17(b) special class process to type certificate AAM aircraft. The United Kingdom, however, has adopted the European Union Aviation Safety Agency (EASA) SC-VTOL airworthiness standards, which have some significant differences compared to FAA's C.F.R. § 21.17(b) special class approach. In general, EASA has taken a more prescriptive approach, designating specific requirements across certification projects.

As the first AAM eVTOLs approach certification, continued collaboration among authorities is needed to achieve streamlined validation in support of global commercialization. ICAO has recognized that the "rapidly evolving AAM ecosystem requires a globally harmonized operational and regulatory framework and guidance."[48] In 2024, ICAO hosted its first Advanced Air Mobility (AAM 2024) Symposium. This event brought together key governmental and industry stakeholders to discuss global harmonization and interoperability. Additionally, in 2025, the FAA and civil aviation authorities from Australia (Civil Aviation Safety Authority), Canada (Transport Canada Civil Aviation), New Zealand (Civil Aviation Authority), and the United Kingdom (Civil Aviation Authority) released a "Roadmap for the Advanced Air Mobility Aircraft Type Certification." The Roadmap aims to align airworthiness and certification standards for AAM, including through data sharing. The core objectives of the NAA Network initiative are to foster collaboration, promote technological advancement, and streamline the validation processes across the network. Participating authorities signed the Declaration of Intent, committing to the Roadmap principles, including a tiered approach that focuses TC efforts proportionately to operational risk. As the first AAM aircraft achieve certification in these jurisdictions, these principles will be put to the test to enable AAM aircraft that can operate globally.

5.2.1 China: The First Country to Type Certificate a Civil eVTOL

China is expected to be one of the early adopters of eVTOL technologies. In 2023, the Civil Aviation Administration of China (CAAC) issued a type certificate to EHang's EH216-S, making it the first eVTOL aircraft in the world to receive such approval. The EH216-S is an autonomous, uncrewed two-passenger eVTOL designed for low-altitude urban air taxi and logistics services.

Little is publicly known about the CAAC's certification process for eVTOL aircraft, but EHang's experience offers some insights. Unlike in the United States, China mainly focuses its early certification efforts on autonomous or remotely piloted aircraft. Specifically, China anticipates logistics as a key early use case, helping the industry to mature its technologies. EHang stands out as one of the few, if not the only, manufacturers launching with autonomous passenger transport.

EHang officially applied for CAAC TC in 2021. In 2022, CAAC issued "special conditions" for the EH216-S, covering requirements for flight performance, structures, design and construction, propulsion systems, systems and equipment, data link, and ground control station. The aircraft reportedly went through a lengthy testing campaign lasting over 30 months, including flight tests and demonstrations, leading to an approval in 2023.[49]

47 A CAA is the government body that regulates a country's civil aviation, such as the US's FAA.
48 FAA Advanced Air Mobility Implementation Plan, V. 1, pg. 3.
49 "EHang Successfully Obtains Type Certificate for EH216-S Passenger-Carrying UAV System Issued by Civil Aviation Administration of China;" EHang News Release, October 13, 2023.

5.3 Ethical Considerations

Introducing any new technology can raise ethical considerations, and AAM is no exception. Many concerns, such as those related to safety, autonomy, affordability, and privacy, are not unique to AAM but rather common to emerging technologies. Others are more specific to the unique operational context and integration requirements of AAM.

For example, AAM has the potential to offer decentralized access to aviation, connecting entirely new markets by air for the first time. In the United States, for example, there are fewer than 500 commercial airports, but over 19,000 existing landing facilities. Decentralization raises important and multifaceted questions about access and social equity—specifically, how the benefits and any negative impacts of AAM (such as aircraft noise, which historically has disproportionately affected lower-income communities) will be distributed.

Equity questions also arise in the context of access to services. Although AAM has the potential to expand access to transportation and critical medical services to underserved communities and create new employment opportunities for underrepresented groups, there are open questions regarding implementation while the industry continues to mature. For instance, it remains unclear how existing accessibility requirements will specifically apply to AAM aircraft and associated facilities.

Ultimately, public trust in AAM will be key to its widespread adoption and success. Some ethical considerations may be addressed through regulatory frameworks, while others may be managed by policy or market forces and community education and engagement.

5.4 Summary

Aviation is one of the most regulated industries in the world. Not only is aviation regulation complex, but it has also proven highly successful—in some places, aviation has achieved unprecedented levels of safety compared to other modes of transportation. Many of the current aviation regulations, however, were not written with AAM aircraft or operations in mind.

As countries prepare for the introduction and integration of new AAM aircraft and operations, civil aviation authorities are evolving aviation regulations for AAM. This chapter examined these challenges and opportunities through the lens of the world's busiest and most complex airspace: the United States.

In the United States, the FAA has the sole and exclusive authority to regulate civil aviation safety, including AAM. Because AAM vehicles are aircraft, their design approval falls under the FAA's TC process. Several AAM manufacturers are seeking TC, but these aircraft include one or more novel design features not fully addressed by the FAA's existing TC regulations and processes. Even without these complexities, TC is typically an intensive, multi-year process. Currently, the FAA is primarily focused on certifying piloted powered-lift aircraft, including eVTOLs.

The decision to categorize certain AAM as "powered-lift" also necessitated new rules for pilots and operations. In 2024, the FAA issued the Powered-Lift Part 194 SFAR interim rule to provide a foundation for initial commercial operations once the first aircraft are certified.

Regarding airspace management, the FAA anticipates that early AAM operations will utilize existing ATC rules and tools. As the industry scales and moves toward increased autonomy, new procedures will likely be required, informed by data gathered from early operations.

Although civil aviation safety is the purview of the federal government, state and local governments play significant roles in AAM regulation and community engagement. In particular, these authorities manage licensing, policy, and infrastructure for landing facilities, and frequently fund or operate these sites.

As the first eVTOLs approach certification, continued collaboration among global authorities is essential for streamlined validation and commercialization. Differences in regulatory development have already emerged. In 2025, the FAA joined authorities from Australia, Canada, New Zealand, and the United Kingdom to release a "Roadmap for the Advanced Air Mobility Aircraft Type Certification" aimed at international harmonization.

Beyond technical and regulatory, social acceptance is a key element of AAM introduction and integration. Introducing new technologies often raises ethical considerations, and AAM is no exception. Although some of these may be addressed through regulatory frameworks, others may be managed by policy or market forces. Ultimately, public trust will be key to AAM's widespread adoption and success.

6

Operational Considerations

Advanced air mobility (AAM) encompasses a wide range of mission types, and each imposes distinct operational demands on aircraft, airspace organization, automation architecture, and ground infrastructure. Two broad operational subsets that capture much of this diversity are urban air mobility (UAM) and regional air mobility (RAM). *UAM* refers to short-range operations conducted within dense metropolitan environments, including passenger, cargo, and service missions. UAM is characterized by confined airspace, high traffic density, stringent acoustic expectations, and limited diversion flexibility. By contrast, *RAM* refers to medium-range missions that extend beyond metropolitan cores into suburban and intercity corridors and is characterized by longer trip legs, greater weather exposure, and increased routing flexibility. Passenger transport occupies roles in both regimes, but UAM provides the clearest example of high-density AAM operations in constrained airspace and serves as the first operational archetype examined in this chapter.

UAM requires precise trajectory control and continuous conformance monitoring within confined 3D corridors, along with community-compatible acoustic profiles. Because UAM missions accumulate more frequent vertical climb and descent cycles than RAM missions, and given the power-intensive nature of these segments, UAM often operates with tighter mission energy budgets due to repeated high-power vertical segments. In comparison, regional point-to-point mobility supports both passenger and cargo transport and extends these missions across longer legs where range requirements, weather exposure, and reserve energy constraints shape the operational strategy. The introduction of cargo missions removes passenger-centric constraints on ride quality and routing and reduces sensitivity to passenger-driven scheduling constraints. A third operational domain within AAM encompasses *public service* missions, characterized by time sensitivity, priority access, and landing-site variability as primary operational drivers. Public service operations, including medical response and emergency deployment, require immediate readiness, dynamic routing outside established corridors, and reliable landing performance in improvised or minimally prepared environments. Beyond transport-centric missions, surveillance, inspection, environmental monitoring, and disaster logistics introduce additional operational considerations, each defined by unique combinations of response time, altitude requirements, sensor payload, endurance, and airspace integration requirements.

Despite this operational breadth, all AAM missions are governed by a common set of fundamental constraints that recur across all operational contexts and mission types. These constraints stem from the same underlying system drivers that shape the preceding mission classes, namely propulsion architecture, system-scale coordination requirements, and airspace integration. Electric propulsion and vertical flight impose tight energy budgets that bound achievable mission profiles.

System-level scalability depends on predictable trajectories and continuous conformance monitoring. As AAM operations mature, pilot-centered deconfliction gives way to supervisory oversight supported by automation, making automation a structural requirement for separation assurance and coordinated flow management. Vertiport throughput, charging power availability, and maintenance resources directly couple ground operations to achievable mission tempo. Lastly, airspace integration further requires that AAM function as a structured flow system defined by predictable trajectories and coordinated access to shared airspace volumes.

Because the full range of AAM mission types is too broad to be fully covered, this chapter develops three representative operational archetypes that capture the diversity of constraints across urban, regional, and public service domains. Each archetype reveals a distinct operational logic that emerges from its specific environmental and performance drivers. First, intracity passenger UAM illustrates high-density automated flight operations in constrained airspace. Next, regional cargo demonstrates medium-range routing in environments with reduced human factors constraints and greater temporal flexibility. Finally, medical-response missions demonstrate how AAM behaves when time sensitivity, landing-site variability, and priority access dominate system requirements. These three cases provide a structured basis for understanding how AAM operates as an integrated transportation system. In Chapter 8, we will examine the economic drivers and implications associated with these mission classes.

6.1 Urban Passenger Mobility Concept of Operations

Intracity passenger mobility occurs within the most spatially constrained and operationally demanding environment of the AAM ecosystem. Dense metropolitan areas impose complex geometric, atmospheric-flow, acoustic, and airspace conditions that affect every aspect of mission planning and execution. The built environment introduces irregular wind fields and thermal gradients, and obstacles dictate climb and descent profiles. Population density imposes stringent acoustic-exposure constraints, and community acceptance depends on operational regularity and on limiting the visual and noise footprints of overflights. Limited diversion options across urban cores require deliberate reserve policies and predefined contingency trajectories to ensure safe outcomes during off-nominal conditions. In the event of reduced ceiling or degraded visibility, the smaller leg length associated with intracity operations reduces opportunities to route around weather and can stymie overall operations. Operations adjacent to busier Class B or Class C airspace introduce controlled airspace boundaries, arrival and departure paths for conventional aviation, and radar surveillance constraints associated with coverage, accuracy, and update performance at low altitude. The combination of these elements produces a 3D operating space that demands highly structured and predictable trajectories. These considerations motivate the operational architecture of intracity passenger missions.

Within the broader context of urban passenger UAM, integration with Class B airports introduces a distinct operational environment defined by high levels of airspace structure, coordination, and procedural discipline. Along with demand aggregation, these characteristics make this environment particularly attractive for early commercial deployment. Class B airspace concentrates arrivals and departures for conventional aviation and establishes strict speed, altitude, and communication requirements that create predictable access and exit conditions for UAM aircraft (Figure 6.1). UAM operations are able to interface with established arrival and departure flows, comply with tower-directed sequencing, and maintain predictable containment within transition corridors that

Figure 6.1 Notional AAM arrival and departure corridors developed at Orlando International Airport (KMCO) during southbound operations [1]. Dark gray shading represents AAM departing and arriving corridors; intermediate shading represents conventional air traffic departures; and light gray represents conventional air traffic arrivals.

connect the airport to the broader urban network. This level of coordination reduces operational ambiguity and supports early high-reliability operations by enforcing consistent behavior across all airspace users. However, the high traffic density and structured flow of conventional operations require that UAM approach and departure profiles minimize interactions with runway flows and that designated UAM transition gates remain laterally or vertically separated from primary instrument procedures. Further, ground operations introduce additional coordination requirements because UAM must integrate with airport surface operations, passenger-processing, and energy restoration processes without disrupting conventional gate activity. Despite these constraints on interaction with conventional aviation operations, the structured airspace and concentrated origin and destination demand within this environment make airport-city missions an attractive early use case, allowing UAM to leverage existing airport infrastructure, established procedures, and supervisory oversight while demonstrating reliable point-to-point service. Airport-city missions thus emphasize disciplined integration and coordinated operations, distinguishing them from less structured urban-only environments and preserving strong operational appeal.

An intracity passenger mission in the urban core begins at a vertiport where ground handling, passenger flow, and charging availability determine departure timing. Departure times must align with corridor access windows that dictate when an aircraft may safely enter structured airspace, and these windows depend on upstream conformance, sequencing constraints, vertiport throughput, and passenger arrival patterns from the surface transportation network. Departure profiles are informed by obstacle clearance requirements, local wind conditions, and noise exposure criteria that govern the transition from hover to climb. Together, these constraints impose high control effort and narrow allowable deviation margins during the earliest phase of flight. These demands increase pilot workload during early operations and inform supervisory oversight and automation requirements as the system matures.

Passenger transport levies additional operational limits that extend beyond aircraft performance and airspace structure (Figure 6.2). Ride-quality constraints restrict allowable pitch, roll, yaw, and linear acceleration rates. These limits determine how rapidly an aircraft can correct for urban wind disturbances or trajectory deviations and prohibit aggressive maneuvering, even when control authority and vehicle performance permit it. Perceived proximity to buildings constrains lateral

Figure 6.2 Joby S4 flying in New York City. *Source*: With permission of Joby Aviation.

positioning within bounded corridors and limits the magnitude of heading changes near obstacles. This reinforces the requirement for stable, predictable trajectories throughout the en route segment.

At the system level, intracity passenger mobility depends on the stability of traffic flows across a network of interconnected corridors and vertiports. Flow stability also depends on the regularity of passenger arrival patterns from the surface transportation network as variability in landside access can propagate into the departure sequence and affect corridor accessibility. Therefore, buffer time must be incorporated to absorb minor deviations without introducing travel-time volatility that undermines service reliability. Conformance monitoring can identify deviations that could affect separation or disrupt scheduling, and supervisory oversight can apply corrections to preserve the temporal structure of the flow. Altogether, intracity UAM functions as a strategically managed flow system where predictable trajectories, stable timing, and coordinated node-to-node transitions are fundamental to maintaining throughput and service reliability.

Arrival operations mirror the structure of departure and must ensure a stable and predictable descent profile that remains compatible with passenger comfort limits while respecting obstacle clearance, local wind conditions, and acoustic-exposure criteria. Access to the arrival corridor depends on upstream sequencing and the availability of pad resources at the destination vertiport; these dependencies require flow coordination to prevent airborne holding and the rapid depletion of onboard energy reserves. During the descent transition, the proximity of urban structures increases motion sensitivity, so changes in vertical descent rate or heading must be limited and smoothed. Final approach paths must align with the touchdown and lift-off (TLOF) areas to minimize lateral deviations and maintain a stable attitude during landing in confined spaces.

Once on the pad, passenger egress determines how quickly the aircraft can be cleared for repositioning, inspection, and energy restoration. The time required for passengers to exit the cabin, collect personal items, and move into vertiport processing areas introduces variability into dwell time. These passenger- and aircraft-driven ground processes determine how rapidly an aircraft can reenter the departure queue and, ultimately, the reliability and schedule stability of the intracity network. The mission concludes only when the aircraft reaches a confirmed ready state with adequate energy reserves, verified system health, and alignment with its next scheduled departure window. For the passenger, operational success also depends on an efficient, intuitive transition to the vertiport's surface transportation network or immediate arrival at their final destination.

6.2 Regional Cargo Concept of Operations

Regional cargo operations represent a distinct operational archetype within the AAM ecosystem because the payload removes passenger-ridership and ride-quality considerations, and the governing constraints shift from occupant comfort to logistics efficiency, routing flexibility, and energy allocation across longer leg lengths. The absence of motion-perception thresholds, proximity-driven discomfort, and ride-smoothing requirements permits trajectory design to prioritize aerodynamic performance, endurance, and network timing over occupant experience. As such, regional cargo missions operate within a relatively broader temporal and spatial envelope than intracity passenger flights, with operational flexibility driven less by passenger synchronization and more by logistics-driven timing, routing adaptability, and network-level scheduling priorities. These characteristics produce an operational environment characterized primarily by payload readiness, aircraft energy state, logistics scheduling, and coordination among geographically distributed nodes. This defines a mission profile fundamentally different from that of urban passenger mobility.

A regional cargo mission begins with payload staging, mass verification, and loading into the aircraft (Figure 6.3). These ground processes exhibit less temporal uncertainty than passenger flow because items can be prepositioned, transferred with mechanical assistance, and validated through automated inventory systems. Consequently, loading and verification produce more consistent dwell times and establish a predictable baseline for departure sequencing. Departure readiness hinges on pad availability, aircraft energy state, and the opening of routing windows that support

Figure 6.3 The Beta Technologies CX300 with cargo. Courtesy of Beta Technologies.

efficient point-to-point movement across suburban or intercity regions. Because cargo does not require synchronization with passenger arrival patterns or modal transfers, departure windows can be aligned with logistics schedules and network-level flow patterns.

The departure segment for regional cargo missions prioritizes energy efficiency and obstacle clearance without requiring acceleration conditioning to maintain occupant comfort. Aircraft can achieve recommended climb gradients and transition profiles based solely on performance limits, local atmospheric conditions, and sequencing requirements within the regional network. This can reduce time spent in power-intensive regimes and improve overall energy availability for longer missions. Departure profiles remain structured and predictable for separation assurance, but they incorporate greater flexibility in climb rate and transition dynamics because motion sensitivity no longer constrains maneuvering.

The en route segment defines the principal distinction between regional cargo and intracity passenger missions. Regional cargo traverses medium-range routes that extend well beyond the dense urban core into areas characterized by lower obstacle density, lower population exposure, and more uniform atmospheric-flow conditions. These characteristics allow broader lateral and vertical routing envelopes and support dynamic adjustments that optimize flight time, avoid adverse weather, or enhance network-level efficiency. Although route containment remains essential for separation assurance, it can be defined over wider corridors, allowing greater latitude for altitude changes or minor heading adjustments. Cargo flights can incorporate direct routing whenever feasible, adjust cruise altitude based on wind patterns, or accept moderate rerouting to balance flow conditions across the network. Longer trip leg lengths allow cruise segments to operate near optimal aerodynamic efficiency to improve endurance and reduce the influence of climb and hover power on total energy consumption.

Energy management for regional cargo missions focuses on balancing cruise efficiency, reserve requirements, and the energy demands associated with longer leg lengths. Cruise performance dominates mission energy consumption, and the stability of the cruise environment enables more predictable energy modeling than the turbulence-prone urban core. Reserve policies must still account for potential deviations due to weather, node congestion, or temporary routing constraints. However, greater routing flexibility allows aircraft to deviate around weather, congestion, or airspace constraints without significantly increasing flight time or energy consumption. Because cycle frequency is lower than in high-frequency urban missions, regional cargo operations experience fewer cumulative thermal cycles and slower degradation patterns across the battery life cycle. These characteristics enable deeper charge cycles, longer dwell periods for energy restoration, and more accurate energy usage forecasting across multi-stop routes.

Automation plays a vital role in regional cargo operations by shifting supervisory oversight from individual aircraft to network-level coordination. As automation assumes a greater share of trajectory execution and conformance management, oversight increasingly operates at the network level, where coordination, flow optimization, and conflict resolution dominate system performance across distributed regional routes. Accurate trajectory execution, separation assurance, and contingency management remain foundational requirements, but these functions are increasingly enforced through automated systems operating within defined performance and separation bounds. These capabilities enable the use of automated lateral repositioning, dynamic rerouting, and altitude assignment strategies to optimize throughput and mitigate adverse weather impacts across the regional network. Within regional cargo networks, supervisory oversight focuses on managing interactions among distributed nodes, verifying conformance at scale, and resolving flow conflicts that emerge from network coupling and scheduling interdependencies.

System-level performance for regional cargo depends on coordination across a network of geographically distributed nodes and differs fundamentally from the tight synchronization required for intracity passenger mobility. Logistics demand can be aggregated into batches, and flights can be assigned flexibly to routes based on payload availability, node congestion, or fleet readiness. Aircraft can be redeployed dynamically to meet shifting supply-chain needs, and temporal precision can be relaxed because cargo demand exhibits lower sensitivity to minute-scale travel-time variability than passenger transport. This enables a scheduling model that optimizes energy efficiency, fleet utilization, and throughput across multiple nodes without enforcing a uniform passenger-driven cadence. Conformance monitoring remains essential for separation assurance, but deviations typically impose fewer downstream penalties because temporal variability can be absorbed at node interfaces rather than within tightly constrained en route corridors and vertiport schedules.

Arrival operations for regional cargo missions mirror departures in their emphasis on efficiency and performance-based maneuvering, with governing constraints defined by vehicle performance and operational efficiency. Descent profiles can utilize recommended rates and geometric transitions that minimize hover time, manage local wind conditions, and align cleanly with the TLOF. Although urban structures are less prevalent in many regional environments, arrival containment remains essential for separation management and predictable flow. Once on the pad, operational focus shifts to unloading, payload verification, and preparation for the next mission. These processes are less variable than passenger egress and can be automated or mechanized to enhance throughput and enable precise forecasting of dwell times. Because cargo operations can span extended operating windows, arrival sequencing can be adjusted to align with logistics cycles, energy availability, or node-level workload balancing.

Regional cargo operations demonstrate how AAM aircraft behave when mission performance is defined by logistics integration, flight efficiency, and routing flexibility within an operational envelope driven by non-passenger payload requirements. This operational archetype highlights AAM's strengths for medium-range distribution networks and supports a well-aligned early commercial application by leveraging existing logistics networks and regional airport capacity.

6.3 Medical Response Concept of Operations

Medical-response missions are a unique operational category within the AAM ecosystem because clinical urgency serves as the primary determinant of operational performance. Mission feasibility depends on launch readiness, routing flexibility, and access to constrained or improvised landing sites. These missions may involve transporting a patient, moving a medical team to a point of care, or transferring critical supplies such as blood, pharmaceuticals, specialized equipment, or even human organs. Each variant imposes unique constraints on vehicle preparation, ground coordination, and operational tolerances during flight; collectively, they constitute the most time-sensitive class of AAM operations.

A medical-response mission begins with a dispatch request that initiates an immediate assessment of aircraft readiness, crew availability, and operational feasibility. Preflight verification is more stringent than in routine operations because the mission may require integration of medical equipment, specialized restraint systems, or unique cabin environmental conditions. Energy reserves must exceed routine requirements because limited diversion options and potential destination changes can lead to unplanned flight extensions. Unlike urban passenger missions that synchronize departures with corridor access windows, medical-response operations prioritize launch immediacy and request expedited corridor access when the network structure allows. This shift in operational priority introduces additional coordination between supervisory oversight, vertiport personnel, and ground medical teams.

The departure and climb phase reflect this elevated urgency. Trajectories must comply with noise limits, obstacle clearance criteria, and local atmospheric-flow conditions, generally following the most direct path compatible with those constraints. Reduced visibility or low ceilings complicate the launch decision because the need for rapid patient transport must be balanced against restricted routing and limited diversion options under degraded weather conditions. This tension places significant importance on predefined "go" or "no-go" thresholds that balance clinical needs with operational risk.

Once airborne, early-phase maneuvering authority may be used more assertively than in routine passenger missions, although acceleration limits must still protect patient stability when a patient is onboard. The en route segment emphasizes stable, time-efficient progression across the network. Routing flexibility is greater than in scheduled passenger operations because medical missions receive priority handling and are not necessarily bound by the fixed corridor timing or node-to-node sequencing that governs routine passenger flows. When transporting a patient, acceleration limits, vibration exposure, and cabin environmental stability constrain heading changes, altitude adjustments, and airspeed variation. When transporting medical personnel, these constraints are relaxed, and routing behavior resembles a more flexible variant of intracity UAM operations. Continuous coordination with receiving facilities ensures that arrival timing aligns with handoff procedures, clinical or logistics team readiness, and downstream transfer or staging requirements. These operational requirements create a feedback loop among aircraft position, patient condition (when applicable), and destination facility status.

Arrival procedures are one of the most variable components of the mission profile. Medical-response aircraft may land at dedicated hospital heliports, designated emergency landing zones, or improvised sites that satisfy minimum surface, obstacle, and wind criteria (Figure 6.4). Approach paths must be predictable to ensure obstacle clearance and ground-team safety and concomitantly flexible enough to accommodate rooftop pads or other confined landing zones. Containment

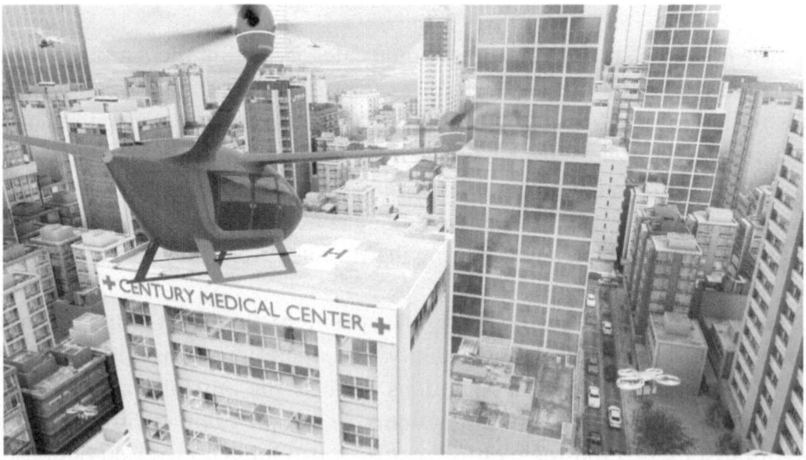

Figure 6.4 Concept graphic showing how an AAM vehicle could aid in medical response by transporting patients to a hospital. Courtesy of NASA/Public domain.

accuracy is crucial because opportunities for lateral adjustment diminish as the aircraft descends into aerodynamically disturbed environments. Ground readiness further dictates arrival timing because handoff actions, deployment activities, or time-critical transfers must occur without delay. Therefore, dispatch and ground coordination extend through the landing phase to ensure alignment between aircraft arrival profiles and destination-side readiness.

At the network level, medical-response operations introduce episodic and unpredictable demands that must be absorbed without destabilizing energy-constrained, high-frequency routine AAM traffic flows. Supervisory oversight assigns priority movement when required, and these assignments propagate into corridor sequencing, node availability, and verification that the aircraft's energy state supports the revised routing. Fleet readiness must include provisions to maintain one or more aircraft in a rapid-launch state, with sufficient energy reserves and verified system health. Because medical-response missions frequently interrupt routine operations, their integration into the broader AAM network depends on effective prioritization logic, accurate conformance monitoring, and the ability to reoptimize flows after the mission concludes.

6.4 Security and Cybersecurity Considerations

AAM security spans both physical and digital domains and must be designed to support a new aviation paradigm characterized by short dwell times at vertiports and high turnover of aircraft, passengers, and operational assets across a distributed network of heterogeneous facilities. This security framework must be integrated into this new environment without eroding the efficiency that defines AAM's value. Physical security establishes foundational controls governing how passengers enter, move through, and access operational areas within vertiports that lack the centralized queuing architecture and spatial buffers of conventional airports. These controls must scale with the service model and corresponding risk profile of each operation. Cybersecurity extends these requirements to the digital infrastructure that coordinates aircraft, vertiports, and supervisory systems. Because AAM depends on persistent data exchange for trajectory management, conformance monitoring, and energy-system management, the security framework must protect the confidentiality, integrity, and availability of these information pathways from initial system design. The combined physical and digital security architecture defines the protection boundary within which AAM must operate and determines whether the system can sustain the cadence, reliability, and trustworthiness required for commercial viability.

6.4.1 Passenger and Facility Security

Screening protocols must protect aircraft, crew, and passengers while preserving the advantages of efficient, seamless, on-demand access to aerial mobility that distributed terminals enable. To accomplish this, screening intensity varies across service models and focuses on *proportionality*. Services that deliver passengers into airport sterile areas require security measures commensurate with those in conventional terminal environments, so that individuals entering secured zones maintain compliance continuity. Without such continuity, time savings from the AAM segment would be lost due to prolonged security reentry procedures at the airport terminal. This requires harmonized standards for identity verification, bag screening, prohibited items lists, and personnel training. In contrast, purely urban point-to-point services operating outside these sterile zones use streamlined screening protocols that balance risk reduction with throughput efficiency.

Distributed vertiports lack the centralized infrastructure and queue-driven screening architecture of major airports. Passenger flow at vertiports is characterized by short dwell times, and this rapid turnover reduces opportunities for extensive secondary screening. This results in most security functions occurring upstream of the boarding area and during passenger movement between the vertiport's entrance and entry into the boarding zone. To accomplish this, spatial organization within the vertiport layout must be designed to ensure that passengers remain within monitored zones throughout the short preflight interval and cannot bypass screening checkpoints.

Within these constraints, reduced-intensity screening protocols for urban point-to-point services rely on a combination of identity and authorization assurance, proportional physical screening, and access control and facility protection (Figure 6.5). *Identity and authorization assurance* establish *who* the passenger is and whether the individual is cleared to travel via a digitally verifiable identity credential. This credential may include a digital ID, mobile driver's license, or biometric match to the reservation record. Authorization assurance relies on reservation confirmation and watchlist screening, both of which can be completed within upstream digital workflows.

Proportional physical screening calibrates the screening protocol to the operational risk profile to assess carried items and identify prohibited materials. For urban point-to-point services with a lower baseline risk, hand-carried items can be screened using compact imaging systems with short cycle times. Restricted-items policies should be defined by the prevailing threat environment, aircraft operating context, and vertiport characteristics associated with the service model, and not defined by standards developed for large commercial terminals.

Distributed networks of relatively small vertiports increase the number of access points compared to a traditional airport and elevate the importance of controlled perimeters, surveillance coverage, and credential-based entry systems. *Access control and facility protection* prevent unauthorized entry into boarding areas and operational zones. This security layer protects both the vertiport facility and aircraft. It determines *who* may enter sensitive areas and incorporates controlled access to boarding areas and credential-controlled entry gates, restricts ramp and charging

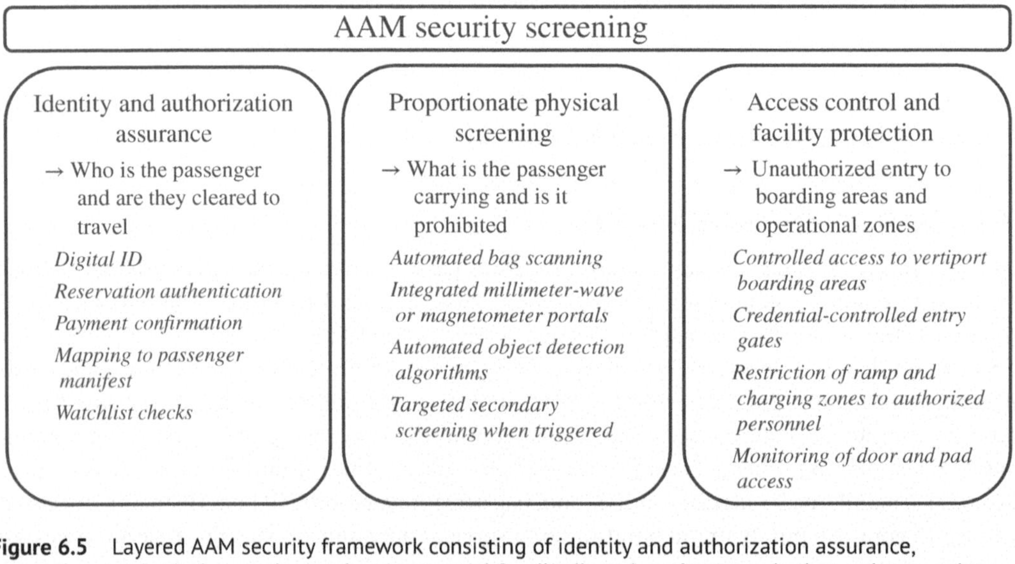

Figure 6.5 Layered AAM security framework consisting of identity and authorization assurance, proportionate physical screening, and access control for distributed vertiports and urban point-to-point passenger services.

zones to authorized personnel, and monitors door and pad access. Ground personnel require role-specific authorization that limits unsupervised access to operationally sensitive zones, such as high-voltage charging areas. Because many vertiports operate in close proximity to public spaces, security design must reduce opportunities for unauthorized entry and still enable the rapid ingress and egress of passengers, crews, and aircraft required for high-cadence AAM service. Collectively, these complementary security layers establish a defensible security posture without compromising the time savings that define AAM's utility.

6.4.2 Cybersecurity

Cybersecurity is a foundational requirement for AAM operations because every mission depends on authenticated commands, verified system states, and uninterrupted data exchanges across vehicles, vertiports, and supervisory systems. AAM aircraft rely on digitally coordinated trajectories, remote status and trajectory monitoring, high-bandwidth communication links, and networked charging and maintenance-diagnostic systems. Operational viability requires that the confidentiality, integrity, and availability of all safety-critical and mission-critical data streams, along with non-repudiation for command, status, and maintenance actions, remain protected against manipulation, degradation, or unauthorized access.

Beyond in-operation flight activities, cybersecurity risk in AAM extends to pre-operational and lifecycle vulnerabilities that directly affect operational integrity. Supply-chain compromise, third-party software dependencies, cloud service availability, firmware integrity, and over-the-air update mechanisms represent critical attack surfaces that can introduce latent vulnerabilities before an aircraft ever enters service. Because these pathways influence software behavior, sensor trust, and system state estimation, lifecycle cybersecurity is a prerequisite for safe and reliable AAM operations and must be integrated from initial system design through sustained operational use.

AAM introduces a wide set of cyber-exposed system interfaces across both airborne and ground-based components. Vehicle avionics, flight-control computers, battery management systems, and navigation sensors must maintain integrity because corrupted or spoofed data can alter trajectory tracking or degrade energy-state estimation. Command-and-control links, supervisory oversight channels, and data exchanges with fleet-management platforms must remain authenticated to prevent unauthorized trajectory modification or false status reporting. Vertiport systems further expand the attack surface through scheduling platforms, pad-access controls, charging hardware, and passenger-processing interfaces that depend on networked communication. These systems link operational, physical, and digital domains, and a compromise at any node can propagate operational instability across the entirety of the network.

Because AAM vehicles operate at low altitude and within close proximity to terrestrial infrastructure, radio-frequency (RF) and navigation-layer threats represent a particularly important class of cybersecurity risk. Threats such as Global Navigation Satellite System (GNSS) jamming, GNSS spoofing, and RF interference can directly degrade navigation integrity, timing synchronization, and trajectory conformance. Mitigating these risks requires a combination of authenticated navigation sources, sensor cross-checking, resilient timing architectures, and hardware-based security mechanisms that protect trusted execution and measurement pathways.

Trajectory integrity is particularly important for AAM because separation assurance depends on predictable motion within structured flow corridors. A corrupted trajectory, altered waypoint, or manipulated conformance report can directly lead to loss of separation, sequencing disruptions, or incorrect conflict-resolution assignments. Supervisory oversight depends on accurate, authenticated inputs from each aircraft, and cybersecurity protections must ensure that the data used to maintain flow stability cannot be modified, delayed, or injected by unauthorized actors. The integration of automated strategic and tactical functions intensifies this requirement because automation increases the reliance on continuous digital inputs.

As automation, autonomy, and AI-enabled functions assume greater roles in trajectory management, conflict resolution, and system optimization, cybersecurity must also address threats specific to learning-enabled and autonomous behavior. These threats include unintended behavior drift, training-data poisoning, adversarial sensor inputs, and manipulation of learned decision boundaries. As a result, security assurance extends beyond correct outcomes to include the ability to explain, audit, and bound autonomous decisions, ensuring that decision rationale and associated risk remain observable to supervisory systems.

Beyond expanding the cyber-attack surface, vertiports' digital infrastructure introduces direct operational risk when compromised. Scheduling systems govern pad assignment, gate access, and passenger throughput, and a compromised system can produce cascading delays or ground lockouts that propagate into departure sequencing and corridor availability. High-voltage charging stations depend on distributed controllers that manage power flow, verify connector compatibility, and record energy usage. Unauthorized modification of these processes could degrade energy availability, misreport state of charge, or interrupt energy restoration cycles. Meanwhile, maintenance records, system health data, and digital logbooks must also all remain protected because these datasets inform airworthiness decisions, predictive maintenance actions, and fleet-level readiness assessments.

Fleet-management platforms connect individual aircraft and vertiports into a coordinated regional system, and their cybersecurity requirements extend beyond vehicle safety. These platforms execute routing assignments, manage fleet utilization, schedule energy cycles, and coordinate recovery after off-nominal events. Even in the absence of immediate safety effects, a cyber intrusion that alters resource allocation, suppresses alerts, or manipulates performance data can degrade network-level efficiency. As AAM networks scale, operational sensitivity to delayed or incorrect information intensifies, and cybersecurity becomes vital to maintaining a stable mission tempo and predictable service levels.

Cybersecurity protections must be embedded in the design of AAM systems at the architectural level as real-time control, navigation, communication, and energy-management functions rely on tightly coupled digital pathways that continuously share data with minimal latency. This requirement reflects the principle of *security by design*. This principle holds that tightly integrated, safety-critical digital architectures cannot be effectively protected by post hoc fixes or purely external defensive measures. *Authentication* and *encryption* are necessary across command, control, navigation, and maintenance datasets. *Network segmentation* must separate safety-critical pathways from business-critical software, reservation systems, and passenger-facing applications. *Fault-mode management* requires that aircraft and vertiports detect anomalous data patterns, revert to verified states, and maintain safe operational modes when communication integrity is uncertain. Supervisory oversight must incorporate automated *anomaly detection*,cross-checking across independent data sources, and predefined procedures for isolating compromised nodes while preserving flow stability across the network.

Cybersecurity protections must be embedded into the design of AAM systems.

Cybersecurity assurance in AAM includes protection, detection, response, and recovery as integral system functions. Effective incident response depends on comprehensive logging, robust time synchronization across vehicles and infrastructure, and fleet-level behavior baselines that enable rapid identification of abnormal states. Recovery strategies must support graceful degradation, controlled reconfiguration, and timely restoration of trusted system states without destabilizing ongoing operations.

Because cybersecurity failures can induce unsafe system states, safety and security form a coupled assurance challenge for AAM engineers. Cyber threats must be evaluated in terms of their potential to compromise trajectory integrity, energy management, separation assurance, and ground coordination. As vehicle density and operational tempo increase, scalable cybersecurity architectures must preserve system-wide trust, maintain situational awareness, and support coordinated recovery across fleets operating at high density.

6.5 Community and Public Acceptance

Community acceptance is the ultimate determinant of whether AAM can function as a sustained transportation system. Local expectations govern where and when aircraft may operate, how vertiports integrate into the community fabric, and which routes achieve durable demand. Perceptions of safety, acoustic and visual acceptability, community benefit, and institutional trust serve as the criteria through which communities evaluate whether AAM operations meet locally defined expectations. These criteria are directly coupled to vehicle design, airspace organization, vertiport placement, and service reliability, and they establish an operational boundary as consequential as energy limits, airspace constraints, or meteorological conditions.

Community acceptance begins with confidence in operational safety. Residents evaluate AAM less through technical metrics and more through *observed consistency*, *visible professionalism*, and *transparent communication* about routine and off-nominal events. Predictable schedules, stable operational patterns, and clear explanations of supervisory oversight and system monitoring contribute to a sense of controllability and reduce uncertainty about when, where, and how often aircraft will operate. Institutional performance carries weight equal to aircraft reliability and operator conduct, through the consistency, transparency, and responsiveness of the organizations that govern and oversee AAM operations. Communities assess more than aircraft and operators; they also evaluate the reliability of municipal partners, regulators, and infrastructure owners, who collectively shape the system's overall safety. Trust is strengthened when these institutions respond promptly to concerns, disclose operational deviations, and demonstrate adherence to commitments.

Acoustic and visual exposure define the daily perception of AAM and play a pivotal role in public acceptance. As we'll explore in Chapter 7, communities evaluate more than sound levels; they also weigh regularity, predictability, and duration of noise exposure. Abrupt changes in operational cadence or unanticipated deviations draw disproportionate attention relative to expected behavior. Similarly, visual presence is assessed by frequency, proximity, and pattern consistency instead of by isolated events. When operators communicate clearly about expected activity levels and maintain consistent usage patterns, residents can form reliable expectations, reducing perceptual intrusion.

Community benefit must also be evident for acceptance to persist. Communities evaluate AAM based on its contribution to broader mobility goals, such as expanded transportation options in the immediate area, improved regional access, support for emergency response, or new transportation options in underserved areas. Land-use compatibility and equitable access should inform siting decisions through visible alignment with municipal priorities and community development objectives. AAM services that demonstrably reduce surface congestion, provide economic opportunities, or enhance travel convenience tend to garner more substantial support.

Governance, responsiveness, and transparency determine whether communities view operators and public agencies as credible stewards of AAM. Acceptance depends on institutions demonstrating that community interests remain central to decision-making. This requires visible channels for feedback and mechanisms that enable communities to understand how concerns translate into operational changes. Over time, acceptance becomes contingent on the reliability and consistency of these governance processes as ongoing determinants of system legitimacy. Communities judge long-term viability by observing how operators and public agencies sustain their commitments, adapt to community evolution, and communicate the rationale behind operational decisions.

Public acceptance forms a broader layer of societal evaluation that extends beyond the immediate communities hosting AAM operations. Individuals outside flight corridors still shape the social legitimacy of AAM by assessing its overall safety record, perceived usefulness, environmental impact, and economic fairness. A lack of familiarity, especially in early operations, with electric propulsion, automation, and supervisory oversight heightens sensitivity to safety and operational reliability, and willingness to fly grows as the system demonstrates consistent performance. Although public acceptance does not directly govern local operational frameworks, it influences the political, regulatory, and market environment that AAM seeks to expand.

Community and public acceptance act as coupled determinants of meaningful AAM success. Community response decides where, when, and how AAM can operate, and public perception determines the pace at which AAM earns legitimacy. Together, they define the societal boundary within which AAM must design its concepts of operation, scale its services, and demonstrate its long-term value.

6.6 Summary

Effective AAM concepts of operation require mission objectives to be coherent with constraints imposed by energy availability, airspace structure, automation capability, and ground-handling capacity. Intracity passenger mobility, regional cargo transport, and medical-response missions demonstrate how payload, spatial constraints, timing demands, and operational urgency drive distinct routing strategies, launch criteria, and integration requirements with the surface transportation system. Together, these use cases illustrate how AAM adapts system architecture to specific operational contexts.

Cybersecurity and public acceptance further mold AAM operations by influencing how digital systems, physical infrastructure, and community interfaces are designed and managed. Cybersecurity requirements govern the protection of communication, navigation, vehicle health monitoring, and supervisory automation against intrusion. These protections must integrate with high-cadence operations without degrading throughput or conformance performance. Public acceptance influences deployment as communities assess AAM through demonstrated safety, acoustic and visual compatibility, transparent governance, and equitable access. Collectively, these factors determine how AAM services integrate into existing urban and regional mobility systems and the extent to which operators can scale services while maintaining trust, reliability, and system performance.

Reference

1 Fraticelli Rivera, V., Thomas, R., Alberto Castro-Pena, C., and Kuba, S. (2025). Assessment of advanced air mobility vehicle integration at Orlando International Airport. *Aerospace* 12 (391). https://doi.org/10.3390/aerospace12050391

7

The Environmental Operating Context of AAM

The promise and challenges of advanced air mobility (AAM) extend beyond technical feasibility; successful implementation depends on how AAM systems interact with and operate within environmental conditions that affect safety, reliability, and public acceptance. This new aviation paradigm will be evaluated on safety, efficiency, and compatibility with broader societal goals such as environmental sustainability and, at a minimum, preservation of quality of life for the public and the communities served. Our consideration of the environment is framed through two operationally relevant perspectives: first, the atmospheric environment, with its variability and hazards, and second, the human environment, where sustainability and noise strongly influence public acceptance.

Weather is the most immediate natural constraint, dictating when and where AAM operations can safely occur. Sustainability reflects a deeper challenge of providing mobility benefits without creating disproportionate environmental or societal costs. Noise, meanwhile, stands at the interface of technology and community impact. Even if AAM is safe and sustainable, its adoption will falter if the public perceives operations as intrusive or disruptive. This chapter examines these three intertwined environmental challenges.

7.1 AAM and the Weather Environment

As described in Chapter 1, AAM operations take place primarily between the surface and roughly 1500m. Conventional aircraft typically traverse this portion of the atmosphere only during ascent and descent, and do so very quickly. In contrast, many AAM vehicles operate exclusively within the lowest 1500m. This region, which can stretch to approximately 2km in altitude, is termed the atmospheric boundary layer (ABL) and is the portion of the atmosphere that most closely overlaps with the AAM operational environment. The ABL is a highly dynamic portion of the atmosphere. Exchanges of heat, momentum, and moisture between the surface and atmosphere create an environment best defined by microclimates and turbulence. With the partial exception of helicopters, which routinely operate in this region and have long contended with the ABL's challenges and limited observational coverage, most traditional aviation operations occur outside the ABL. As AAM introduces sustained, high-tempo flight operations into this region at greater scale, the ABL becomes a primary operational environment instead of a transient one.

Within this operating context, it is reasonable to expect that AAM vehicles will need to operate safely across the same range of weather conditions encountered by conventional airline operations.

When these two paradigms operate concurrently, familiar weather hazards, including winds, thunderstorms, icing, reduced visibility, and low cloud ceilings, will present many of the same operational challenges to AAM. However, the sustained exposure of AAM operations to the ABL fundamentally alters how these hazards are encountered and managed, reflecting differences in vehicle flight characteristics, short-range high-frequency mission profiles, sustained low-altitude exposure, and the proximity of operations to urban environments with modified wind fields and urban heat island (UHI) effects.

Electric vertical takeoff and landing (eVTOL) aircraft, therefore, present a different set of operational trade-offs when contending with low-altitude atmospheric hazards. Operating at lower altitudes does not inherently reduce exposure to icing and, in many regions, may increase it, since supercooled liquid water (SLW) is often concentrated in shallow layers associated with stratus clouds, freezing fog, and freezing drizzle common in northern and coastal climates. As discussed in Chapter 2, electric propulsors can deliver short bursts of high power, but this capability does not eliminate icing risk and may impose unplanned power demands and reduced energy margins if climb or maneuvering becomes necessary. Such responses remain conditional on available vertical clearance, remaining energy reserves, vehicle performance limits, and the spatial extent of the icing layer.

Similarly, although eVTOLs can operate at reduced approach speeds comparable to helicopters, this capability does not fundamentally relax visibility constraints, particularly in environments characterized by fog, low ceilings, or complex urban lighting. Electric propulsion avoids the power degradation experienced by internal combustion and turbine engines at high density altitude, but this advantage primarily preserves available thrust rather than expanding overall performance margins. Taken together, these vehicle and operational characteristics indicate that eVTOL operations do not enjoy systematically improved weather tolerance. Instead, they encounter a different balance of constraints that must be managed through vehicle design, energy-aware operations, and conservative dispatch decisions.

Thus, the dynamic nature of the ABL presents unique weather challenges for AAM and motivates new approaches to weather observation and prediction to ensure both safety and effective commercial integration.

7.1.1 The ABL: A New Environment for Sustained Aviation Operations

The *ABL*, sometimes referred to as the planetary boundary layer (PBL), is the lowest portion of the atmosphere and is directly influenced by the Earth's surface. While this region typically extends from the surface to about 1–2 km on average, its exact height varies in both time and space. The ABL can be more precisely defined by its key characteristics, including (1) the influence of the underlying surface, (2) the presence of turbulence, and (3) its diurnal cycle. A standard way to define the ABL is as the portion of the atmosphere directly influenced by surface forcings, such as frictional drag, evaporation, surface heating and cooling, pollutant emissions, and terrain-induced flow modification, on short timescales. Another defining feature of the ABL is turbulence. Unlike the *free atmosphere* above, the ABL is often turbulent through much of its depth, with turbulent eddies transporting heat, momentum, moisture, and pollutants. Finally, the ABL evolves on a diurnal cycle (the characteristic 24-hour rhythm driven by solar heating), a hallmark that distinguishes it from the free atmosphere aloft (Figure 7.1).

The exact height of the ABL varies according to atmospheric stability. This occurs because a stable boundary layer resists vertical air motion, while an unstable one promotes it. At night, radiational cooling of the Earth's surface often produces a shallow, stable boundary layer capped by a temperature inversion. Here, outgoing thermal (long-wave) radiation is emitted upward from the

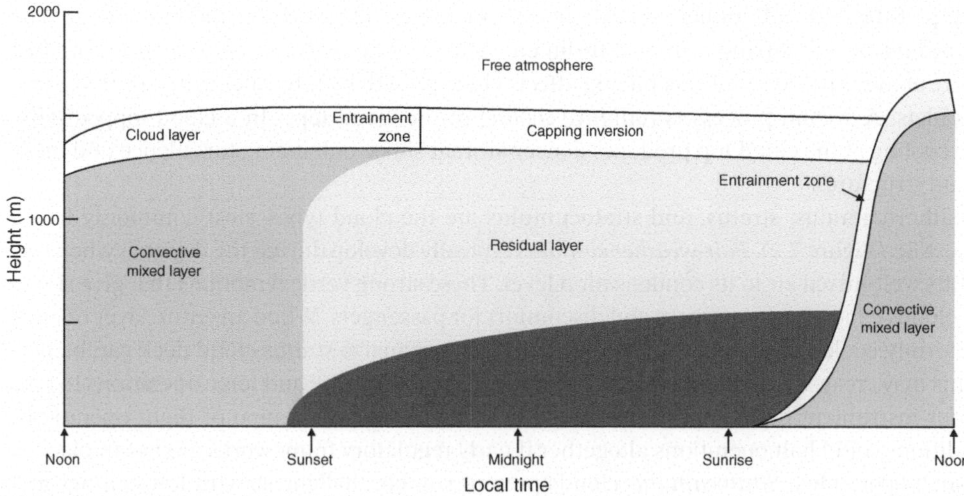

Figure 7.1 Typical diurnal evolution of the ABL. *Source*: Paul Markowski et al., 2010 / with permission of John Wiley & Sons.

surface without compensation from incoming solar (short-wave) radiation, causing the ground to cool. The air in direct contact with the surface cools by conduction and, on still nights, this chilled layer steadily deepens as the night progresses. This creates a vertical temperature gradient where colder air lies beneath warmer air. Such a profile, known as a *temperature inversion*, suppresses vertical mixing, confines turbulence close to the surface, and can trap pollutants and moisture. Under these conditions, the nighttime boundary layer may extend only a few hundred meters above the ground.

After sunrise, solar heating reverses the near-surface temperature profile. The ground warms, the air above the surface heats through conduction, convection redistributes this heat upward through the boundary layer, and the temperature once again decreases with height. This unstable gradient encourages vertical motion, as buoyant thermals rise and turbulent eddies mix the air. The result is a deepening daytime mixed layer that can grow to altitudes of 1–2 km under strong surface heating. While this convective environment disperses fog and pollution, it also generates gusts, shear, and turbulence. By evening, cooling resumes, stability increases, and the cycle begins again. In this way, the ABL "breathes" daily, alternating between shallow, stable nights and deeper convective afternoons. Its depth is further modulated by larger-scale influences, such as passing high- and low-pressure systems. This diurnal cycle is illustrated in Figure 7.1.

Above the ABL lies the free atmosphere, which is only weakly influenced by the surface and more closely resembles the operational environment of conventional aircraft. Regardless of the atmospheric stability regime, the AAM operational environment is primarily within the ABL.

ABL Characteristics

Although the ABL can be described by its structure and diurnal evolution, it is the specific processes and phenomena within it that will directly affect AAM operations. Cloud formation in the ABL arises from exchanges of heat and moisture between the surface and the air above. For clouds to form, moist air must be lifted upward. Turbulence in the lower ABL stirs the air, allowing water vapor to rise, cool, and condense into droplets. Without surface fluxes of heat and moisture, or the mixing provided by turbulence, the ABL remains stratified and lacks the support needed for cloud development.

At the top of the ABL, additional processes can influence cloud formation and persistence. The first is *entrainment*, the mixing of moist, turbulent ABL air with warmer, drier air from the free atmosphere above. The extent of this mixing affects cloud growth by introducing dry air that evaporates droplets. A second process is *radiative cooling* of the cloud tops. In a cloud-topped ABL, longwave cooling at the cloud top produces denser air that sinks, enhancing turbulence and modifying cloud structures.

Fair-weather cumulus, stratus, and stratocumulus are the cloud types most commonly found within the ABL (Figure 7.2). *Fair-weather cumulus* typically develop during the daytime when turbulence lifts well-mixed air to its condensation level. These strong vertical motions that give rise to cumulus clouds can cause bumpiness and discomfort for passengers. When an entire layer of air is more uniformly cooled to the point of condensation, an expansive *stratus* cloud deck can be produced. This may create a low ceiling that can severely restrict visibility and force operations to take place under instrument flight conditions. While this may reduce the tempo of flight operations, such conditions could halt operations altogether if early regulatory frameworks are not in place to support such operations. *Stratocumulus* clouds present similar challenges, with low ceilings and reduced visibility, thus impeding visual flight operations. These clouds can develop when stratus layers fragment or when cumulus clouds spread horizontally. As previously noted, strong radiative cooling at the top of stratocumulus layers can promote their persistence, and this persistence can continue to shade the surface, alter the surface energy budget, and influence the dynamics of the entire ABL.

Visibility is strongly modulated by boundary-layer processes. With nighttime radiational cooling and stable conditions, weak mixing permits moisture accumulation and aerosol stratification. This can create favorable conditions for *fog*, which can be considered a stratus cloud in contact with the ground. Fog processes are sensitive to the delicate balance between turbulence and stability. During the day, convective turbulence disperses fog and pollutants, improving horizontal visibility, though this same mixing may also loft particulates or dust.

While thunderstorms are not strictly an ABL phenomenon (they form above it), they strongly modify the ABL through precipitation and intense vertical motions. *Thunderstorms* can ventilate boundary-layer air via strong updrafts or stabilize it through precipitation-induced surface cooling. Convective downdrafts generate pools of cold air and gust fronts as air spreads outward after striking the surface. This dynamic can drastically alter surface wind fields. Such modifications occur over short timescales, from minutes to tens of minutes, and present hazards for low-altitude operations by introducing turbulence, sudden wind shifts, and, potentially, sharp reductions in vehicle performance margins.

Within the ABL, *winds* are governed by the combined effects of the pressure gradient force, frictional forces, and the Coriolis effect. Pressure gradients initiate motion, the Coriolis effect deflects the wind, and friction slows the flow. Within the immediate surface layer, typically extending

Figure 7.2 Cumulus (left), stratocumulus (center), and stratus clouds (right).

from the ground to tens of meters above the surface depending on terrain roughness, surface drag reduces wind speed toward zero. With height, the surface influence diminishes, and wind speed increases as frictional drag weakens, gradually adjusting toward a balance between the pressure gradient force and the Coriolis effect. In idealized conditions of steady state, horizontal homogeneity, and neutral stability, the vertical wind profile, U, within the surface layer follows a logarithmic increase with height, z,

$$U \propto \ln(z) \tag{7.1}$$

This sharp increase reflects the diminishing drag from roughness elements, such as vegetation and the built environment, with height. Above this surface layer, empirical approximations are often applied, with power laws approximating the growth of mean wind speed. Empirical exponents, α, typically range from 0.1 to 0.3 depending on surface roughness, atmospheric stability, and other thermal or mechanical mixing parameters, but a value of $\alpha = 1/7$ is most commonly used,

$$U \propto z^{\alpha} \tag{7.2}$$

In the upper portion of the ABL, airflow increasingly approaches balance between the pressure gradient force and the Coriolis effect. Neither the logarithmic nor the power-law approximations account for the Coriolis effect. Therefore, at these higher altitudes, wind speeds gradually approach free-atmosphere values through an exponential-like adjustment,

$$U \propto 1 - e^{-z/L} \tag{7.3}$$

Here, L represents a characteristic adjustment depth, describing how quickly the influence of surface drag diminishes with height. Figure 7.3 illustrates these conceptual models in a stacked manner, but the reader should note that these transitions are not precise and that vertical wind profiles are highly variable, changing with terrain, surface roughness, atmospheric stability, and other local conditions.

While the mean wind describes the general structure of airflow, the actual conditions experienced by vehicles are far less orderly. The atmosphere is inherently turbulent. *Turbulence* refers to the irregular, chaotic motions superimposed on the mean flow (Figure 7.4). Mathematically, the instantaneous wind, $u(t)$, can be decomposed into a mean component, $\overline{u}$, and the turbulent component, u',

$$u(t) = \overline{u} + u' \tag{7.4}$$

These turbulent fluctuations are a defining feature of the ABL and become especially pronounced in urban environments. For AAM, the way vehicles respond to turbulence will directly influence stability, energy use, and passenger comfort.

Turbulence is generated by both mechanical and thermal processes. *Mechanical turbulence* arises when air is forced to flow around or over obstacles, or when velocity differences (wind shear) create instabilities in the flow. In cities, roughness elements such as buildings, trees, and bridges impose drag on the flow, creating wakes, vortices, and shear zones that fragment the mean wind into irregular motions. These eddies generally scale with the size of the roughness elements: they are smaller for trees or low-rise structures and are much larger in areas with tall and densely packed buildings. Stronger background winds amplify this process, feeding more kinetic energy into the turbulent field. For AAM operations, mechanical turbulence is particularly important in and around the urban canopy, where recirculating flows around structures can challenge vehicle stability during takeoff, landing, and low-altitude cruise. These dynamics will be more thoroughly explored in our discussion of the urban boundary layer (UBL).

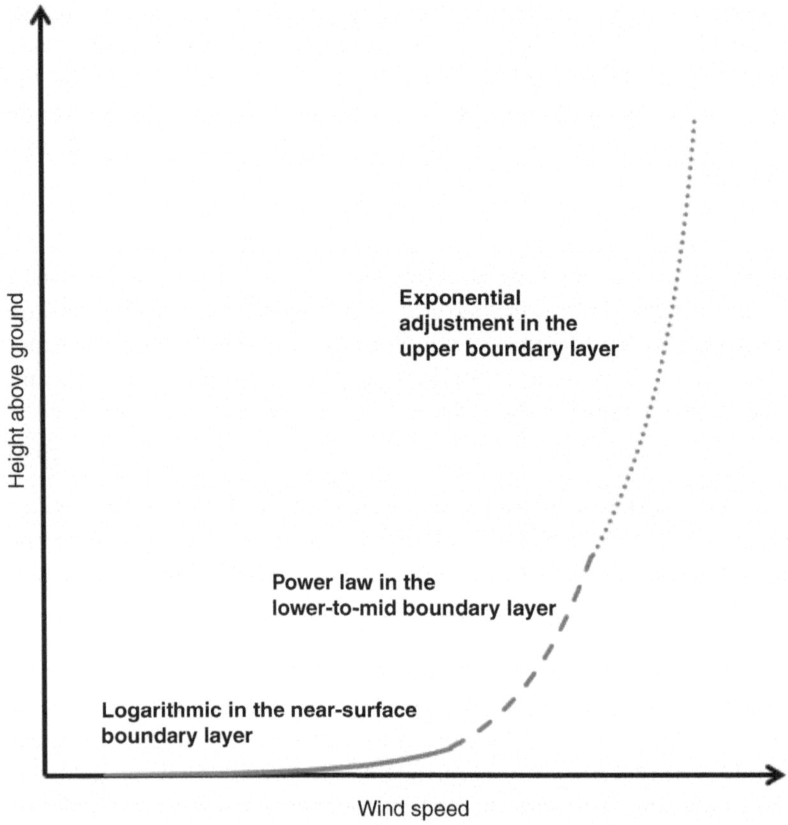

Figure 7.3 Conceptual wind speed profiles through the ABL. Flow near the surface can be described logarithmically, a 1/7 power law provides an empirical extension through the mid-ABL, and an exponential adjustment illustrates the gradual approach to flow in the free atmosphere aloft. The transitions are chosen for clarity and do not represent exact physical boundaries. Actual wind profiles vary with atmospheric stability, surface roughness, and ABL depth.

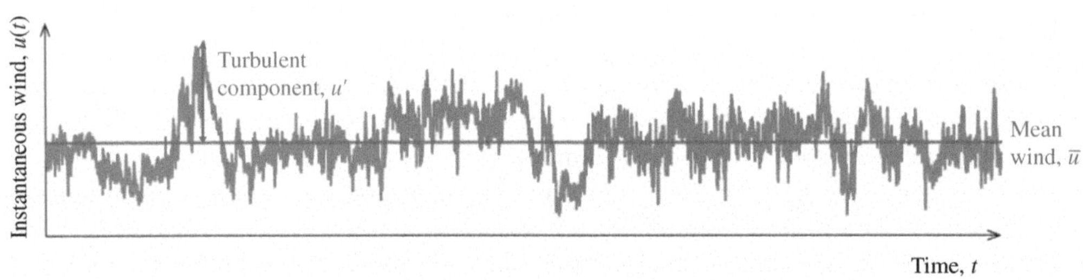

Figure 7.4 The instantaneous wind, $u(t)$, can be decomposed into the mean wind, $\overline{u}$, and the turbulent component, u'.

Thermal turbulence is generated by buoyancy when differential heating of the surface produces rising plumes of warm, less dense air and compensating subsidence of cooler, denser air. Strong solar heating of rooftops, asphalt, and bare ground produces vigorous thermals by day, while cooling at night tends to suppress these vertical motions. The scale of thermal eddies depends on the depth of the ABL. Near the surface, they may be only a few meters across, while in strongly convective daytime conditions, they can extend hundreds of meters or more. When large thermal eddies dominate, these fluctuations can persist much longer than the relatively smaller mechanical eddies. For AAM, buoyant turbulence can degrade ride quality, challenge vehicle stability, and increase control demands, particularly during hot conditions or mid-afternoon operations when thermals are strongest.

A quasi-frequent but often invisible nighttime wind feature is the *nocturnal low-level jet* (LLJ). As surface heating diminishes after sunset, turbulence within the ABL weakens. This reduction in mixing allows faster-moving air aloft, less affected by surface drag, to decouple from the ground. The result is a stream of accelerated winds, typically located between 100 and 300 m above the surface but occasionally extending up to 900 m. Wind speeds in these jets commonly reach 10–30 m/s. While sometimes appearing as a narrow stream of fast-moving air, LLJs can span hundreds of kilometers in width and extend a thousand kilometers or more in length. The sharp vertical gradients in wind speed associated with LLJs create strong shear, and this can affect eVTOL stability and increase power demand during ascent, descent, and low-level cruise.

Another distinct hazard for AAM vehicles operating in the ABL is icing. *Icing* is a critical hazard for AAM vehicles because, unlike conventional transport aircraft, AAM vehicles operate at low altitudes within portions of the ABL where shallow SLW-rich layers are commonly encountered. SLW is most common and concentrated in shallow layers associated with stratus clouds and fog, where liquid water content is relatively high despite the layer's limited vertical depth. Aircraft flying within these layers often encounter rapid and locally severe icing conditions, in contrast to the more gradual accumulation seen in deeper layers associated with synoptic systems. Accordingly, AAM vehicles confined to lower altitudes face an elevated risk of severe icing when they remain embedded within SLW-rich environments for extended periods instead of briefly transiting through them.

Helicopters and small drone operations have demonstrated how rapidly ice can accrete on exposed lifting and propulsive surfaces, leading to a loss of aerodynamic efficiency, sharp increases in drag, and, in some cases, hazardous imbalances from uneven ice shedding. Unlike fixed-wing aircraft, where ice accretion tends to occur more uniformly along lifting surfaces, rotorcraft and propeller-driven vehicles are especially vulnerable to unsteady accretion because their rotating blades continually sweep through varying temperature and moisture fields. However, for eVTOLs, electric propulsion does offer the advantage of motors capable of delivering very high levels of power for short durations until thermal limits are reached. This capability provides the potential for short-term high rates of climb intended to escape shallow icing layers or the application of emergency thrust before significant accretion compromises safety. Nevertheless, even in regions not traditionally considered icing-prone, the nature of icing in the ABL poses a significant hazard for AAM operations.

7.1.2 The UBL and the Built Environment

As AAM vehicles transition from rural to urban environments (Figure 7.5), they may encounter zones of wind shear and turbulence generated by abrupt changes in surface roughness and heating characteristics. This flow adjustment develops just above the surface and deepens with distance downwind. Near the city's edge, only the lowest portion of the flow reflects urban conditions, while the wind aloft retains predominantly rural characteristics. With increasing downwind distance, however, the influence of the urban surface expands upward until the entire boundary layer takes on a distinctly urban profile.

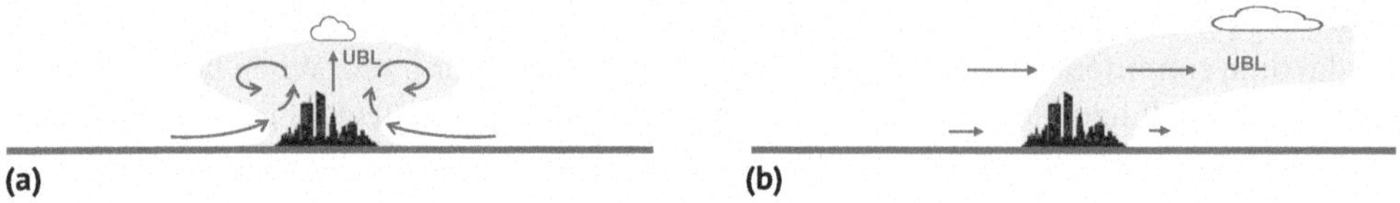

Figure 7.5 ABL and UBL structures in: (a) a calm regional flow regime, and (b) a moderate regional flow scenario.

The UBL is the portion of the ABL that is modified by the presence of a city. Accordingly, the UBL's structure and properties can differ significantly from a surrounding rural environment. Whereas the rural ABL is affected primarily by natural surface processes, the UBL reflects the additional influence of the built environment. As illustrated in Figure 7.5, in calm conditions, the UBL can appear as a dome of modified air over a city; under prevailing winds, it develops as an internal boundary layer that grows with downwind distance and may grow until it fills much of the ABL. As discussed in Chapter 6, while regional AAM services will extend well beyond urban environments, a substantial fraction of operations will take place within the UBL, especially takeoffs and landings, making detailed understanding of the UBL particularly important for AAM operations.

The structure and behavior of the urban climate are fundamentally shaped by the properties of the surface beneath it, the city's metabolism (its exchange of energy and materials with the surrounding environment), and the structures that populate it. In cities, the surface is composed of a mosaic of materials, or *fabrics*, such as asphalt, concrete, brick, glass, vegetation, and water. Each fabric type has distinct thermal, radiative, and moisture properties that determine how it absorbs and releases heat, and stores or sheds water. An urban area's *surface cover* describes how these fabrics are proportioned. Together, urban fabric and surface cover dictate the exchanges of heat, moisture, and momentum that drive turbulence and mixing in the UBL. For example, extensive impervious cover amplifies surface heating and raises sensible heat fluxes, while vegetated patches promote evaporation and localized cooling. Thus, the unique patchwork of urban surfaces gives rise to the distinctive microclimates of each city and operating environments that AAM vehicles must accommodate.

The urban ecosystem relies on external inputs of energy, water, and raw materials, procured to sustain human activities. These resources are processed and transformed and ultimately returned to the environment, often in altered or degraded forms, such as waste heat, water or water vapor, and particulate and gaseous emissions. This cycle constitutes the *urban metabolism*, through which energy and material by-products released into the atmosphere modify thermal, radiative, and visibility conditions within the UBL. Particulate matter, liquid droplets, and other aerosols, including secondary products of gaseous emissions, can reduce visibility via the scattering of shortwave radiation in the visible portion of the spectrum. Reduced visibility can pose challenges for AAM operations by limiting a pilot or sensor's ability to detect obstacles, other aircraft, and vertiports during low-altitude flight. Persistent haze or fog can impede reliance on visual flight rules (VFR), forcing greater dependence on instruments, navigational aids, and air traffic management (ATM) systems to maintain safety and reliability.

We have considered how an urban environment's surface properties, metabolic outputs, and structures can alter its character relative to an adjacent rural area. Collectively, these factors can create the UHI. The UHI refers to the characteristic temperature difference between urban and surrounding nonurban areas, where cities are typically warmer, especially at night. This arises because urban materials with low albedo and high heat capacities absorb and store solar radiation during the day, releasing it slowly after sunset. The effect is exacerbated by the dominance of impervious surfaces that limit evaporative cooling, the geometry of urban canyons that trap long-wave radiation, and the direct addition of anthropogenic heat from the built environment. Under these conditions, the UBL often remains several degrees warmer than nearby rural air, with the strongest contrasts occurring during clear, calm periods.

The UHI is a critical consideration for AAM. By elevating air temperatures within the UBL, the UHI alters the thermodynamic environment within which vehicles must operate. Higher temperatures reduce air density, effectively increasing density altitude. This can diminish vehicle performance and require greater power during hover, takeoff, and climb. UHI effects are most noticeable in densely built city centers where vertiports are most likely to be located.

By altering the urban thermal environment, the UHI also modifies the stability and turbulence characteristics of the urban atmosphere. By reducing nocturnal cooling and enhancing daytime heating, it shifts the lower atmosphere toward more neutral or unstable conditions, promoting stronger vertical mixing and turbulence. For AAM, this translates into ride quality variability, increased control demands, and potentially higher energy consumption for low-altitude cruise. The UHI can also degrade air quality by fostering conditions conducive to pollutant accumulation and smog formation, thereby reducing visibility.

Vertical Structure and Urban Flow

Understanding the vertical structure of the UBL is important because it defines the environment that AAM vehicles must traverse during every urban flight operation. This vertical structure can be subdivided into three layers: the urban canopy layer (UCL), roughness sublayer (RSL), and inertial sublayer (ISL) (Figure 7.6).

Urban Canopy Layer

The UCL extends from the ground up to the mean building height. Within this layer, airflow is fully 3D in nature as it is deflected laterally, overturned vertically, and imbued with vortices and up- and downdrafts. For AAM vehicles, this is the zone for takeoff and landing and where eVTOLs must negotiate sharp wind gradients.

Gaining a comprehensive understanding of the 3D flow around buildings is a formidable task, but consideration of general flow characteristics around basic buildings with varying spacing can develop useful insight. Figure 7.7 shows three basic UCL flow regimes around buildings. Which regime develops depends on the relationship between building height (H) and the spacing between them (W).

In *isolated roughness flow* ($H/W < \sim 0.35$), buildings or other roughness elements are widely spaced so that each generates its own wake with limited interaction from neighboring structures.

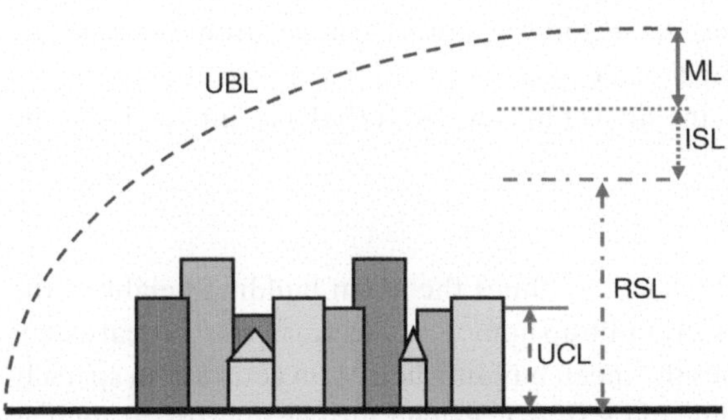

Figure 7.6 Structure of the UBL.

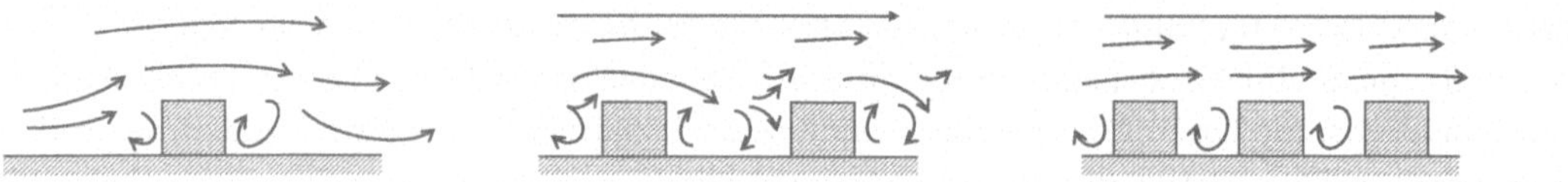

Figure 7.7 Basic UCL flow around an isolated square building (left), in a wake interference regime (center), and a skimming flow regime (right).

The airflow around each element resembles that of a solitary obstacle with flow separating at sharp edges, producing counterrotating vortices on either side of the building's base and a 3D recirculation zone in the lower-pressure lee of the building, before gradually reattaching farther downwind. Because the perturbed flow around an isolated building does not significantly overlap with wakes of other elements, turbulence is relatively localized, and the mean flow is only marginally perturbed. This regime is more representative of suburban or low-density development than dense urban cores. For eVTOL operations, isolated roughness flow implies that vehicles may encounter localized but intermittent gusts when maneuvering around individual obstacles such as warehouses or isolated high-rises. The overall environment is less turbulent than in denser districts, but variability in wind direction near wakes and cavities may complicate low-altitude maneuvering.

Wake interference flow ($0.35 < H/W < {\sim}0.65$) occurs when buildings are moderately spaced such that the recirculation zone downwind of one element overlaps with the windward face of the next. This arrangement causes the recirculating flow from the upstream building to reinforce vortices at the base of the downstream building, leading to stronger turbulence and greater vertical motions. The flow is less organized than in isolated roughness, with alternating gusts, eddies, and flow reversals filling the spaces between structures. This regime presents a more challenging environment for ascent and descent through the urban canopy. Vehicles may encounter enhanced shear and turbulence between buildings, increasing energy demand for stability and creating greater variability in ride quality. Piloting through these conditions requires anticipating rapid changes in the flow over short distances.

When buildings are packed densely together, a *skimming flow* ($H/W > {\sim}0.65$) regime develops. Here, the overlying airflow effectively "skims" across the rooftops, with little direct penetration into the canyons below. Within leeward cavities, recirculating vortices dominate and exchanges with the overlying flow are limited. The shear zone at roof level becomes the key driver of turbulence where downward intrusions of higher-momentum air ("sweeps") and upward expulsions of slower-moving air ("ejections") intermittently mix canyon air below with air above the roof level. This regime is typical of downtown cores with high aspect ratio (high H/W) canyons.

For AAM operations, skimming flow creates a sharp contrast between relatively strong and coherent rooftop-level winds and highly recirculating canyon winds below. Vehicles transitioning through rooftop level must contend with strong shear, and vertiports located within canyons may be exposed to stagnant, polluted air with unpredictable gusts. Low-altitude cruise above rooftops may benefit from more uniform flow but remains subject to elevated turbulence intensities at the shear interface. This is the domain of the RSL.

Roughness Sublayer

The RSL typically extends upward to approximately 2–3 times the mean building height of the urban area. In skimming flow regimes over densely built-up homogeneous areas, the vertical extent of the RSL may be reduced to as low as 1.5 times the mean building height. In contrast, in sparsely built-up areas with isolated roughness elements, the RSL may extend further upward, as high as 4 times the mean building height. Regardless of depth, wind speeds are still reduced in magnitude within the RSL, but mechanical turbulence is intensified and fully 3D in character, responding to individual roughness elements. For AAM, transitions through the RSL during climb or descent are among the most demanding phases, with stability and control most directly impacted. A region of particular concern is the rooftop shear zone where velocity gradients are strong as the surface and roughness elements below absorb the flow's momentum while the flow above remains less impeded.

Inertial Sublayer

Above the RSL, the atmosphere becomes more horizontally homogeneous, and mean vertical motions dissipate. This ISL is no longer characterized by the influence of individual roughness elements as their effects blend and the microclimates present in the RSL disappear. While wind speed is still stymied here by the urban area, turbulence in the ISL is more homogeneous. This extends from the airflow to other atmospheric properties as well. Like wind speed, temperature and humidity become horizontally consistent at the neighborhood scale. Thus, the ISL offers relatively steady conditions useful for corridor planning within urban districts as a comparatively stable low-level cruise ride experience can be expected here.

As eVTOLs transition through the ISL and climb higher, the influence of buildings and other roughness elements continues to weaken through the remaining depth of the UBL. The overlying *mixed layer* becomes increasingly dominated by buoyant motions, with mechanical turbulence associated with surface drag playing a diminished role. Neighborhood *climate zones* characteristic of the ISL blend into a city-wide atmospheric character. Vehicles climbing into the mixed layer encounter fewer gusts associated with individual structures but experience greater exposure to large, thermally driven eddies. With these conditions shaping stability, ride quality, and power demand during ascent and low-altitude cruise, the mixed layer serves as the broader atmospheric backdrop for urban and regional AAM operations.

It is important to remember that the flow patterns described here are abstract; they merely provide a conceptual framework to describe and better understand the general effects of the urban environment. While this insight may assist in developing useful intuitive models, more detailed numerical models are required to capture true flow patterns over any particular urban locale with high fidelity. These models, and the observations required to run and validate them, are discussed next.

7.1.3 Spatiotemporal Weather Observations

Among all regions of the atmosphere, the ABL remains the least observed. The ABL sits in an observational "blind spot"; it is too high for most ground-based towers, too low for satellites and weather radars, and too variable to be represented by the sparse surface networks and twice-daily balloon launches that underpin conventional meteorology (Figure 7.8). The very features that define the ABL, such as its rapid evolution, its strong diurnal cycle, and its sensitivity to local terrain and the built environment, make the region especially challenging to monitor. Observational networks have historically been designed for synoptic weather prediction and high-altitude aviation, leaving little incentive to resolve the turbulent flows of the lowest kilometers. Thus, the atmospheric region most critical to AAM remains the least measured.

Limitations of Current Systems

Because meteorological observing systems have historically not been designed with the ABL in mind, existing observing infrastructure leaves major observational gaps for AAM. While surface

Figure 7.8 Common meteorological observational instrumentation. Doppler weather radar (left), MetOp second-generation A-type weather satellite (top center), ASOS (bottom center), AWOS (top right), and radiosonde (bottom right). *Source*: Eric Phelps / Wikimedia Commons / CC BY 3.0.

observation networks provide reliable and highly accurate observations, these measurements tend to be widely spaced point measurements. Automated Surface Observing System (ASOS) units, typically deployed at larger airports, and Automated Weather Observing System (AWOS) units, typically deployed at smaller airports, are excellent examples. However, the spatial separation of these systems can often leave tens to hundreds of kilometers without any direct measurements. Broad-area observing systems, including radar, satellites, and radiosondes, offer minimal resolution in the lowest kilometer where AAM vehicles operate. Collectively, these sensors provide a patchwork of information that can describe broad synoptic conditions, but they still fail to resolve the fine-scale winds, visibility, and cloud height variability that govern safe and efficient flight near the ground. For decades, helicopter emergency medical services have contended with this "weather gap," where the lack of localized ceiling and visibility data has led to higher risk or canceled missions. Similar challenges appear in road weather networks, where sparse coverage limits the detection of hazards, such as freezing fog or localized icing.

Remote sensing systems, such as radar and satellites, provide wider coverage but bring their own limitations. Doppler radars are optimized for precipitation monitoring versus clear-air returns, a limitation for observing low-level winds on clear days. With radar beams tilted to sample large storm volumes, the lowest beam often leaves the critical first few hundred meters, where AAM vehicles fly, essentially unsampled. Urban buildings further complicate radar returns, degrading the signal in some of the areas of greatest interest to AAM operations. Satellites have been major drivers of forecasting improvement over the last several decades and are an invaluable tool for global coverage and synoptic monitoring. However, they remain limited in the lowest kilometer of the atmosphere, where they cannot detect sharp vertical gradients and are too coarse to resolve building-scale flows and microclimates.

Radiosondes provide high-quality in situ-based vertical profiles of temperature, humidity, pressure, and winds, but their limitations are significant for AAM. Outside of special circumstances, radiosondes are only launched twice daily and at sparsely spaced sites. This leaves large gaps in both space and time and makes them unsuitable for supporting the rapid update cycles demanded by high-tempo, and especially urban, flight operations. Further, the lowest kilometer is often under-sampled due to their rapid ascent rates and the balloons drifting with the wind. Consequently, resulting profiles may not represent the exact area of interest. Despite the labor-intensive nature of radiosonde launches and their lack of direct turbulence or aerosol measurements, radiosondes remain invaluable for synoptic forecasting but are poorly suited for the spatiotemporal demands of AAM nowcasting.

Because numerical weather prediction (NWP) models depend on these same observations for initialization, observational gaps in the ABL directly translate into shortcomings in forecasting model performance. As a result, fine-scale features of the lower atmosphere often remain underrepresented, even in the most advanced high-resolution weather models. While NWP at 3-km grid spacing has become routine, such models cannot capture the sharp gradients, building wakes, and turbulent structures that characterize the AAM operational environment. These gaps motivate the development of new sensing strategies specifically tailored to AAM.

Emerging Needs for AAM

Large commercial air carriers have long benefited from dense and highly reliable meteorological observation networks at airports. In contrast, many AAM operations will occur at hubs and vertiports that lack comparable instrumentation and historical weather characterization. This gap is especially consequential because AAM vehicles are inherently more sensitive to low-altitude atmospheric variability than traditional transport aircraft. Smaller vehicle mass, lower inertia, and tighter thrust margins increase susceptibility to rapidly changing winds, turbulence, and thermal gradients. In addition, AAM missions allocate a greater fraction of total flight time to takeoff, approach, hover, and landing within the ABL, where flows are strongly influenced by surface roughness and the built environment.

Within this context, improved characterization of low-altitude wind fields becomes a foundational operational requirement for AAM. Enhanced observation and prediction capabilities support safety, reliability, and energy efficiency by enabling routing strategies, stabilized approach profiles, and vertiport layouts aligned with prevailing and building-modified flow patterns. This alignment reduces unnecessary power expenditure during power-intensive flight segments, particularly during approach, hover, and landing where the available performance margins are most constrained.

New sensing paradigms are beginning to address these observational needs. The emergence of ASTM International's Weather Information Provider (WIP) standard is an example of one new model and represents a fundamental shift in how aviation weather data is defined, sourced, qualified, and delivered. The WIP framework addresses this need by advancing from a platform-based system, where only certain certified sensors are recognized, to a performance-based system where any source can contribute as long as its data quality is characterized. Under this model, diverse observations, whether from a Doppler wind light detection and ranging (LiDAR) or sound detection and ranging (SoDAR) at a vertiport, an adjacent Internet of Things (IoT)-enabled rooftop weather station, or a telemetry stream from an eVTOL on approach, flow into a common reporting structure that forms a Vertiport Weather Observation System (VWOS). Each dataset is tagged with metadata about its location, time, accuracy, and performance tier. High-precision Tier 3 data may be used to define launch and landing limits, while Tier 1 and Tier 2 data, when accompanied by well-characterized uncertainty, can support operational risk management decisions by pilots and automated systems. In such cases, knowledge of expected error enables operators to apply appropriate safety margins relative to aircraft performance thresholds. In this way, the framework does not exclude unconventional sources but integrates them transparently, allowing their operational value to be realized without overstating their reliability.

Approaches such as WIP open the door for an immense expansion in weather sensing. Cyber-physical architectures present the opportunity for these stationary sensors to be integrated with mobile sensing nodes, with observational support from traffic cameras, smartphones, surface vehicles, small uncrewed aircraft systems (sUAS), and larger AAM aircraft themselves. At the same time, the growth of IoT and smart city infrastructure potentially allows for dense, low-cost sensing of temperature, humidity, visibility, and precipitation across the urban landscape into a type of Neighborhood Weather Observation System (NWOS). Fused into a coherent operational picture, these systems would combine heterogeneous data streams into an Integrated Weather Observation System (IWOS) with certified products for pilots, dispatchers, and automated ATM systems. Still, these observations alone are not enough; they must feed into nowcasting and forecasting frameworks capable of rapid updates at urban scales.

7.1.4 NWP for AAM

NWP uses computer simulations based on mathematical equations to predict future weather conditions. NWP thus serves as the basis for operational forecasting worldwide. These predictions are accomplished on a 3D grid of points where the numerical forecast is solved (Figure 7.9). Global and regional NWP models routinely produce accurate forecasts of synoptic and mesoscale phenomena with grid spacings on the order of a few kilometers, and data assimilation systems capable of ingesting millions of observations each cycle. A model with 3-km grid spacing can represent, or resolve, weather systems on the order of several kilometers or larger; smaller-scale features are

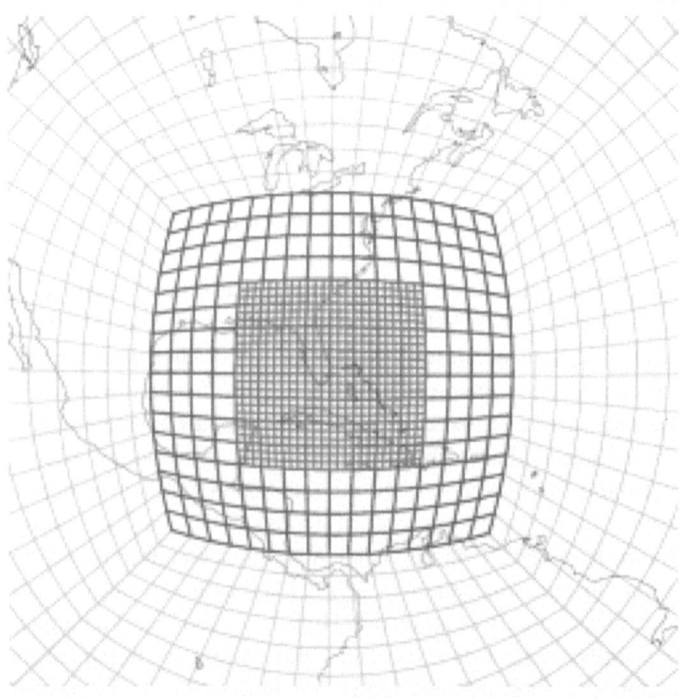

Figure 7.9 A nested grid which covers the Southeastern United States with finer resolution than the coarser "parent" grid. Courtesy of NOAA. Public domain.

approximated, or "parameterized." These systems, exemplified by convection-permitting models such as the High-Resolution Rapid Refresh (HRRR) in the United States or the Applications of Research to Operations at the Mesoscale (AROME) in France, are widely used to provide short-range guidance on convective storms, ceilings, visibility, turbulence, and icing in support of conventional aviation. However, these capabilities are poorly aligned with the spatial and temporal scales that define AAM operations.

The ABL, especially its lowest levels, remains the most difficult regime for NWP to resolve. Kilometer-scale models can explicitly resolve organized convective systems, such as thunderstorms and frontal structures, but do not resolve the turbulent wakes of buildings, sharp shear layers, or localized thermals that dominate the urban atmosphere. Even when run at finer spacings, forecast skill remains constrained by limited near-surface observations. The current assimilation systems for these models are designed around satellite radiances, radiosondes, and aircraft reports, all of which provide limited information in the ABL, and especially the UBL. As a consequence, fine-scale boundary-layer features are systematically under-represented, and forecast skill for low-altitude winds, visibility, and turbulence tends to plateau even as model resolution increases.

Outside of coarse resolutions and limited surface observations, the temporal refresh rate of the model output is another significant limitation. Operational forecast cycles typically update every 1–3 hours, with products issued for time horizons of 6–24 hours or longer. While suitable for synoptic-scale aviation planning, these cycles do not match the timescales of boundary-layer evolution, where changes can occur within minutes. Nowcasting systems provide forecast guidance spanning approximately 0–6 hours and, in practice, focus on the first 1–2 hours where observational data dominate predictive skill. These systems extrapolate radar, satellite, and surface data to provide near-term hazard awareness, but they similarly struggle within the lowest kilometer where radar beams overshoot, satellite retrievals are coarse, and surface stations remain too sparse to resolve urban-scale variability. The result is a persistent gap in time-critical, low-altitude weather intelligence precisely where AAM vehicles operate.

Bridging the gap between current capabilities and the requirements of AAM presents challenges in several interconnected areas. To represent microscale atmospheric processes, which typically occur at spatial scales ranging from tens of meters to approximately 2 km, forecast models must be executed with grid spacings of tens to hundreds of meters. Supporting parameterizations must also be suited to both open terrain and urban canopies. Such models will need to be nested within coarser systems to inherit boundary conditions but configured to explicitly resolve variability at scales relevant to infrastructure elements, including vertiports and urban air corridors. As discussed, these models cannot function without new observational input strategies, and assimilation frameworks must be adapted to handle heterogeneous data streams, weighting them by quality and incorporating metadata to ensure transparency.

Forecast update cycles must also accelerate to reflect operational tempos. Rapid-update cycling, where models are refreshed frequently as new data become available, is better aligned with the demands of urban air mobility than the fixed hourly cycles of current global and regional models. Probabilistic approaches, using ensembles of rapid-update forecasts, are essential to capture the uncertainty inherent in the dynamic ABL and UBL. Such ensembles will need to provide probabilistic forecasts along with confidence metrics for hazards such as low-level wind shear or rooftop turbulence.

Equally important to forecasting skill and updating is how model output translates into operational decision tools. Although designed for convective hazards at larger scales, the Federal Aviation Administration's (FAA) Shake (Simplified Hazard Area for Convective Environments) and Sharp (Simplified Hazard Area for Rapidly Developing Phenomena) program is an example of a simplified hazard depiction model that distills complex meteorological data into intuitive hazard "footprints" that pilots and dispatchers can easily act upon. In the AAM context, the same philosophy must be applied at boundary-layer scales. Successfully integrating these features would produce a rapid-update urban model with dense observations fused into probabilistic hazard depictions that delineate no-launch sectors, rooftop shear zones, and transient gust fronts, each tagged with confidence levels and times of validity. Such products would provide a clear interface between meteorology, procedures, and autonomy, allowing operators and automated systems to act on explicitly quantified and operationally relevant risk instead of unprocessed meteorological fields.

Artificial intelligence and machine learning (AI/ML) are increasingly explored as complements to, not replacements for, physics-based atmospheric models. However, at boundary-layer and urban scales, the application of AI/ML is fundamentally constrained by the availability and fidelity of observational data needed for training, calibration, and validation. Many ML approaches aim to emulate computationally expensive turbulence schemes, correct systematic model biases, or downscale coarser forecasts to vertiport-scale resolutions. Yet these approaches implicitly assume the existence of representative "ground truth" at the relevant spatial and temporal scales, an assumption that is often not satisfied within the ABL or UBL.

In urban environments, critical microscale variability in winds, visibility, low stratus clouds, and turbulence occurs at spatial scales that remain below the resolution of conventional observational networks. When training datasets rely on model analyses or reanalyses as proxies for truth, there is a risk of reinforcing existing model deficiencies instead of correcting them, particularly when the underlying physics is under-observed. This limitation is far less pronounced at global or synoptic scales, where dense satellite coverage and long historical records support robust AI/ML performance, but becomes more critical at sub-kilometer scales relevant to AAM operations.

Nonetheless, AI/ML can offer value when applied within clearly defined bounds, such as accelerating selected components of physics-based models, providing probabilistic corrections conditioned on quantified uncertainty, or assisting with pattern recognition in well-observed phenomena. However, at microscale resolutions, AI/ML effectiveness remains tightly coupled to advances in boundary-layer sensing, especially in urban areas where microclimates are poorly sampled. Hybrid approaches that merge first-principles physics with data-driven components offer the opportunity to extend forecast skill, but only when observational strategies are sufficient to support meaningful validation and continuous recalibration.

High-performance computing developments, including GPU-accelerated large-eddy simulation (LES) frameworks such as NCAR's FastEddy, illustrate a parallel pathway in which physical realism is preserved and enhanced through computational efficiency instead of relying solely on statistical inference. These approaches demonstrate how real-time or near-real-time microscale modeling may become increasingly operationally feasible for capturing building wakes, rooftop shear layers, and other urban flow features at tens-of-meters resolution. For AAM, physics-anchored models, supported by dense and purpose-built observations, remain the most defensible foundation for microscale weather prediction, with AI/ML serving a supporting role rather than a substitute for unresolved physics.

Conceptually, the trajectory of NWP for AAM is moving toward a hierarchical, multi-scale architecture with information flowing between all model scales. Global and continental models will continue to provide boundary conditions, and regional models at kilometer scale will refine mesoscale features such as fronts and convection. Urban nests, informed by dense observations, will resolve city-scale circulations, and microscale LES domains will explicitly simulate turbulence and flows around vertiports. These layers must be linked by assimilation systems capable of continuously ingesting distributed observations and by decision frameworks that translate forecasts into hazard-focused products. The final architecture is an integrated system specifically designed to represent the scales of motion most relevant to AAM operations.

7.2 Sustainability

Our modern understanding of sustainability is most often traced to the Brundtland Report published by the United Nations' World Commission on Environment and Development [1]. The report defines *sustainability* as "meeting the needs of the present without compromising the ability of future generations to meet their own needs." This definition has since guided both sustainability thought leadership and policy that balances economic activity with environmental stewardship and social well-being.

Sustainability is one of the defining measures by which AAM will be judged. While electrified flight is often touted as "zero-emissions aviation," the reality is far more nuanced. True sustainability requires looking beyond the aircraft in flight; it requires consideration of the full life cycle of materials and batteries, the carbon intensity of the supporting electricity grid, and the integration of AAM vehicles into the broader transportation system. A central concept here is the idea of a *carbon footprint*, a measure of the total greenhouse gas (GHG) emissions associated with an activity, product, or organization, often expressed as carbon dioxide equivalent (CO_2e). In practical terms, a carbon footprint accounts for all emissions linked to producing and using energy, including the upstream emissions associated with the fuel's extraction and refining, as well as emissions associated with the operation itself. For electric aircraft, the carbon footprint is tied directly to the carbon intensity of the electricity grid. Against this backdrop, our examination of sustainability and AAM considers the topic from two perspectives: (1) the sustainability of AAM operations themselves and (2) the role of AAM in supporting broader sustainability goals, particularly sustainable urban development.

7.2.1 Operational Sustainability in AAM

When assessing the sustainability of AAM, it is tempting to focus exclusively on the vehicle in flight and the absence of emissions. However, as we have already established, true sustainability is far broader in scope; it encompasses the entire life cycle of the aircraft and its components, the energy systems that support it, the operational practices that prescribe its use, and the frameworks companies employ to measure their environmental and social performance. In other words, sustainability in AAM operations cannot be separated from the supply chains, facilities, and governance structures that underpin the sector.

Electric propulsion enables eVTOL aircraft to fly without in-flight GHG emissions. However, this does not equate to zero emissions. The total impact depends on the carbon intensity of the grid that supplies the charging power, the materials and manufacturing that go into producing batteries and composite structures, and the recycling or disposal pathways available at the end of life. Other propulsion pathways under consideration for sustainable aviation, such as the hybrid-electric and hydrogen systems discussed in Chapter 2, come with similar caveats. Hybrid-electric aircraft, aircraft that combine combustion engines with batteries, improve efficiency and reduce emissions compared to conventional aircraft. Nonetheless, they still have the same concerns as purely electric flight, plus the direct burning of fuel and, consequently, a "tailpipe" carbon footprint. Hydrogen-powered aircraft promise water vapor as the only in-flight by-product when hydrogen is used in a fuel cell; however, overall sustainability depends on whether the hydrogen is produced using renewable or fossil-based energy sources. In all cases, even in the absence of a "tailpipe" emission, upstream energy production remains a determining factor and is today still dominated by coal-, natural gas-, and petroleum-based sources. This illuminates a broader truth: no propulsion technology is automatically zero-carbon.

This reality mirrors lessons from the automotive sector. Battery electric vehicles (BEVs) and hybrid electric vehicles (HEVs) have reduced average carbon intensities compared to gasoline-powered cars, but their carbon intensity varies depending on grid composition, manufacturing, and operational use. While eVTOLs can fly point-to-point, the additional energy required for vertical lift can reduce their competitiveness relative to electric surface transportation. Worked example 7.1 illustrates these ideas in concrete terms by placing eVTOL operations alongside internal combustion cars and BEVs on a per-passenger-kilometer emissions basis. Note how assumptions about grid carbon intensity and occupancy change the sustainability picture.

Worked Example 7.1:

Estimate how carbon intensity compares across transportation modes for a 100 km trip.

Given:
Representative vehicle-kilometer (vkm) GHG intensities (well-to-wheel) for a 100 km trip (single occupant):

- Internal combustion engine vehicle (ICEV): 0.20 kg-CO_2e per vkm[a];
- BEV: 0.09 kg-CO_2e per vkm[b];
- eVTOL: 0.15 kg-CO_2e per vkm (representative mid-grid, cruise-dominated mission assumption).[c]
  - The vkm considers emissions per kilometer traveled by a vehicle, regardless of how many passengers it carries.
  - GHG intensities refer to the amount of GHGs emitted, expressed in grams of CO_2e, per unit distance. "CO_2e" expresses the impact of all GHGs in terms of the amount of CO_2 that would have the same warming effect.
  - Well-to-wheel (WTW) takes into account everything stretching from energy production to use in the vehicle (Figure 7.10).
  - A 100 km trip (single occupant) assumes a typical 100 km journey, which is long enough for cruise flight to dominate an eVTOL mission but short enough to be comparable with car commutes.

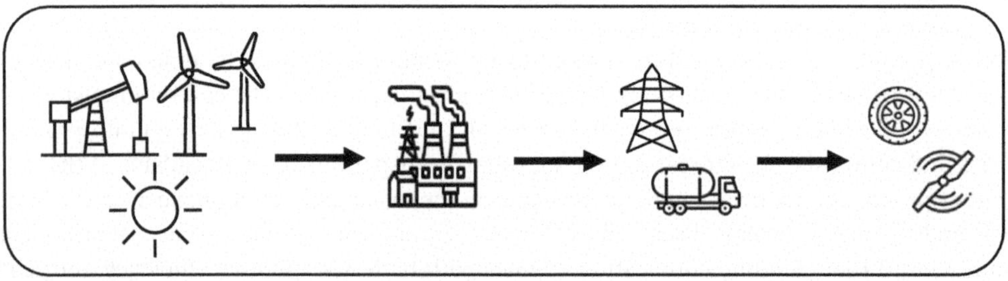

Figure 7.10 WTW includes upstream energy production, delivery, and in-vehicle operation.

Find the Answer:
Convert vkm to passenger-km.

$$\text{PKT} = \text{VKT} \times \eta \tag{7.5}$$

where PKT and VKT represent vkms traveled and passenger-kilometers traveled, respectively, and η is the number of passengers. VKT represents the total distance traveled by a vehicle, regardless of how many passengers are inside; and PKT represents the total distance traveled by all passengers combined. Therefore, if a vehicle travels 100 km with four passengers, it has a PKT of 400 km.

Case A: One occupant in each vehicle:

ICEV: 0.20 kg-CO_2e per vkm/1 = 0.20 kg/km = 200 g/km
BEV: 0.09 kg-CO_2e per vkm/1 = 0.09 kg/km = 90 g/km
eVTOL: 0.15 kg-CO_2e per vkm/1 = 0.150 kg/km = 150 g/km

Case B: Illustrative higher-utilization occupancies (cars: 1.5 passengers; eVTOL: four passengers):

ICEV: 200 g/km/1.5 = 133 g/km
BEV: 90 g/km/1.5 = 60 g/km
eVTOL: 150 g/km/4 = 0.050 kg/km = 38 g/km

Exposition:

AAM should be evaluated in the same carbon conversation as other transportation modalities and should be compared on a per-passenger basis. Electrification reduces emissions but does not eliminate them; actual carbon emissions depend on how the electricity is supplied. Occupancy strongly affects sustainability outcomes. With a single passenger, an eVTOL is outperformed by BEVs but with higher occupancy, an eVTOL can outperform both ICEVs and BEVs.

---

a U.S. Environmental Protection Agency. (2024, November). The 2024 EPA Automotive Trends Report: Greenhouse Gas Emissions, Fuel Economy, and Technology Since 1975 (EPA-420-R-24-022). https://www.epa.gov/automotive-trends

b U.S. Energy Information Administration. (n.d.). How Much Carbon Dioxide Is Produced Per Kilowatt-hour of U.S. Electricity Generation? Frequently Asked Questions (FAQs), U.S. Department of Energy. https://www.eia.gov/tools/faqs/faq.php

c Mission-phase energy estimates for eVTOLs (hover, climb, cruise, descent, and reserve) are based on NASA and FAA concept of operations modeling [2]. These assumptions were applied in Kasliwal et al. [3], who reported ~0.15 kg-CO_2e per vkm for a 100 km mission. Subsequent work by Liu et al. [4] confirmed the mission-phase breakdown and extended the scope to a full LCA, reporting ~0.24 kg-CO_2e per passenger-km under Chinese grid conditions.

The adoption of life-cycle analysis (LCA) extends sustainability assessment beyond operational emissions by including upstream material extraction, manufacturing energy, operational use, and end-of-life disposal or recycling. These interconnected stages are commonly represented in a life-cycle diagram (Figure 7.11), depicting how an aircraft's environmental footprint begins long before its first flight and continues well after it leaves service through disposal or material recovery. The diagram shows how materials move from extraction through design, production, and years of operation, before ultimately ending in either disposal or recycling. Considering this complete sequence distinguishes LCA from approaches that focus solely on operational emissions.

For electric aviation, LCA must pay particular attention to battery production, an energy- and resource-intensive process involving critical minerals such as lithium, cobalt, and nickel. Battery production directly demonstrates how sustainability in AAM operations is inseparable from

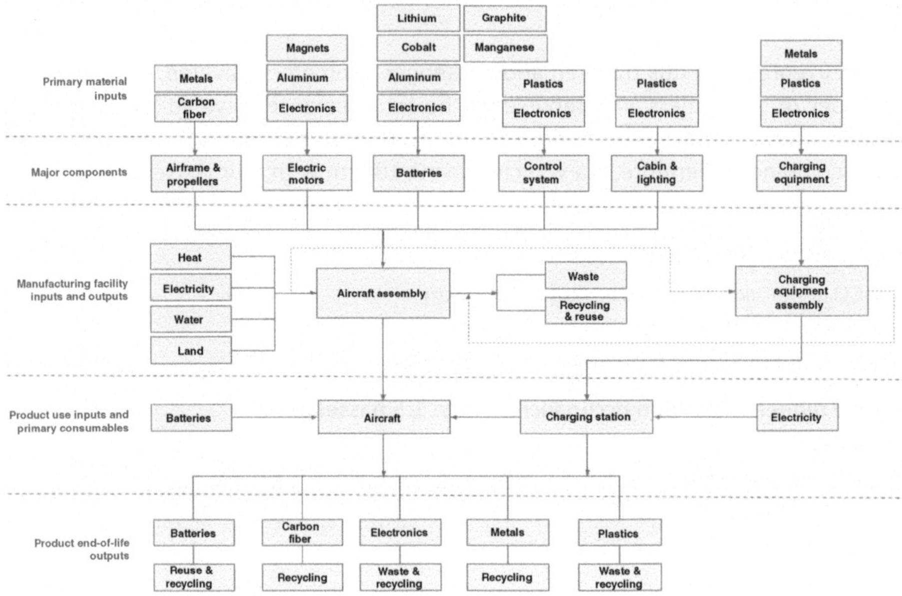

Figure 7.11 Life-cycle stages of an electric aircraft and its associated charging equipment. Courtesy of Beta Technologies.

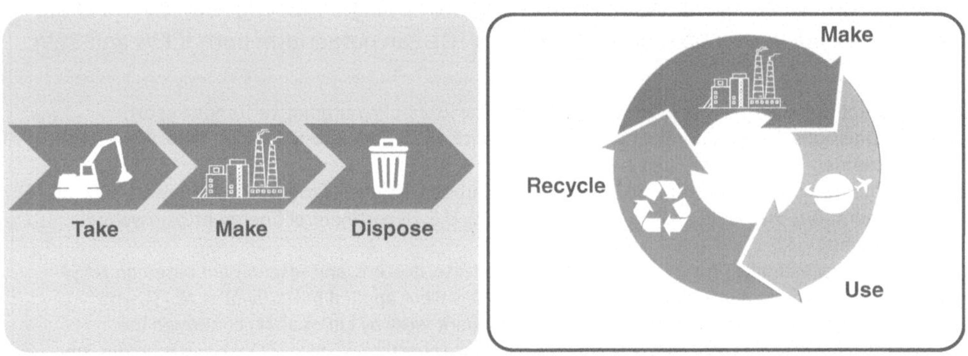

Figure 7.12 Comparison of a linear economy (left), where materials flow one-way from extraction to disposal, and the circular economy (right), where items are designed for reuse, recycling, and recovery. Circular practices reduce waste and the carbon footprint of supply chains.

ethical and transparent supply chains. For batteries, ethical sourcing requires that such minerals be mined and processed under fair labor conditions and without environmental harm, while transparency demands that companies track and disclose where inputs come from so that regulators, investors, and the public can verify good-faith practices. Beyond sourcing, sustainability also requires attention to circular economy practices that move away from the traditional linear "take–make–dispose" model. Under a circular framework, materials are designed for durability, reuse, recycling, and recovery, enabling batteries, composites, and other components to re-enter the production cycle instead of becoming waste (Figure 7.12).

While LCA widens the lens on sustainability considerations, sustainability *performance* can be more thoroughly evaluated using an *Environmental, Social, and Governance (ESG) framework.*

An ESG framework allows sustainability to become both an engineering objective *and* a business necessity. Emissions inventories now extend to include facility operations, supply chain logistics, and employee travel. In some cases, companies might account for "avoided emissions," such as emissions reductions achieved by replacing fossil-fuel ground trips with electric aerial trips. While these accounting practices vary in rigor, they are an important step toward embedding sustainability within corporate strategy.

Within the ESG framework, the environmental dimension addresses issues such as emissions and resource use; the social dimension emphasizes workforce considerations, community impacts, and the inclusiveness of AAM services; and the governance dimension encompasses company ethics, transparency, and operational accountability. This perspective requires attention to how jobs are created and sustained across manufacturing and operations, how flight corridors and vertiports interact with surrounding neighborhoods, and whether services are accessible to a broad cross-section of the population instead of serving only affluent users. Together, these categories remind us that sustainability encompasses energy and emissions as well as how manufacturers and service providers behave and engage with the communities they serve.

7.2.2 AAM and System-level Sustainability

As we have seen, sustainability in AAM cannot be evaluated at the aircraft level in isolation. Most large cities around the world have adopted ambitious climate targets, many aiming to reach net-zero emissions by mid-century. Surface transportation is a major challenge for meeting these commitments, as private automobiles remain the largest source of urban transport emissions. AAM is often presented as a solution, but its actual contribution to a net-zero pathway depends heavily on context. Urban systems are complex, with interconnected goals for climate, infrastructure, and quality of life. This makes the urban integration of AAM one of the most vital dimensions of sustainability.

The sustainability value of AAM will depend on vehicle efficiency, the types of trips displaced, the degree of passenger loading achieved, and whether operations lead to a net reduction in total travel demand. In some modeled networks, additional ground trips to reach vertiports, coupled with inefficient routing or low utilization, can actually produce higher overall emissions than conventional alternatives [5]. This demonstrates that, while electrification might improve the emissions profile of a mode, it does not guarantee sustainability at the system level.

One of the most important dimensions of sustainable urban development is whether new transportation options complement or compete with existing transit. Rail and bus systems move the largest numbers of passengers with the lowest emissions per passenger-kilometer. If AAM draws passengers away from these systems, it risks undermining their financial viability and environmental performance. Conversely, if AAM is designed as a complementary mode, it can fill gaps within these existing systems and extend the reach of sustainable transport networks. In practice, this means siting vertiports at multimodal hubs: railway stations, bus terminals, or other major transit corridors. A passenger arriving at a central business district by train, for instance, could transfer to an eVTOL for the final 10 km of a trip. Here, AAM could act as a mobility amplifier for transit rather than a competitor. It is interesting to note that this is the inverse of what is sometimes proposed for AAM: where the eVTOL provides middle-mile service and requires complementary first and last mile transportation. In this case, the eVTOL itself provides this first or last mile connection.

The siting of vertiports is also an important determinant of system-level sustainability. As discussed in Chapter 4, vertiports require scarce urban space, and poorly located facilities can reinforce patterns of sprawl and car dependence. When vertiports are located far from dense urban cores or established transit corridors, long access trips can be induced, eroding potential emissions benefits at the trip level. Strategic placement within dense, multimodal nodes, by contrast, can strengthen the role of vertiports within the broader transportation system. Moreover, reusing underutilized infrastructure, such as repurposed structures or industrial land, can minimize land consumption and environmental impact. Vertiport designs that emphasize modularity and flexibility, allowing infrastructure to scale incrementally with demand, further support sustainable integration by avoiding excessive initial buildout.

The carbon intensity of the underlying charging grid has already been discussed. However, grid stewardship and management can either mitigate or exacerbate the impacts of a carbon-intensive grid. Charging eVTOL fleets requires high-capacity power delivery that often occurs in bursts aligned with peak travel times. If these demand periods cannot be met through renewable sources and must, instead, be met by fossil-based peaking plants, the emissions reductions of electric flight may be offset or erased. By contrast, if vertiports are designed with smart-grid integration and renewable sourcing, they can become drivers of urban decarbonization. A smart-grid monitors and manages electrical supply and demand on a minute-by-minute basis, allowing utilities to smooth peaks and avoid overloads and coordinate batteries, microgrids, and other electrical storage so that renewable power can be stored and dispatched when needed. Consequently, with thoughtful planning, AAM can encourage investments in renewable generation, distributed storage, and microgrids. Under these conditions, AAM offers a new mobility option and serves as an impetus for energy transition.

As we have seen, the sustainability of AAM is not defined solely by the aircraft or charging infrastructure; it is also defined by the communities into which it is introduced. Properly managed, AAM can reinforce climate commitments, expand mobility options, and support renewable energy transitions. Managed poorly, AAM risks adding complexity, inefficiency, and environmental burden without delivering commensurate benefits.

7.3 Aeroacoustic Considerations for AAM

Would a fully autonomous eVTOL without any passengers flying over a barren desert make a sound? This question is a twist on the more familiar question, “If a tree falls in a forest and no one is around to hear it, does it make a sound?” While people may answer this question differently, we will avoid debating a strict definition of the word “sound” and, instead, first focus on the nature of sound and then on the sources of sound from AAM aircraft.

Acoustics is the study of how sound is generated, propagated, and received. When this sound is generated by aerodynamic forces and propagates through air, it is termed *aeroacoustics*. The related concept of *noise* is an important topic in AAM, purporting a leading role in the differentiation of this type of air travel from conventional aviation. To garner public acceptance for AAM, the noise produced by AAM vehicles will have to be at a level deemed acceptable by the underlying community, presumably a noise level and acoustic character that blends, to some extent, into the background. For many, noise will be the first way they experience AAM.

AAM aircraft and operations introduce distinct acoustic challenges. The distributed propulsion architecture discussed in Chapter 2, with multiple closely spaced rotors, produces complex interactions, while differences in rotor speed, blade pitch, and asynchronous rotor operation can further complicate the acoustic signature of these vehicles. Even with quieter vehicles, lower operating altitudes and a higher frequency of daily flights may still increase community noise exposure. This section examines the physical mechanisms of sound generation in AAM and the operational implications.

7.3.1 Origin of Sound

Most sounds in the air are produced by vibrations. The strings inside a piano or strung across a violin, for instance, vibrate through the deliberate action of the pianist or violinist, creating a sound. This primary vibration creates a secondary vibration in the surrounding air. This larger, secondary, vibration results in regions where air molecules are closer together, known as a local area of *compression*. In adjacent regions, where air molecules are further apart, regions of reduced pressure, known as *rarefactions*, are formed. The disturbance of air molecules, and the resulting areas of high and low pressure, is illustrated in Figure 7.13.

It is important to note that the air molecules are not traveling from the sound source to the receiver; the energy driving the pulse is what makes its way through the air to our eardrum. The affected air molecules simply move in a to-and-fro manner within a localized area. When this energy makes its way to our ear, our eardrum is likewise set into motion like the vibrating string. This wave of energy is categorized as a longitudinal wave.

The pressure fluctuations that the human ear can sense, via a vibrating eardrum sending electrical signals to the brain, range from 20 micropascals (μPa) to 20 Pa. This vast range represents the ear's astounding ability to detect pressure variations across six orders of magnitude. Because of this vast range, and to keep the numbers manageable, the *decibel scale* (dB) is employed when discussing sound (Figure 7.14). A *decibel* (dB) is a logarithmic unit for pressure fluctuations and, therefore, provides a pragmatic approach for sound level in lieu of an impractical linear scale. This logarithmic strategy for sound level helps keep the numbers manageable and, as we'll see, best aligns with how our physiology deals with this vast range of sound levels.

Sound intensity level (SIL), with units of dB, is defined as:

$$\mathrm{SIL} = 10\log_{10}\frac{I}{I_0} \tag{7.6}$$

where I represents sound intensity in W/m^2 and I_0 represents a sound intensity reference value of $1 \times 10^{-12}\,\mathrm{W}/\mathrm{m}^2$, often taken to be the threshold of human hearing. Inspecting the units for I and I_0,

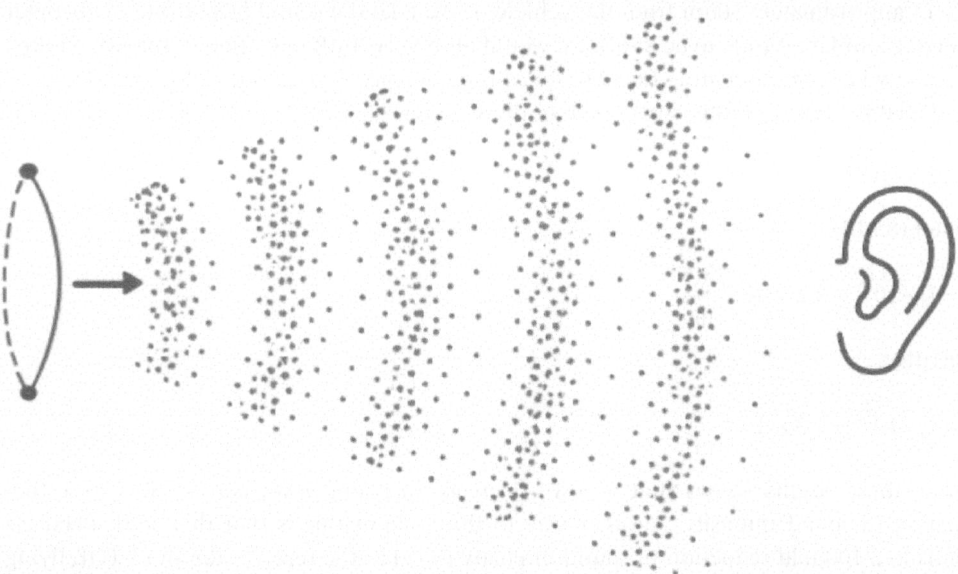

Figure 7.13 Longitudinal sound wave with compressions and rarefactions created by a vibrating string.

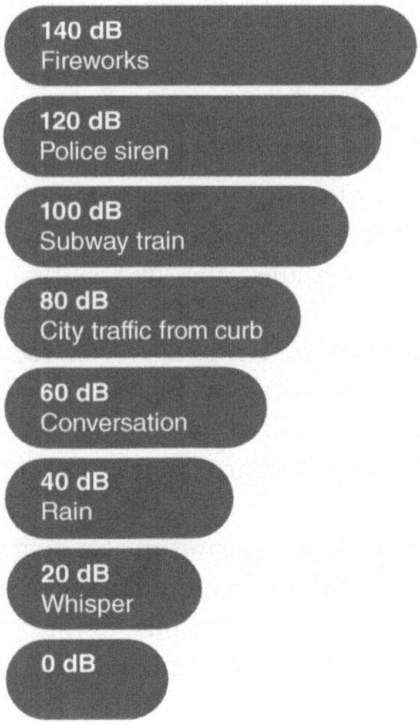

Figure 7.14 The dB levels of various sound sources.

we see that SIL is tied directly to acoustic power per unit area. Equation (7.7) shows that by taking I to be 2 times the reference value, $I_0\left(\frac{\Delta I}{I_0}=2\right)$, results in a 3 dB SIL increase. Accordingly, as demonstrated in Eq. (7.7), every 3 dB increase corresponds to a doubling of sound intensity,

$$\Delta \text{SIL} = 10\log_{10}(2) = 3 \text{ dB} \tag{7.7}$$

During cruise flight, some eVTOL developers estimate sound levels of approximately 65 dB at a distance of approximately 500 m from the vehicle. If we take the sound level from a helicopter flying over at a similar altitude to be 85 dB, we would observe a 20 dB reduction in the SIL emitted from such an eVTOL. Manipulating Eq. (7.6) for sound intensity, I, yields Eq. (7.8). Equation (7.8) enables calculation of sound intensity at both of these dB levels,

$$I = I_0 \bullet 10^{\frac{\text{SIL}}{10}} \tag{7.8}$$

Helicopter (85 dB):

$$I = I_0 \bullet 10^{\frac{85}{10}} = 3.2 \times 10^{-4} \frac{\text{W}}{\text{m}^2}$$

eVTOL (65 dB):

$$I = I_0 \bullet 10^{\frac{65}{10}} = 3.2 \times 10^{-6} \frac{\text{W}}{\text{m}^2}$$

Inspecting these results shows that a 20 dB decrease in sound level corresponds to a 100-fold reduction in sound intensity. Another way to think about this is that this 20 dB decrease corresponds to a 100-fold reduction in sound intensity [W/m^2] at the receiver for an eVTOL flying overhead compared to a helicopter overflight at the same altitude. This does not, however, make the eVTOL "100 times quieter," as might be assumed. Loudness, a distinct psychoacoustic concept, will be discussed later within the context of noise perception and measurement.

7.3.2 Atmospheric Attenuation

As noise propagates away from the source, such as an aircraft, the atmosphere plays an important role in how that noise is perceived on the ground. The most fundamental effect is spherical spreading. Recall that SIL represents acoustic intensity, defined as acoustic power spread out across an area:

$$\mathrm{SIL} = \frac{P_{\text{acoustic}}}{A_{\text{sphere}}} \tag{7.9}$$

Assuming the vehicle to be a point source of noise, with time, this (relatively) small sphere of acoustic energy radiates outward in all directions. This causes the initial acoustic energy to be spread out over an ever-increasing spherical area with time. Taking the surface area of a sphere to be: $A_{\text{sphere}} = 4\pi r^2$, and substituting this into Eq. (7.9), yields Eq. (7.10). Inspecting Eq. (7.10), we see that sound intensity decreases in proportion to $1/r^2$ (Figure 7.15). This "inverse square law" explains why an observer far from an aircraft perceives much lower noise levels than one directly beneath the flight path.

$$\mathrm{SIL} = \frac{P_{\text{acoustic}}}{4\pi r^2} \tag{7.10}$$

Beyond geometric spreading, the atmosphere itself absorbs and scatters sound energy. Molecular absorption, particularly by water vapor, preferentially attenuates high-frequency components of noise. Accordingly, noise components at higher frequencies decay more rapidly with distance than low-frequency components. This frequency dependence is why aircraft noise heard at long distances is often dominated by a low-frequency rumble. Thus, humidity strongly influences noise propagation and makes local weather conditions an important factor in noise propagation.

Temperature and wind gradients add to this weather dependency. The speed of sound is proportional to the square root of absolute temperature. Accordingly, under a typical temperature lapse rate, where air temperature decreases with altitude, sound waves near the surface travel faster than sound waves aloft. Consequently, this temperature profile, common during daytime convective conditions, refracts acoustic energy upward, decreasing the noise footprint on the ground. The inverse of this temperature profile, a temperature inversion, occurs when warmer air overlays

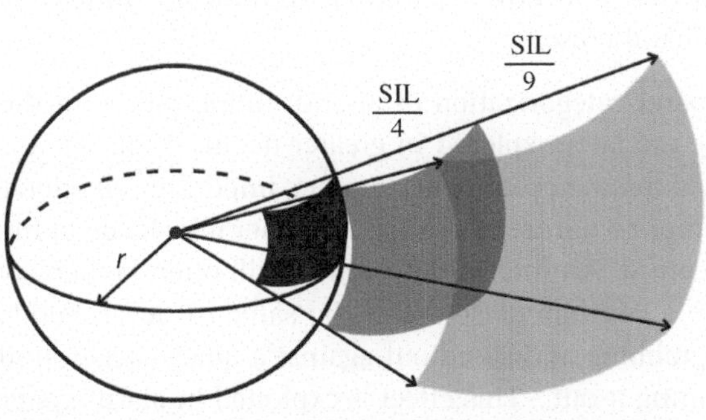

Figure 7.15 The spreading of acoustic energy follows the "inverse square law" where the sound intensity decreases in proportion to $1/r^2$. Courtesy of Kayla D. Taylor.

cooler air near the surface. This is most likely found at night but, regardless of when it occurs, acoustic energy is bent downward, intensifying ground-level noise. This is why you can often hear sounds from much further away at night or during clear and calm atmospheric conditions. Analogously, nominal vertical wind shear in the atmosphere, where wind speed increases with height, can refract sound downward in the direction of the wind while shielding areas upwind. These atmospheric effects mean the same aircraft can have very different community noise impacts depending on the time of day, season, or prevailing weather.

Adjacent to the atmosphere, the ground itself interacts with sound waves. Energy is absorbed by the ground, with the amount of absorption depending on surface type. Soft ground, such as grass or soil, provides greater attenuation, while hard surfaces, like concrete or asphalt, reflect more energy. These reflections can combine with direct sound waves, leading to areas of sound amplification or partial cancellation. Collectively, atmospheric and ground interactions demonstrate that aircraft noise exposure depends on both source characteristics and the propagation path between the aircraft and the observer.

7.3.3 The Qualitative Nature of Sound and Noise

Our discussion of sound must consider more than just its quantitative power; we must also address its qualitative characteristics. This brings us to a consideration of *noise*. The difference between sound and noise is somewhat subjective, but the distinction often relates to the sound's perception, purpose, and organization. Noise stems from an unwanted or annoying sound (*perception*). This is often the result of the sound not having a specific intent or message (*purpose*). With the attribute of intent missing, many sounds that individuals find unpleasant are frequently random in nature (*organization*). Perhaps these differences bring us back to consideration of the musician's skill that opened this discussion.

Recall that sound is the propagation of small pressure disturbances through the air. The character of these disturbances depends on the nature of their source. For aircraft, the resulting sound can generally be categorized as either tonal or broadband. Each category uniquely contributes to the overall acoustic signature of an aircraft and must be considered when designing vehicles and assessing their community impact. These two fundamental types of noise differ in their origin and perception:

Tonal noise consists of sound waves concentrated at discrete frequencies. Such sounds typically originate from periodic mechanical or aerodynamic processes, such as a rotating rotor or propeller. Tonal noises are often described as a "whining," "humming," or "buzzing" sound.

Broadband noise, in contrast, is distributed across a wide range of frequencies. Broadband noise is produced by stochastic, turbulent flow phenomena that lack a repeating pattern, such as boundary-layer turbulence and flow separation. A "hiss" or diffuse rush of air is commonly associated with broadband noise.

The tonal and broadband categorization of aircraft noise will be further developed when the sources of aircraft noise are later explored in greater detail. While the tonal or broadband categories describe a sound's frequency content, another important characterization of sound is its temporal nature. The temporal nature of a sound, whether it is steady or intermittent, can strongly influence how it is perceived. An intermittent sound will often be perceived as more noticeable and intrusive relative to a continuous sound at the same dB level. Sudden changes in a sound can be more attention grabbing and stand out against a quiet background, preventing the brain from habituating or "tuning it out." This effect is exploited in alarm systems that, typically, emit intermittent signals instead of continuous tones. The accumulation of these discrete sounds can actually contribute to higher levels of psychological stress than an analogous continuous sound.

In addition to frequency and temporal characteristics, the perception of noise is influenced by its acoustic context. One important phenomenon relevant here is *masking*, occurring when the presence of one sound reduces the ability to perceive another. An aircraft passing overhead may mask the sound of the breeze through a neighborhood's trees or a conversation between two individuals at ground level. This can alter the perceived loudness of the overflight and cause it to be thought of as louder than it actually is because the aircraft's sound contrasts with the quieter background elements. At night, when ambient noise levels are lower, masking may work in reverse. In this circumstance, in the absence of background sounds, an aircraft overflight stands out more distinctly. This can result in nighttime overflights seeming more intrusive. Masking illustrates how different communities may experience (and view) AAM noise. When compared to a major urban area, such as New York City, where the typical background noise on busy city streets and commercial districts may be 70 dB, a suburb, where the background ambient noise level might be 50 dB, may find AAM overflights much more intrusive.

7.3.4 Measuring Noise

As discussed in Chapter 2, AAM aircraft utilize unique aerodynamic and propulsion strategies, and these characteristics produce noise signatures that differ from those of conventional fixed-wing aircraft and helicopters. Accurate noise assessment for AAM operations requires metrics that extend beyond a single, frequency-neutral dB value. To evaluate the potential impact of AAM noise on communities, acoustic metrics must account for sound level (amplitude), spectral content (frequency), temporal variation, and contextual factors such as time of day. Below are several key acoustic metrics that can be used to characterize both continuous and transient noise generated by AAM aircraft (Table 7.1).

Sound Pressure Level

As previously discussed, sound is a longitudinal pressure wave consisting of alternating regions of compression and rarefaction. Sound pressure is the pressure exerted by these waves. Sound

Table 7.1 Summary of sound metrics.

Metric	Captures	Use
SPL	Physical pressure	Raw measurement
dBA	Human hearing	Environmental noise
MSL	Peak sound level	Snapshot
SEL	Total sound energy	Compare events of varying duration
SENEL	Total energy within a targeted sound window	Aircraft overflight
LEQ	Average energy	Long-term exposure
DNL	LEQ with a nighttime (2200–0700 h) penalty	Community/airport
CNEL	DNL with evening (1900–2200 h) penalty	Community/airport
PNdB	Human perception	Aircraft tone-based
EPNdB	PNdB	Aircraft certification

pressure level (SPL) is defined as the ratio of the root-mean-square (RMS) sound pressure to a reference pressure and is expressed in dB. SPL correlates with perceived loudness and is given by:

$$\text{SPL}\ [\text{dB}] = 20\log_{10}\left(\frac{p}{p_0}\right) \tag{7.11}$$

where p represents the RMS sound pressure in Pa and p_0 represents a reference sound pressure value of 2×10^{-5} Pa, often taken to be the threshold of human hearing. The logarithmic nature of Eq. (7.11) again reflects the physiological response of human hearing that perceives changes in sound pressure on a logarithmic, versus linear, scale. A useful rule of thumb here is that for mid-range sounds, a 10 dB increase in SPL is perceived as roughly twice as loud. Conversely, a 10 dB decrease in SPL is perceived as about half as loud. Revisiting our previous eVTOL and helicopter comparison, the eVTOL that is 20 dB lower than the comparable helicopter is consequently perceived to be about 4 times quieter (½ × ½ = ¼).

A-weighted dB

As discussed earlier, the human ear is sensitive to a wide range of frequencies, spanning approximately 20 Hz–20 kHz, with peak sensitivity between 1000 and 4000 Hz. The A-weighted dBA applies a frequency-dependent weighting filter to SPL measurements to approximate the ear's sensitivity. The A-weighted scale de-emphasizes low- and high-frequency ranges where the human ear is less sensitive, while leaving mid-range frequency components largely unchanged. For this reason, dBA better accounts for the effect of noise exposure on humans. Figure 7.16 shows the A-weighting filter applied to measured SPLs to produce A-weighted levels. Note the 0 dB adjustment in the ear's most sensitive range between 1000 and 4000 Hz and the reduction given to observed sounds at lower and higher frequencies.

Maximum Sound Level

The maximum sound level (MSL) is simply the highest A-weighted sound level observed during a sound event. The MSL is particularly useful when considering how the noise level can change during an aircraft overflight, increasing as the aircraft approaches and decreasing as it flies away.

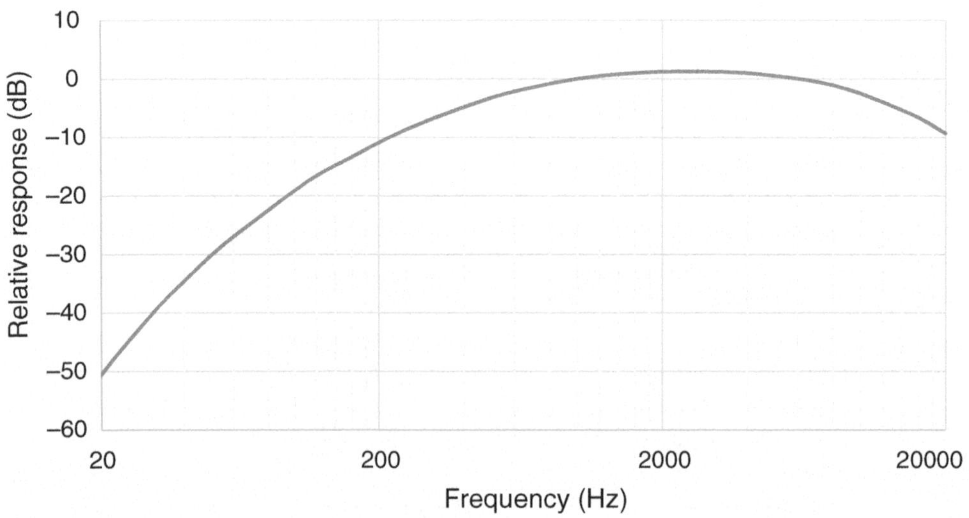

Figure 7.16 A-weighting curve.

However, while the MSL is useful for quantifying peak noise exposure, MSL does not capture the duration of the event, such as the entirety of the overflight or the cumulative exposure of multiple overflights.

Sound Exposure Level

To account for the duration of a sound event, the sound exposure level (SEL) is used. SEL expresses the total acoustic energy of a noise event as if it were compressed into a 1-second period. SEL is calculated via Eq. (7.12). Here, $p(t)$ is the time-varying RMS sound pressure between times t_1 and t_2. Inspecting Eq. (7.12), we see that $p(t)$ is squared. This results from acoustic energy being proportional to the square of the pressure. If the customary 1-second time interval is used, $t_2 - t_1$ is simply set to 1second. It is important to note that if SPL measurements in dB are used, the SPL observations must first be converted into linear pressure values. Taking a step back and further inspecting, we see that SEL considers both the total energy and temporal characteristics of a sound event. It also makes for an easier comparison between sound events of varying duration,

$$\mathrm{SEL} = 10\log_{10}\left(\frac{1}{(t_2 - t_1)}\int_{t_1}^{t_2}\frac{p^2(t)}{p_0^2}\mathrm{d}t\right) \tag{7.12}$$

Single-event Noise Exposure Level

The single-event noise exposure level (SENEL) is a more practical implementation of the SEL and is especially well suited for aircraft noise analysis. SENEL considers only the portion of a sound event above a specified threshold, often taken to be within 10dB of the MSL. This focus yields a more targeted evaluation of the most acoustically significant portion of a sound event. Equation (7.13) is used to calculate SENEL. The expression, $\max\left[p^2(t) - (p_{\mathrm{th}})^2, 0\right]$, considers only the portions of the sound event exceeding the threshold, p_{th},

$$\mathrm{SENEL} = 10\log_{10}\left(\frac{1}{(t_2 - t_1)}\int_{t_1}^{t_2}\max\left[p^2(t) - (p_{\mathrm{th}})^2, 0\right]\frac{1}{p_0^2}\mathrm{d}t\right) \tag{7.13}$$

Figure 7.17 graphically portrays these sound metrics for a representative aircraft overflight. Here, the MSL is concurrent with the aircraft being directly overhead. The sound gradually increases as the aircraft approaches and gradually decreases as the aircraft recedes.

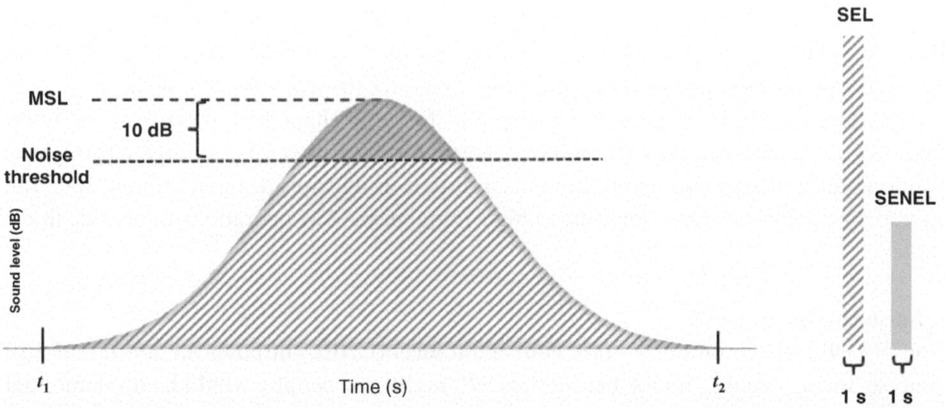

Figure 7.17 MSL, SEL, and SENEL during a representative aircraft overflight.

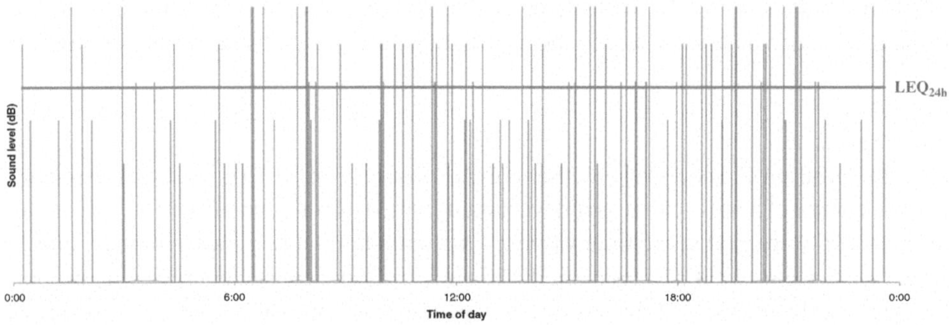

Figure 7.18 The LEQ provides the time-averaged sound energy of a fluctuating sound over a specified time period.

Equivalent Continuous Noise Level

The equivalent continuous noise level (LEQ) represents the constant sound level that, over a given period, *T*, would contain the same total acoustic energy as the actual time-varying sound (Figure 7.18). Although 24-hour periods are common, LEQ use is also appropriate for 12-hour daytime or nighttime periods.

LEQ is calculated via Eq. (7.14), where *T* is the length of the specified time period. It is important to note that the LEQ is not the same as the arithmetic mean of the SPLs (in dB) over the time period of interest. Unlike a simple arithmetic mean of SPL values, LEQ is computed from averaged energy, not dB values directly. Neglecting this ignores the logarithmic nature of the dBA and results in the under-representation of high-intensity events. Therefore, when starting with dB data, each observation must first be converted into linear form. From here, observations can be averaged, with the final value converted back to dB,

$$\text{LEQ} = 10\log_{10}\left[\frac{1}{T}\int_0^T \frac{p^2(t)}{p_0^2}\,\mathrm{d}t\right] \tag{7.14}$$

Day–Night Average Sound Level

The day–night average sound level (DNL), typically expressed in dBA, is a 24-hour LEQ-based metric that applies an additional 10dB penalty for nighttime noise. This accounts for an individual's increased sensitivity to noise during the overnight hours, typically understood as 2200–0700 hours. Calculated similarly to LEQ, the penalty is applied to nighttime events before averaging. DNL is an appropriate metric for assessing long-term community exposure to environmental noise.

In some regulatory frameworks, particularly those historically shaped by airport-centric noise management, DNL has emerged as the dominant metric for land-use compatibility, although its emphasis on long-term energy averaging limits its ability to fully capture the temporal, spectral, and operational characteristics most relevant to high-frequency AAM operations in diverse international contexts.

Community Noise Equivalent Level

An extension of DNL, the community noise equivalent level (CNEL) imposes an additional 5dB penalty for noise during evening hours, before the DNL nighttime penalty would be implemented (Figure 7.19). Typically applied between 1900 and 2200 hours, CNEL reflects the higher potential for disturbing individuals in their home during the evening.

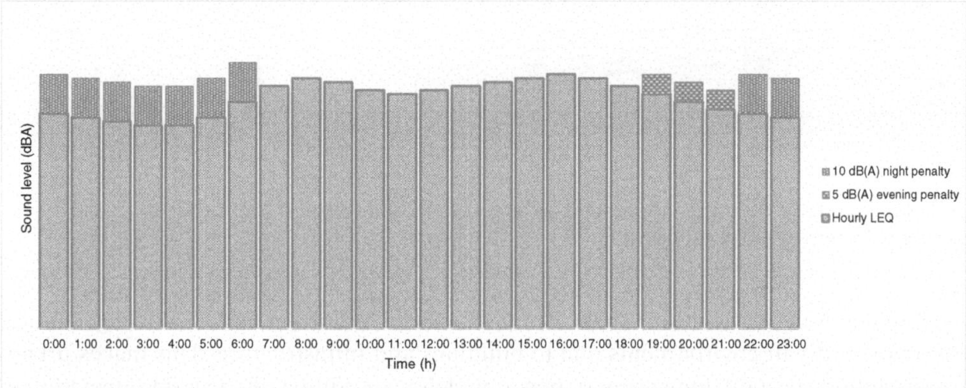

Figure 7.19 Day–evening–night noise level metric. Hourly LEQ values have a 5 dB penalty applied during evening hours and a 10 dB penalty applied during nighttime hours.

Perceived Noise Level

The perceived noise level (PNdB) accounts for the frequency content, spectral bandwidth, and tonal characteristics of a sound without relying on the A-weighting of dBA measurements. This allows PNdB to capture tonal noise elements that do not readily blend into background sound and are often perceived as more annoying. Because A-weighting may under-represent these tonal components, PNdB is well suited for assessing AAM vehicle noise.

Effective Perceived Noise Level

The effective perceived noise level (EPNdB) extends the PNdB by incorporating both the duration of the noise event and its time-varying characteristics, such as the variation in loudness and tonal content throughout an aircraft overflight event. With PNdB providing an instantaneous perceived noise level, EPNdB integrates the PNdB in an analogous manner to how SEL integrates SPL. EPNdB is the standard noise metric used in aircraft noise certification by the International Civil Aviation Organization (ICAO), as it provides a comprehensive measure of the human-perceived noise impact of an entire overflight event.

7.3.5 Noise Contour Maps

Any of the noise metrics discussed thus far can be spatially represented around an aerodrome, be it an airport or vertiport, using a noise contour map. These maps connect points of equal noise exposure to illustrate how noise levels vary across a geographic area. The resulting contours represent constant values of a selected metric, such as LEQ or DNL, and the spacing between contours reflects the local noise gradient, with contours more closely spaced indicating a more rapid change in noise level. Figure 7.20 shows representative noise contour maps for a traditional airport and a vertiport.

The shape of a noise contour map is primarily determined by the noise signature of the overflying aircraft and their approach and departure paths. Hence, the vertical takeoff and landing, or short takeoff and landing, profiles of AAM vehicles result in more compact and radially symmetric contours, removing the typical elongation found along runways at conventional airports. This geometry reflects a more concentrated noise footprint near the vertiport itself. However, asymmetric contours around vertiports can still exist due to the presence of approach and departure corridors.

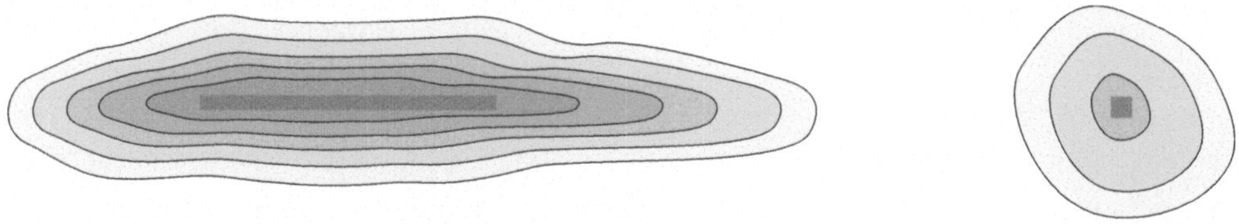

Figure 7.20 Notional noise contour maps around an airport (left) and vertiport (right). Darker colors represent areas of increased noise exposure.

While terrain can influence contour shape through acoustic reflection and shadowing, similar effects can occur in urban environments due to buildings and infrastructure. This makes urban topology especially important for vertiport noise modeling. Additionally, variable operational demand, such as reduced activity at night, can cause significant differences between day and night noise contour maps. Such maps are important tools in land-use planning for AAM and are useful for evaluating zoning compatibility, environmental impact, and regulatory compliance.

7.3.6 Sources of Noise

While noise contour maps comprehensively capture all sound stemming from an aerodrome operation, including ground operations, the primary contributor to noise around airports is the aircraft themselves. The same is expected to hold for vertiports. Noise from aircraft originates from several physical mechanisms. These can generally be classified as either mechanical sources or aerodynamic sources associated with unsteady airflow around the vehicle and its propulsion systems. As many would expect, the dominant contributor to aircraft noise stems from the propulsion system, but a second contributor arises from airflow interacting with the airframe itself. This factor is often overlooked; however, observations of large transport-category aircraft during approach and landing demonstrate that substantial airframe noise persists even during significant reductions in engine power. One common example is the noise generated by unsteady flow interacting with deployed landing gear. A useful way to frame the sources and character of noise for eVTOL aircraft is through comparison with conventional rotorcraft. Helicopters have long served as a benchmark for vertical flight, and they are notoriously loud!

Conventional helicopters produce a complex mix of engine, gearbox, and rotor (both main and tail rotor) noise. In contrast, eVTOLs, powered by electric motors, substantially reduce noise associated with combustion engines and conventional transmissions, focusing attention more heavily on the rotating blades of distributed electric propulsion (DEP) systems. Even outside of electric motors, the DEP architecture fundamentally alters the perceptual qualities of rotor noise, producing an acoustic profile distinct from that of a helicopter. Like all aircraft, secondary noise contributions still come from air making its way over airframe elements, such as pylons, landing gear, and wings, as turbulent energy is converted to acoustic energy. However, these contributions are generally smaller in magnitude for eVTOLs than for conventional aviation counterparts, although they can become significant at higher speeds.

Understanding these components and their directional characteristics is essential to designing "neighborly" aircraft and operations for AAM. To improve our understanding, it is helpful to further categorize the main sources of noise for an eVTOL. There are four primary eVTOL noise sources to consider: (1) *thickness noise*, (2) *loading noise*, (3) *blade–vortex interaction (BVI) noise*, and (4) *broadband noise*. Each source contributes to a varying extent depending on the flight condition and the configuration of the AAM vehicle. We will investigate the first three of these noise sources as tonal noise sources before discussing broadband noise.

Tonal Noise

Thickness Noise

Much like an oar displacing water, thickness noise results from the displacement of air by a blade as it rotates. Recall the association between periodicity of mechanical or aerodynamic events and tonal sounds from Section 7.3.3. This periodic displacement results in thickness noise being tonal in nature, dominated by the blade-passing frequency (BPF) and its harmonics. The BPF is the rate at which a blade passes a given point during rotation (Eq. 7.15), showing how thickness noise depends on the number of blades, the geometry of the blades, and the overall rotor speed. With the highest blade speed at the rotor tip, the strongest pressure pulses are generated at the rotor tip,

$$\text{Blade passing frequency} = \text{Number of blades} \times \text{Rotational speed of the rotor} \tag{7.15}$$

Example 7.2: Determine the BPF for a four-bladed rotor rotating at 950 rpm:
Solution: $4 \times 950\ \frac{\text{rev}}{\text{min}} \bullet \frac{1\ \text{min}}{60\ \text{s}} = 63.33\ \text{Hz}$

Inspecting Eq. (7.15), we see that increasing the number of blades shifts tonal energy toward higher frequencies and distributes aerodynamic loading across more blades, often reducing the prominence of individual tonal peaks. For this reason, DEP designs often employ higher blade counts than conventional helicopters, in part to reduce loading per blade and improve acoustic characteristics. With pressure fluctuations more pronounced near the leading and trailing edges of the blade, thickness noise is primarily emitted in the plane of the rotating blades.

Loading Noise

While thickness noise depends on a blade's shape and motion, it is not tied to the blade's aerodynamic performance. Loading noise arises from the generation of time-varying aerodynamic forces produced by a blade, namely lift and drag. The resulting fluctuations in air pressure, the essence of aerodynamic "loads," radiate out and produce acoustic waves. With the aerodynamic performance of the blade responsible for its existence, loading noise is also tonal in nature and dominated by the BPF. However, its directivity differs with a propagation pattern both perpendicular to the blade surface and the plane of rotation. This is why loading noise propagates both through the rotor disk and outward. Lowering blade loading through design, such as using higher aspect ratio blades or distributing thrust across more rotors, can reduce loading noise.

BVI Noise

BVI noise is caused by the interaction of a rotor blade with a vortex shed from a previous blade's passage. As each rotor blade generates lift, strong vortices form near the blade tip (much like wing tip vortices form at the tip of a wing) as the higher-pressure air behind the blade attempts to flow toward the lower-pressure air in front of the blade. As these vortices are shed, subsequent blades encounter these vortices. When subsequent blades encounter this dynamic airflow, abrupt fluctuations in aerodynamic loads generate acoustic impulses that, again, radiate both through the rotor disk and outward. For traditional helicopters, these interactions produce the familiar "slapping" sound, often most noticeable during descent.

For AAM vehicles adopting DEP, BVI noise manifests differently but can still loom as a concern for urban operations, especially during descent or low-speed approaches into vertiports. However, many AAM vehicles employ smaller distributed rotors operating with optimized loading and tip-speed characteristics, which can reduce vortex strength and associated BVI noise relative to conventional helicopters. Further, the higher blade counts employed in DEP architectures result in acoustic energy being shifted from sharp impulsive peaks to a broader noise spectrum.

In DEP systems, the use of multiple closely spaced rotors introduces an additional layer of complexity where tonal peaks from each propulsor can add constructively or destructively (Figure 7.21). For example, loading noise can be exacerbated for vectored thrust aircraft employing tiltrotor or tiltwing architecture due to the resulting complex and unsteady aerodynamic loads, as well as across a broader class of AAM vehicles that rely on differential thrust in place of conventional aerodynamic control surfaces for maneuvering. However, *rotor phasing*, defined as the relative rotational offset between adjacent rotor blades, offers one potential means of mitigating these effects by reducing the prominence of any single dominant tone (Figure 7.22).

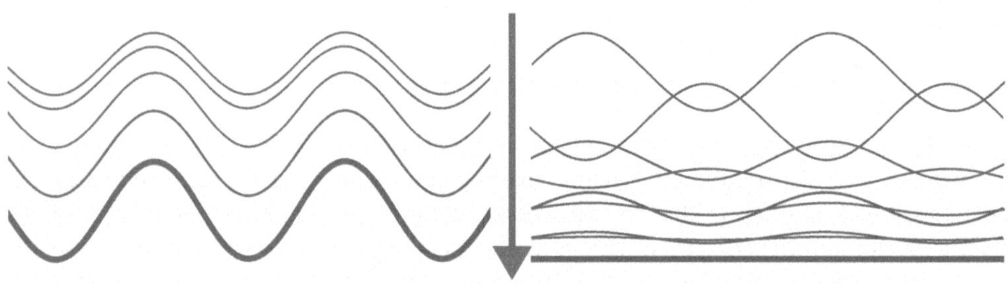

Figure 7.21 Waves in phase constructively interfering (left). Waves out of phase and destructively interfering (right).

Figure 7.22 Example of two three-bladed propellers 60 ° out of phase.

Appropriate phasing strategies, such as fixed offsets of 90° or 120° between neighboring rotors, can distribute tonal energy across a broader frequency spectrum. At the same time, the close proximity of multiple propulsion units increases the likelihood of rotor–wake and rotor–vortex interactions. If the spatial arrangement of propulsors and the associated wake interactions, particularly during descent and low forward-speed flight, are not carefully managed, impulsive noise analogous to BVI can emerge.

Broadband Noise

Broadband noise arises from turbulent flow phenomena that lack dominant frequencies. Here, the noise is instead spread across a wide (broad) frequency band. The incoherent pressure fluctuations associated with the underlying turbulence cause more diffuse noise propagation than the more directional nature of tonal noise. The most common turbulent flow phenomena giving rise to broadband noise are generated by turbulent boundary layers, flow separation, vortex shedding, or wake–structure interactions. Therefore, broadband noise can stem from both rotating blades and the airframe. Here, it is again helpful to further divide broadband noise into three subcategories and consider broadband noise that is predominantly associated with rotating blades and that which is associated with turbulent flow around the airframe.

Rotating Blades

Blade–Wake Interaction Noise Blade–wake interaction (BWI) noise can be classified as a form of leading-edge-generated noise. In general, BWI noise is generated when a rotating blade interacts with the wake generated by a preceding blade. This mechanism was partially addressed in the discussion of BVI. Recall that BVI occurs when a subsequent blade encounters the concentrated tip vortices shed by a previous blade, producing impulsive, tonal acoustic emissions. Under this definition, BVI represents a specific and particularly intense subset of BWI noise. However, BWI noise can also arise when a blade interacts with the inboard wake of a preceding blade, a mechanism commonly referred to as blade–turbulent wake interaction noise. Interactions between a blade and the weaker, more diffuse inboard portions of an upstream wake produce broadband acoustic energy, in contrast to the sharp impulsive noise associated with the concentrated tip vortices characteristic of BVI. In general, BWI is less of a concern for most AAM designs because most of these vehicles do not typically cruise with rotors in edgewise flight. Rather, they accomplish wingborne flight with propellers (axial flight).

Blade Self-noise Blade self-noise is aerodynamic noise brought about by a blade interacting with the surrounding airflow. The aerodynamic phenomena that are most responsible for this are delineated in Figure 7.23. However, overall, a turbulent boundary layer convecting past a sharp trailing edge is most responsible for a blade's self-noise. Therefore, blade self-noise can be thought of as primarily blade trailing-edge generated noise.

Inflow Turbulence Noise The source of inflow turbulence noise (TIN) is easily ascertained by its name. TIN originates from turbulent air being ingested by rotor or propeller blades. This turbulence can arise from various sources, including atmospheric conditions, surrounding obstacles, and other components of the aircraft itself. For AAM vehicles with a DEP configuration, the close spacing of multiple propulsors and airframe elements increases the likelihood of turbulent wakes shed from one component being ingested by another. In urban environments, the presence of buildings, trees, and other roughness elements further increases ambient turbulence levels. As

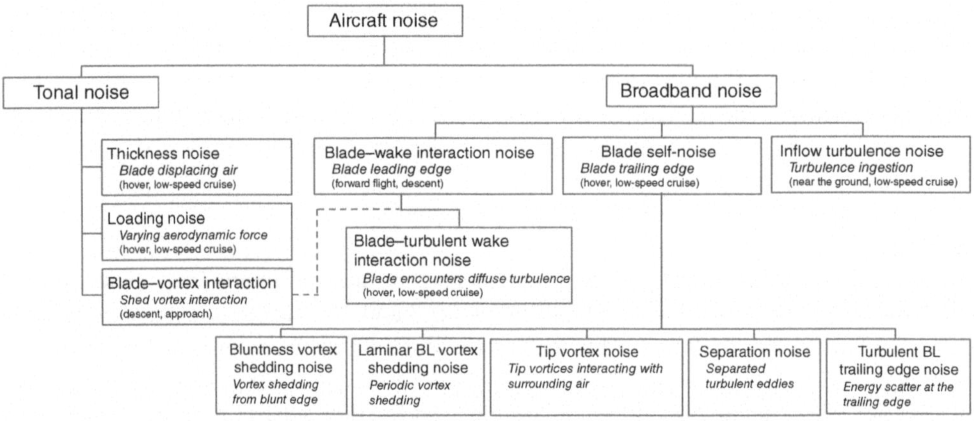

Figure 7.23 Tonal and broadband noise sources associated with rotating blades. Note that tip vortex noise is included here as a broadband self-noise contribution, although its classification can overlap with wake-interaction mechanisms depending on flow coherence and operating condition.

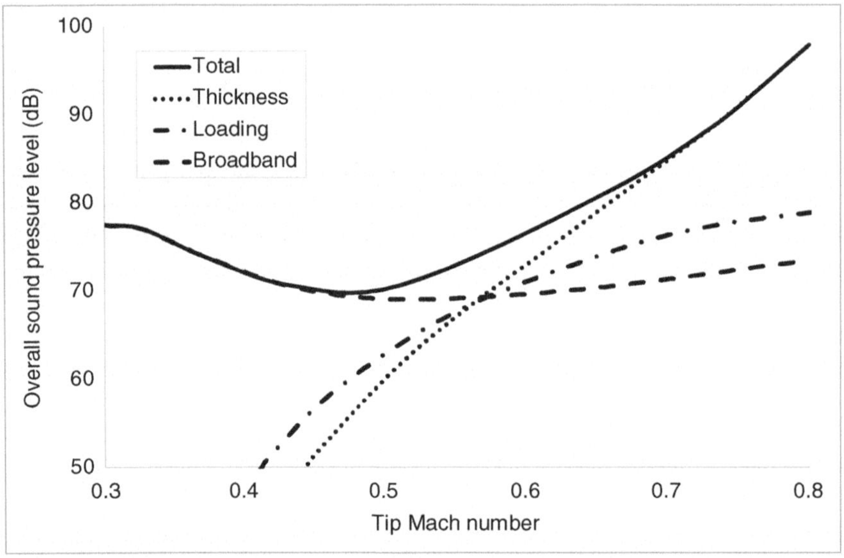

Figure 7.24 Variation in noise components by tip Mach number for a notional eVTOL rotor [6]. Courtesy of Kayla D. Taylor.

rotor blades ingest this turbulent flow, unsteady aerodynamic loads are generated and are then radiated as broadband acoustic energy.

Figure 7.23 organizes the tonal and broadband noise sources associated with rotating blades.

As we have consistently seen, many DEP AAM aircraft use a greater number of smaller blades operating at lower tip speeds than conventional helicopter main rotors. As Figure 7.24 shows, this changes the distribution of tonal and broadband components relative to helicopters. In contrast to a helicopter's rotor tip speeds in the 0.55–0.7 Mach range, an eVTOL's tip speeds are generally designed to be below 0.5 Mach. Figure 7.24 suggests that, for the representative rotor considered, the region near 0.55 Mach marks a transition in the relative importance of broadband and tonal components. Consequently, for lower-tip-speed eVTOL rotors similar to the notional case shown in Figure 7.24, broadband noise can become the dominant contributor to total noise as tonal noise from thickness and loading diminishes.

Example 7.3: Determining a propeller's helical tip speed and Mach number

Consider a propeller with a diameter (d) of 2 m, rotating at 1320 rpm while moving forward at 67 m/s at an altitude of 1000 m.

The helical tip speed (v_{hel}) and Mach number (M_{hel}) come from the vector combination of the blade tip's rotational (v_{tip}) and forward airspeed (v_∞) components.

Solution: First, we must find the blade tip's rotational speed:

$$v_{tip} = \Omega\left(\frac{d}{2}\right), \text{ where } \Omega = 2\pi\left(\frac{\text{rpm}}{60}\right) [\text{rad/s}]$$

$$v_{tip} = 2\pi\left(\frac{1320 \text{ rpm}}{60}\right)\left(\frac{2 \text{ m}}{2}\right) = 44\pi \text{ m/s}$$

The helical tip speed of the blade is given by:

$$v_{hel} = \sqrt{v_{tip}^2 + v_\infty^2}$$

From problem statement, the forward airspeed (v_∞) is 67 m/s.

With this, the helical tip speed of the blade is:

$$v_{hel} = \sqrt{(44\pi \text{ m/s})^2 + (67 \text{m/s})^2} = 153.6 \text{ m/s}$$

The helical tip Mach number is given by:

$M_{hel} = \frac{v_{hel}}{a}$, where a is the speed of sound and is taken to be 336 m/s at 1000 m.

With this, the helical tip Mach number of the blade is:

$$M_{hel} = \frac{v_{hel}}{a} = \frac{153.61 \text{ m/s}}{336 \text{ m/s}} = 0.46$$

Note: By inspecting our equation for v_{tip}, we see how a smaller blade diameter decreases the tip velocity.

Airframe Noise

Airframe noise arises from interactions between the airflow and the airframe components themselves, as distinct from noise generated by rotating blades. As with rotating blades, the turbulent flow structures associated with these interactions also results in noise emissions across a wide range of frequencies. Accordingly, airframe noise is classified as a broadband noise source. While generally quieter than propulsion-based noise, airframe noise can be more prominent during low-power phases of flight, such as descent, or at high speeds. The underlying turbulence can be generated by flow separation over components such as pylons, tilt mechanisms, gear struts, landing skids, and other bluff airframe elements. Although broadband noise sources are often categorized by origin, rotor wakes impinging on the airframe can generate the same turbulence-driven broadband noise mechanisms.

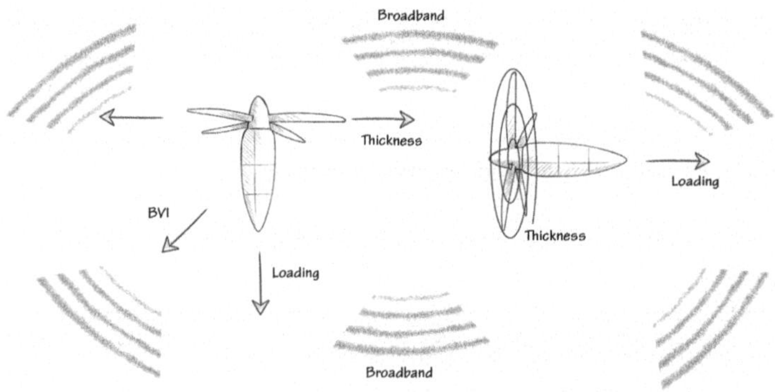

Figure 7.25 Predominant radiation directions for noise for an eVTOL's rotating blades in a helicopter-style lift rotor configuration (left) and a propeller-style thrust configuration (right). Additional insight for noise propagation during transition can be found in [7].

From an operational and perceptual standpoint, broadband noise is most noticeable during flight phases with lower rotor thrust, such as descent or low-power forward flight, when tonal rotor noise is diminished. While broadband noise may sound less intrusive due to its diffuse character, it can still stand out against quieter backgrounds. Consequently, it still poses a significant challenge from a community noise perspective. What's more, it lacks discrete frequencies that could potentially be filtered or masked and often extends into frequency ranges that are most perceptible (and most annoying) to the human ear. Thus, AAM noise mitigation strategies must be developed to address both tonal and broadband noise. We will address this in the next subsection, but before doing so, it must be noted that the directivity of eVTOL noise can be just as important as its magnitude. The predominant propagation directions associated with these noise sources are summarized in Figure 7.25. These directional properties should be considered when designing vertiports, approach and departure corridors, and other AAM operational procedures.

7.3.7 Designing to Reduce Noise

Mitigating noise from AAM aircraft is a multifaceted challenge given its varying sources and tonal and broadband components. Thus, a multipronged approach, spanning aerodynamic design, rotor configuration, control logic, and operational procedures, is needed without compromising performance or safety. As AAM matures, the ability to minimize and control acoustic emissions will be critical to public acceptance and regulatory success.

Mitigating tonal noises often involves spectral shaping strategies and shifting the dominant acoustic tones outside of the most sensitive frequency bands of human hearing. As already discussed, the BPF is tied to the dominant tonal frequencies associated with rotor thickness and loading noise. While thickness noise is largely kinematic and difficult to entirely eliminate, reduced blade tip speeds and rotational rates can decrease both the intensity and dominant frequency content of the radiated acoustic energy. Combined with slender blade profiles or tapered blade tips, this further reduces local pressure disturbances and associated acoustic radiation. These slender blade profiles are made feasible by increasing the number of rotor blades, N, which maintains or raises the *rotor's solidity*, σ, defined as the ratio of total blade area, a_b, to the overall swept disk area, A (Figure 7.26). With greater solidity, the thrust responsibility is distributed across additional blade surface area, allowing each blade to operate at a lower angle of attack and displace less air. While

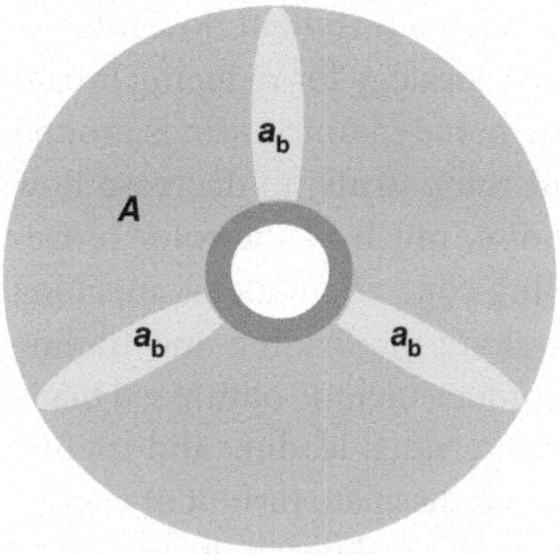

Figure 7.26 Rotor solidity is the ratio of total blade area, a_b, to the swept disk area, A.

the number of blade passages increases per revolution with an increased number of rotor blades, pressure pulses from each passage are weaker with, consequently, the noise energy spread across a larger number of weaker pulses. Accordingly, although the BPF increases with more blade passages, the overall perceived noise often decreases because of the reduced amplitude of each event outweighing the added frequency content. However, this desire for an increased number of blades must be tempered by a consideration of the resulting increased weight of the additional blades, rotor drag, aerodynamic efficiency, and mechanical complexity,

$$\sigma = \frac{Na_b}{A} \tag{7.16}$$

We have seen that increasing the number of rotor blades lowers per-blade loading, reducing the lift requirement of each blade. This reduction in per-blade loading is expected to lessen the intensity of pressure fluctuations and lead to a corresponding decrease in loading noise. Loading noise can be further minimized if the total thrust generated can be distributed across each blade more evenly. Minimizing loading noise also requires careful consideration of thrust modulation and blade pitch changes. This can be of special concern for DEP configurations that rely on differential thrust for transitions and maneuvering. Here, control systems that incorporate thrust rate limiting and smooth transitions between flight modes may have to be considered.

The presence of multiple rotors in close proximity presents a significant challenge for BVI noise. As highlighted earlier, BVI is especially problematic during low-speed descents when vortices tend to remain in proximity to the rotor disk. Here, thoughtful consideration of rotor spacing and phasing must be made. Phasing adjusts the angular position of the blades across multiple rotors so that pressure pulses and vortices do not coincide or reinforce each other in a manner that amplifies tonal noise. This approach of "de-synchronization" spreads the acoustic pulses more evenly in time, perhaps even in a quasi-random manner. It also creates a more continuous and less tonal noise signature. However, this crafting of a less tonal noise must be balanced to avoid creating a harsher broadband acoustic signature. Regardless, passive measures, such as vortex generators, fairings, or winglets, and active methods, must manage wakes, minimize vortex impingement, and prevent constructive interference.

For distributed propulsion systems, as previously discussed, slight but intentional irregularities in rotor spacing, phasing, or speed can smear out strong tonal components, yielding a more diffuse and less intrusive acoustic signature. While these modifications must be made without inflicting penalties in aircraft performance or controllability, the mitigation of diffuse, broadband, noise must also be considered. Remember that broadband noise originates from turbulent flow interactions on the blades or with the airframe. Thus, one easy and effective strategy for reducing broadband noise is minimizing bluff body characteristics and employing aerodynamic fairings around complex geometries, such as landing gears, tilt mechanisms, and rotor struts, to decrease flow separation and vortex shedding. However, aside from airframe noise, much of the rotor broadband noise arises from boundary-layer effects and leading or trailing-edge phenomena, meaning passive and active control measures for the boundary layer and along the leading and trailing edges hold potential. Besides optimized airfoil design, passive techniques include optimized blade tip geometries, porous blades that can include damping layers, and serrated leading and trailing edges. Porous blades reduce turbulent boundary-layer interactions and dissipate turbulent energy. Serrated blade leading edges can serve to break up large coherent turbulent eddies in the incoming airflow and reduce the surface pressure fluctuations, potentially scattering them and resulting in acoustic emissions that partially cancel each other out. This concept draws inspiration from owl feathers, whose fine serrations enable their characteristically silent flight. In a somewhat analogous manner, coherent turbulent structures that might convect past the trailing-edge simultaneously, and thus create a coherent pressure fluctuation, can be broken apart into smaller, less energetic, out-of-phase turbulent eddies that partially destructively interfere with one another by the use of serrated trailing edges. Here, the reduced radiated acoustic energy is also spread across a broader frequency range. This also weakens low-frequency components, often perceived as louder and more intrusive. Examples of serrated leading and trailing edges are shown in Figure 7.27.

Presently, passive noise mitigation strategies are generally more practical for rotor blades than active strategies, owing to their inherent simplicity, reliability, and advantages in weight and energy efficiency. Active systems, by contrast, introduce the potential to adapt in real time to changing conditions, such as inflow turbulence, variations in flight dynamics, or even exceedance of specific noise thresholds. This adaptability could make them particularly attractive for reducing acoustic impact during operations around vertiports. Examples of active noise control concepts include blowing or suction-based boundary-layer control, trailing-edge flaps, blade twist morphing, and variable rotor phasing. Despite these promising possibilities, these systems remain difficult to implement because of their mechanical complexity, added mass, increased power demands, and certification challenges. For these reasons, passive approaches are expected to remain the dominant means of noise mitigation in AAM design for the foreseeable future.

Finally, in conjunction with technical noise mitigation strategies, operational strategies must be considered. This can entail optimizing altitude during cruise to increase the propagation distance but will be especially important around vertiports where increased separation distances are not possible, and the public is likely to be most sensitive to noise. Here, noise abatement procedures

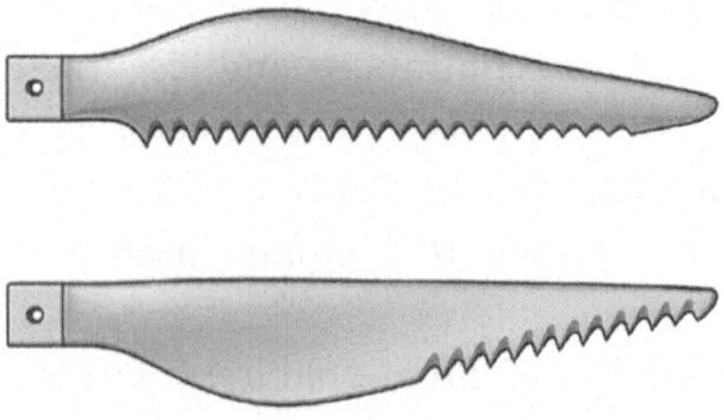

Figure 7.27 Rotor blade with leading edge (top) and trailing edge (bottom) serrations.

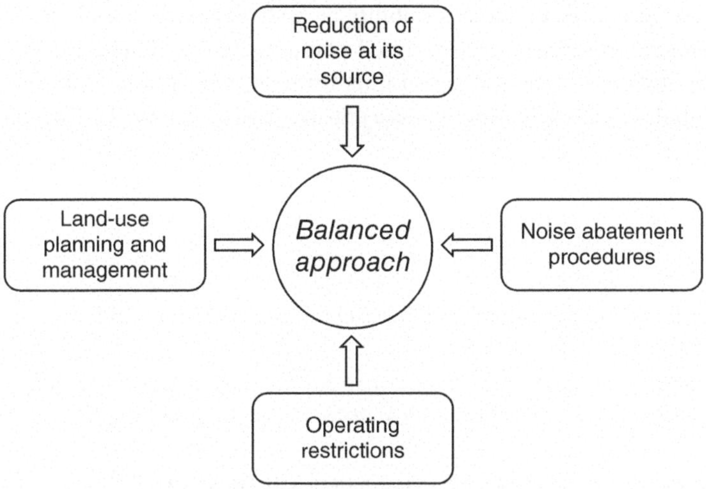

Figure 7.28 The four elements of a balanced approach to aircraft noise management.

may be implemented. The physical design of vertiports, however, will influence the ground-level noise exposure, and the siting of these vertiports will determine where the community intersects with the noise. These ideas are encapsulated in what ICAO calls its "balanced approach" [8] (Figure 7.28).

7.4 Summary

AAM operates within low-altitude and urban environments that impose constraints on operational integration that differ from those of conventional aviation. Sustained operations within the ABL expose AAM vehicles to surface-driven variability in weather. These effects are exacerbated in the urban environment.

The sustainability promise often associated with AAM cannot be evaluated through electrification alone. Emissions outcomes depend on life-cycle impacts that extend beyond the aircraft in flight. Material extraction, battery production, electricity generation, passenger occupancy, and supply-chain governance all influence net environmental performance. At the system level, AAM may reduce emissions in some use cases and offer limited benefit in others. Outcomes depend on trip substitution patterns, grid carbon intensity, and integration with existing transportation networks. As a result, sustainability emerges as a function of system design and operational context, not simply propulsion technology in isolation.

Noise represents a parallel constraint on AAM deployment. DEP enables reductions in peak noise levels compared with conventional rotorcraft. However, frequent low-altitude operations near urban populations could introduce cumulative exposure effects that remain significant. The acoustic signature of AAM aircraft reflects a combination of tonal and broadband sources shaped by rotor design, operating condition, and vehicle configuration. Perception depends on sound level, frequency content, temporal structure, and ambient context. Effective noise management requires coordinated attention to aircraft design, propulsion architecture, control strategies, vertiport siting, and flight procedures.

Together, sustainability and noise demonstrate that the viability of AAM depends on more than technical feasibility. Long-term outcomes depend on integrated planning across vehicles, infrastructure, and operations. Addressing environmental constraints through deliberate design and system-level coordination is essential if AAM is to function as a constructive component of future mobility systems.

References

1 World Commission on Environment and Development. (1987). *Our Common Future*. Oxford University Press.

2 Silva, C., Johnson, W., and Antcliff, K.R. (2018). Design of Urban Air Mobility Vehicles: Mission Analysis and Considerations. AIAA AVIATION Forum; FAA, 2020, Urban Air Mobility Concept of Operations.

3 Kasliwal, A., Furbush, N.J., Gawron, J.H., McBride, J.R., Wallington, T.J., De Kleine, R.D., Kim, H.C., and Keoleian, G.A. (2019). Role of flying cars in sustainable mobility. *Nature Communications* 10 (1555).

4 Liu, M., Hao, Y., Geng, Y., Dou, H., Sun, X., Bai, F., Qu, X., Wang, K., Liu, Z., and Zhao, F. (2025). *Journal of Cleaner Production* 490 (143910). https://doi.org/10.1016/j.fmre.2025.04.002

5 Zhao, Y., Zhang, W., Wu, Y., and Yin, Y. (2022). Environmental impact analysis of on-demand urban air mobility: a case study of the Tampa Bay Area. *Transportation Research Part D: Transport and Environment* 108 (103318). https://doi.org/10.1016/j.trd.2022.103318

6 Greenwood, E., Brentner, K.S., Rau, R.F., and Ted Gan, Z.F. (2022). Challenges and opportunities for low noise electric aircraft. *International Journal of Aeroacoustics* 21 (315–381). https://doi.org/10.1177/1475472X221107377

7 Jamaluddin, N.S., Celik, A., Baskaran, K., Rezgui, D., and Azarpeyvand, M. (2024). Aerodynamic noise analysis of tilting rotor in edgewise flow conditions. *Journal of Sound and Vibration* 582 (118423). https://doi.org/10.1016/j.jsv.2024.118423

8 ICAO. Aircraft noise. https://www.icao.int/environmental-protection/aircraft-noise

8

Business Logic of Advanced Air Mobility

Advanced air mobility (AAM) is often introduced through vehicle technologies, propulsion architectures, or mobility concepts. Yet the long-term viability of AAM depends less on technical capability and more on the business models that structure how these aircraft are deployed and integrated into the broader transportation system. Technology enables AAM, but economics sustains it. Accordingly, this chapter examines whether AAM operations can be sustained under realistic cost, demand, and utilization conditions, with emphasis on durable and repeatable business viability as opposed to abstract measures of broad adoption. The analytical focus shifts away from equipment-centric descriptions and toward the cost structures, utilization patterns, demand conditions, and market contexts that determine whether AAM operations can persist as economically viable services over time.

From an economic perspective, a defining characteristic of AAM is its rebalancing of cost drivers that have historically shaped aviation. In early AAM deployment, economic comparison is most meaningful when framed against the specific transportation modes that AAM is intended to supplement or selectively substitute for within short-range, time-sensitive missions, including surface transportation assets, conventional regional aircraft, and helicopters. Electric and hybrid-electric propulsion alter this economic balance in several fundamental ways. Over comparable mission profiles, energy costs decrease because electrical energy is typically less expensive per unit of delivered shaft power than aviation fuel, and electric motors convert stored energy to mechanical work with higher efficiency than combustion engines. In tandem, maintenance-related operating costs change in both magnitude and structure. Maintenance requirements shift toward condition-based inspection because electric propulsion systems eliminate many thermally stressed, life-limited components that drive time-based maintenance in conventional engines. This shift is reinforced by continuous health monitoring enabled by high sensor density and digital diagnostics. As operating cost and maintenance downtime decrease, economic performance becomes increasingly dependent on dispatch frequency, realizable market demand, and the efficiency with which ground operations support sustained operational cadence. Under operating conditions dominated by short-range, high-frequency missions, cost sensitivity shifts away from cruise efficiency, altitude capability, and range, and toward dispatch reliability, turnaround cadence, and sustained daily utilization supported by sufficient and repeatable demand. Within these operating conditions, AAM economic performance exhibits increased sensitivity to constraints commonly associated with surface-based transportation systems when compared with sensitivity to traditional aviation services.

This reframing has important implications for how AAM markets should be evaluated. Rather than asking whether AAM can replace existing aviation services, a more instructive question is whether AAM can operate competitively alongside surface-based transportation under short-range, high-frequency, time-sensitive conditions. In many early applications, particularly in logistics and public-service missions, the relevant comparison is cost per flight or cost per unit time saved versus cost per mile. When time is explicitly valued, aerial operations that appear expensive on a per-mile basis can become economically competitive or even advantageous at the system level.

For this reason, freight and logistics applications are examined first, followed by passenger transport. Logistics-driven operations tolerate constraints that passenger services do not. Cargo missions are less sensitive to ride quality, public perception, and discretionary pricing, and they place a premium on reliability, speed, and network efficiency. These characteristics allow freight operations to achieve higher utilization earlier, making them a natural test bed for AAM economics. The operational experience, infrastructure investment, and cost discipline developed in cargo applications directly inform subsequent passenger markets, particularly regional air mobility (RAM).

The examination of passenger applications first focuses on regional operations that offer relatively flexible cost and utilization requirements, followed by urban air mobility (UAM), where air taxi and ridesharing concepts operate under substantially tighter constraints on pricing, infrastructure, and operational cadence. Passenger-oriented RAM occupies an intermediate position, inheriting many operational characteristics of logistics while introducing passenger-centric considerations such as consumer-facing schedule reliability and fare sensitivity. UAM, including air taxi and ridesharing concepts, is the most constrained use case from an economic perspective. Urban operations have shorter mission lengths, higher infrastructure complexity, stricter noise and community acceptance requirements, and stronger competition from established ground transportation. These factors do not preclude UAM viability, but they significantly narrow the conditions under which sufficient and sustained profitability is achievable.

Public-service applications form a distinct category within the AAM ecosystem because they are evaluated using fundamentally different performance and value criteria than commercial passenger and freight services. Metrics such as response time, geographic coverage, and mission availability take precedence over revenue generation or throughput. In public-service use cases, including medical transport, emergency response, disaster relief, and public safety operations, economic value is often realized through cost avoidance, improved coverage, or reduced response time instead of direct operating revenue. Consequently, quantitative evaluation focuses on mission availability, fleet readiness, and system resilience versus passenger yield or freight throughput. Despite these differences, the same underlying cost drivers apply, allowing meaningful comparison with conventional helicopters and ground-based emergency assets.

Several economic concepts recur across all sections of the chapter. Asset utilization, taken as the ability to execute revenue-relevant missions at high frequency in response to realizable demand, emerges as a dominant determinant of viability. Infrastructure integration, including the physical layout of vertiports, the availability and coordination of charging or refueling systems, and the sequencing of ground operations, directly influences achievable flight rates by determining turnaround time between flights, thereby linking economic outcomes to vertiport design and operational planning discussed in earlier chapters. Range limitations, driven primarily by energy storage constraints, bound feasible mission profiles and shape where AAM can be deployed effectively. These constraints explain why certain applications scale naturally and others remain niche, independent of technological maturity.

By structuring the discussion around these shared economic principles and applying them across freight, passenger, and public-service domains, the chapter establishes a framework for evaluating where AAM delivers measurable value, where it competes directly with incumbent modes of transportation, and where structural limitations persist. This economic grounding supports the examination of hybrid and emerging business models presented later in the chapter and provides continuity with the longer-term perspectives developed in Chapter 9.

8.1 Freight and Logistics Applications

Freight and logistics uses are the most economically viable entry point for AAM. Unlike passenger transport, logistics operations are governed by objective functions that prioritize reliability, speed, and network efficiency over comfort, perception, or discretionary pricing. These characteristics allow freight applications to tolerate operational constraints that would be unacceptable in passenger markets and, in doing so, expose the structural economic advantages of AAM more clearly than any other use case.

From an economic standpoint, freight operations are fundamentally asset-driven. Vehicles (Figure 8.1) are primarily evaluated through cost per mission, time-normalized cost, and achievable utilization, with traditional performance metrics such as speed, range, or efficiency treated as constraints rather than primary value drivers. For short-haul, high-cadence freight missions, AAM aircraft compete with both conventional small cargo aircraft and surface transportation assets. However, their strongest economic leverage often arises in segments dominated by ground logistics. Thus, ground logistics systems, including medium-duty trucks and delivery vans operating along regional freight corridors, provide the most relevant economic baselines for evaluating early AAM deployment.

Electric and hybrid-electric AAM platforms are particularly well aligned with logistics mission profiles characterized by short distances, high cadence, and time sensitivity. These missions minimize the economic penalty of limited range and maximize the value of rapid transit and frequent dispatch. When energy cost per mission is reduced and maintenance downtime becomes more predictable, the dominant economic variable shifts from cruise efficiency to operational flight tempo. Under these conditions, an aircraft that can complete multiple short flights per day can achieve lower effective cost per delivered unit than a higher-performance platform operating over fewer cycles.

Freight applications also benefit from reduced sensitivity to demand aggregation. Passenger services require sufficient density to fill seats at specific departure times and price points compatible with operating cost. These requirements impose substantial constraints on routing flexibility and schedule design. Cargo demand, by contrast, can be aggregated across time and distributed across a larger set of feasible routes and departure windows. This operating flexibility enables smoothing of demand variability and supports sustained utilization. Higher achievable utilization

Figure 8.1 Elroy Air Chaparral C1 eVTOL cargo aircraft. Courtesy of Vertical Flight Society.

increases the number of economically viable missions that a single vehicle can support within a given operating day, improving revenue stability and reducing exposure to short-term fluctuations in market conditions.

Geography further amplifies the logistics case for AAM. Regions with challenging terrain, dispersed populations, island geographies, or limited surface infrastructure impose high time and cost penalties on ground transport. In such environments, aerial logistics can compete on total system cost even when per-mile expenses exceed those of surface modes. The relevant comparison becomes the cost of delay, inventory buffering, and service unreliability that ground networks incur versus the cost of transport alone.

Despite these advantages, freight-focused AAM remains constrained by several nonnegotiable factors. Range limitations impose hard bounds on feasible route lengths, and payload constraints restrict applicability to specific segments of the logistics market. Infrastructure availability, particularly access to charging or refueling capacity that supports rapid turnaround, directly governs achievable utilization. These constraints do not undermine the logistics case for AAM, but they do sharply define the conditions under which economic viability can be achieved. For these reasons, freight and logistics applications provide a clear foundation for understanding AAM economics. These applications reveal how cost structures shift under electric propulsion, why utilization dominates performance in short-haul operations, and how AAM competes most effectively as a time-critical logistics asset instead of a direct replacement for conventional aviation.

8.1.1 Surface-based Freight as the Primary Economic Baseline

A useful way to frame freight-oriented AAM economics is to compare aerial logistics against the operational reality of ground-based freight systems. Medium-duty trucks dominate regional and last-mile logistics because they offer relatively low capital cost and flexible routing within a mature surface transportation infrastructure (Figure 8.2). Their principal limitations are speed, exposure to congestion, and sensitivity to terrain.

Figure 8.2 A representative medium-duty freight truck. *Source*: Rab, Driver of P300NJB @Grampian Continental / flickr / CC BY 2.0.

AAM platforms invert this cost–time trade space. Capital cost per vehicle is higher, but transit time is dramatically reduced, routing is decoupled from surface congestion, and performance is largely insensitive to geographic obstacles. The result is economic competitiveness that depends on how effectively time savings and utilization offset higher upfront capital investment, and not on absolute cost per mile alone.

This distinction is particularly important in logistics networks where delivery time carries economic value, either through premium pricing for speed or through indirect costs associated with delay and uncertainty. In such cases, the relevant economic metric is total system cost across the logistics chain as opposed to transportation cost alone. AAM can reduce these secondary costs by compressing delivery windows and improving schedule reliability, even when direct operating cost per mile exceeds that of surface transport. This effect is strongest for short- to medium-distance routes, typically on the order of several to a few hundred kilometers, where ground travel time grows disproportionately with congestion, circuitous routing, or terrain constraints.

Asset utilization further differentiates aerial and ground logistics economics. Ground vehicles are often utilization-limited by a combination of driver availability, regulatory duty-hour constraints, and congestion-related delays. Electric AAM platforms operating in short-haul freight missions, by contrast, are utilization-limited primarily by turnaround time and energy replenishment. Because aerial routes are typically more direct, achieve higher effective travel speeds, and avoid many congestion and terrain constraints, individual missions can often be completed in substantially less elapsed time than comparable ground deliveries. When vertiport infrastructure supports rapid charging, concurrent ground handling activities, and preplanned dispatch schedules, a single aircraft can complete multiple missions within a duty cycle that would constrain a ground vehicle to far fewer deliveries. In this circumstance, capital investment recovery is spread across a larger number of revenue-generating missions, narrowing or reversing the cost disadvantages.

8.1.2 Conventional Aerial Freight Economics

The economic implications of lower energy cost, a reduced maintenance burden, and utilization-driven capital cost recovery become clearer when aerial logistics is compared to existing aviation alternatives. Turboprop aircraft, such as the Cessna Caravan shown in Figure 8.3, have long served as workhorses for regional freight and logistics. This aircraft type is optimized for carrying substantial payloads over relatively longer trip lengths with high dispatch reliability. That same optimization leads to penalties for short-haul operations, including fuel consumption associated with takeoff and climb that is largely independent of trip length, greater cycle-driven maintenance burden, and a relatively high minimum viable mission length driven by fixed per-flight costs that dominate total operating expense. AAM aircraft incur many of these same cost categories but at lower absolute levels and with turnaround processes designed to support higher flight rates. This makes economic viability possible over substantially shorter mission distances. As such, turboprop cargo aircraft remain cost-effective within their intended operating regime but are less suited to short-haul, high-frequency logistics roles. AAM aircraft do not replace such platforms across the full spectrum of cargo operations. Instead, they occupy a complementary niche where shorter distances, faster turnaround, and higher flight rates dominate the economics. This distinction explains why AAM logistics often compete more directly with medium-duty freight trucks than with small cargo aircraft, even though the latter are technically capable of serving similar routes.

Figure 8.3 A FedEx Cessna 208B Super Cargomaster, a purpose-designed turboprop used as a platform for regional freight and logistics operations. *Source*: Steve Knight / Flickr / CC BY 2.0.

Figure 8.4 New Zealand's rugged topography, including mountainous terrain, volcanic regions, and dispersed inland communities, imposes significant time and cost penalties on surface logistics. These conditions create favorable operating contexts for AAM, particularly for time-sensitive freight where aerial routing can reduce delay, variability, and total system cost. *Source*: Michal Klajban / Wikimedia Commons / CC BY-SA 4.0.

8.1.3 Network Structure and Routing Economics

Network structure plays a central role in determining whether freight-oriented AAM operations can achieve economic viability. Ground logistics networks are optimized around a hub-and-spoke architecture constrained by road topology, time-varying congestion, and indirect routing imposed by terrain, waterways, and urban infrastructure. These constraints can cause delivery time to grow disproportionately with route length, particularly in regions where surface routes require substantial deviation from direct point-to-point travel. AAM networks, by contrast, operate in a largely unconstrained geometric space, allowing route distance to scale more linearly with straight-line separation. This difference decreases travel time variability and simplifies scheduling, both of which have measurable economic value in logistics planning.

The benefits of aerial routing become most prominent in environments where surface networks impose disproportionate distance-related penalties. Mountainous regions, island geographies, and sparsely connected rural areas exemplify this (Figure 8.4). In such settings, ground transport may require circuitous routing, multimodal transfers, or time-consuming ferry crossings that increase cost and reduce reliability. AAM vehicles can bypass many of these constraints, perhaps enabling same-day delivery across a region that would otherwise be fragmented across a multiday, multimodal surface logistics network. The economic value of reduced delivery time is often matched by the value of predictable delivery timing.

These geographic advantages become even more economically meaningful when combined with high operational cadence. Logistics networks are rarely optimized for a single delivery; they are optimized to sustain predictable, high-volume flow over time through balanced scheduling, asset utilization, and network reliability. AAM platforms can advance this optimization by supporting frequent departures and arrivals that enable logistics operators to move from batch-based delivery to continuous or near-continuous flow. This shift reduces inventory dwell time and lowers the need for buffering at intermediate nodes. From an economic standpoint, these effects reduce working capital tied up in inventory and improve the responsiveness of the logistics system to changing demand conditions and delivery schedules, benefits not captured by per-mile transportation cost metrics alone.

8.1.4 Hybrid Logistics Networks and Selective Aerial Integration

AAM delivers the greatest economic leverage when operations are integrated with surface transportation and assigned to route segments that benefit from aerial routing within a broader multimodal logistics system. Hybrid logistics models selectively insert aerial segments into predominantly surface-based networks to target portions of a route where time savings justify higher vehicle cost. In such designs, AAM can bridge the most time- or distance-constrained route segments, such as the first or last connection between a hub and a remote destination. Constraining aerial use to route segments with the highest marginal economic return reduces sensitivity to range limitations and concentrates higher-cost assets where economic payoff is strongest.

Operational resilience further strengthens the use of AAM for freight. Ground logistics systems remain vulnerable to localized disruptions such as road closures and congestion. Aerial logistics provide an alternative pathway that can bypass such disruptions. However, AAM operations do introduce distinct sensitivities, particularly related to weather and airspace availability. Nevertheless, diversification across transport modes improves overall network robustness. In economic terms, increased resilience lowers the expected cost of disruption by reducing delay-related penalties and service interruptions over time. This effect becomes increasingly important as logistics networks increasingly support time-critical or high-value goods.

8.1.5 AAM as a Timing and Reliability Enabler

A critical distinction between freight-oriented AAM and many passenger concepts of operation lies in how value is realized across the logistics chain. In logistics operations, transportation is rarely the end product. Instead, it is one component within a broader system that includes inventory management, scheduling, warehousing, and downstream distribution. Costs incurred at one stage of the chain often propagate elsewhere, meaning that improvements in transit time or reliability can generate savings that are not completely visible in transportation budgets. Thus, AAM's ability to compress delivery windows and reduce variability produces second-order economic benefits that amplify its apparent value.

These effects are particularly relevant for logistics networks operating under just-in-time or near just-in-time conditions. When delivery timing is uncertain, operators compensate by holding additional inventory, maintaining excess warehousing or transport capacity, or accepting higher rates of missed or delayed deliveries. Each of these responses carries an economic penalty. However, a logistics system that can rely on predictable, rapid aerial transport can reduce inventory dwell time and improve synchronization across network nodes. In this context, AAM functions less as a transportation mode and more as a timing and reliability enabler within the supply chain.

It is important to recognize, however, that the logistics case for AAM is not universal. The economic advantages described here diminish rapidly as mission distance increases beyond the effective range of electric and hybrid-electric platforms. Payload further narrows the value proposition to specific segments of the AAM freight market, concentrating economic viability on high-value, time-sensitive goods versus bulk commodities. Weather sensitivity and airspace access also impose constraints that must be managed through operational planning and network design.

Within short-haul, time-sensitive, high-cadence freight operations, freight-focused AAM occupies a distinct niche between surface transport and conventional aviation. It does not seek to replicate the economies of scale associated with long-haul air cargo, nor does it aim to displace trucks across dense, low-cost surface corridors. Instead, it targets segments where time sensitivity, geography, and cadence interact to create economic friction for surface transport and inefficiency for traditional aircraft. When these conditions align, aerial logistics can deliver competitive or superior system-level outcomes despite higher per-mile transportation costs. This economically beneficial niche explains why freight and logistics applications are widely viewed as a leading use case for AAM viability. The absence of passenger-service constraints allows these operations to achieve sustained economic performance under a wider set of operating conditions.

Worked Example 8.1: Representative Short-haul Freight Trip

Scenario: A time-sensitive, high-value, same-day shipment moves from a regional distribution hub to a destination separated by an approximately 120 km straight-line distance across a coastal bay. Surface transportation must route around the water barrier, resulting in a ground distance of approximately 215 km (Figure 8.5).

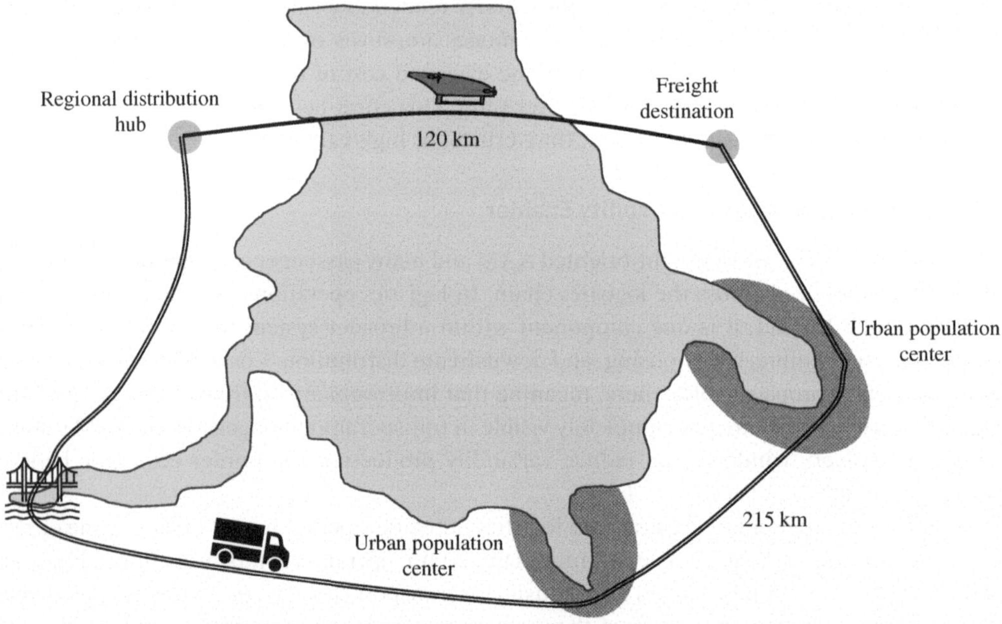

Figure 8.5 Representative short-haul time-sensitive freight trip constrained by a water barrier. Surface logistics routes must detour around the barrier and through urban congestion.

Assumption: For consistency in analysis across transportation modes, a trip is defined as a single one-way revenue delivery between the hub and the destination. Return legs are treated as nonrevenue repositioning segments. This accounting reflects common hub-and-spoke freight operations and allows utilization to be compared on an equivalent basis.

Calculations: The following calculations illustrate how elapsed time and daily asset utilization shape operational economics across trucks and electric AAM cargo aircraft.

1. **Door-to-door elapsed time:** In addition to the calculated flight or drive time, assume an additional 0.67 hours (40 minutes) for combined ground processing at both ends. This time represents loading and dispatch activities that occur concurrently with energy replenishment or refueling.

   Truck: Assume an average line-haul speed of 65 km/h over the 215 km route after taking into account routing intersections, speed limits, and moderate urban congestion,

   $$t_{\text{drive}_{\text{truck}}} = \frac{215\ \text{km}}{65\ \text{km/h}} = 3.31\ \text{h} +\ 0.67\ \text{h} = 3.98\ \text{h}$$

   AAM: Assume an average airspeed of 240 km/h for short-haul operations,

   $$t_{\text{fly}_{\text{AAM}}} = \frac{120\ \text{km}}{240\ \text{km/h}} = 0.50\ \text{h} + 0.67\ \text{h} = 1.17\ \text{h}$$

   Insight: In this scenario, a trip occupies a truck for nearly 4 hours, whereas an AAM aircraft completes the same delivery in just over 1 hour. *This disparity in elapsed time is the first-order driver of utilization advantage.*

2. **Achievable daily trips and utilization effects:** Assume a 10-hour operational day for each asset and that a roundtrip (outbound revenue leg plus return repositioning leg) must be completed before another revenue delivery can occur.

   Truck:

   $$\frac{10\ \text{h}}{(3.98\ \text{h} \cdot 2)} \approx 1$$

   AAM:

   $$\frac{10\ \text{h}}{(1.17\ \text{h} \cdot 2)} \approx 4$$

   Insight: A single truck can support approximately one revenue-generating roundtrip per day, whereas a single AAM aircraft can support up to four. This difference illustrates how reduced elapsed time increases achievable delivery rate, allowing fixed capital costs to be spread across more revenue-generating missions and making *utilization a dominant economic driver in short-haul freight operations.*

8.2 Passenger Transport: Regional and Urban Applications

Passenger transport is the most visible application of AAM, yet it is more economically constrained than freight. The same operational and economic relationships identified in Section 8.1 (per-mission cost structure, turnaround time, asset utilization, and infrastructure integration) also govern passenger operations. As in freight applications, these relationships shift economic sensitivity away from vehicle performance and toward achievable utilization across passenger AAM operations. However, when the payload changes from cargo to people, additional requirements are introduced that materially alter the economic conditions for viable operations.

Figure 8.6 Interior of Vertical Aerospace's Valo. Courtesy of Vertical Aerospace.

Passenger comfort, schedule reliability, pricing sensitivity, public acceptance, and safety perception impose more immediate and stringent constraints on passenger operations than freight service (Figure 8.6). These factors tighten the coupling between operations and demand by making revenue generation directly dependent on reliably filling seats for specific departures. Consequently, shortfalls in utilization, operational disruptions, or infrastructure constraints translate immediately into lost revenue and limit the ability to pool demand flexibly across time, routes, or vehicles. Therefore, AAM passenger service must satisfy a tighter alignment between cost structure, demand characteristics, and operational cadence than is required for logistics-driven applications.

Passenger AAM is best framed as a progression of operating regimes characterized by increasingly restrictive economic conditions, not as a single unified market. The two use cases that dominate this progression are passenger-oriented RAM and UAM. Although often conflated, these categories differ significantly in mission length, demand structure, infrastructure requirements, and the incumbent transportation modes against which they compete. Each applies the same underlying economic framework but does so under distinct constraints that dictate achievable utilization, pricing flexibility, and scalability. We now examine passenger-oriented RAM and UAM in sequence.

8.2.1 Regional Air Mobility

RAM occupies an intermediate economic position between freight logistics and urban passenger transport. On the passenger AAM spectrum, RAM possesses several favorable characteristics drawn from cargo-oriented operations, particularly longer mission lengths, more predictable and concentrated passenger demand across scheduled routes and time windows, and the ability to leverage existing aviation infrastructure. These characteristics reduce sensitivity to fixed per-flight overhead and enable utilization to be planned with greater stability than is achievable in short-duration, highly distributed urban passenger operations.

In practical terms, RAM typically encompasses passenger trips on the order of approximately 80–800 km. These routes often connect secondary cities, regional airports, or suburban population centers to larger metropolitan hubs. Representative examples include shorter regional corridors such as Chicago–Rockford, as well as cross-border European connections such as Paris–Brussels, where straight-line distances fall within the RAM envelope and surface travel times are shaped by congestion, indirect routing, and schedule variability. These trip lengths allow fixed per-departure costs, such as passenger processing, dispatch, and energy replenishment, to be amortized over longer flight times, reducing per-mission cost sensitivity to small variations in turnaround time relative to urban operations. Hence, passenger-oriented RAM aligns more closely with conventional regional aviation than with urban passenger mobility, even when an AAM vehicle platform is capable of serving both regional and urban missions.

Demand structure further differentiates RAM from urban passenger applications. Regional services typically connect dispersed population centers, remote communities, or secondary airports to larger hubs, where passenger demand can be concentrated around fixed, scheduled departures. This temporal concentration improves seat occupancy predictability and supports higher average utilization than is achievable in highly distributed, short-notice urban passenger markets. Although absolute passenger volumes may be lower than in metropolitan cores, concentrating demand at specific times reduces revenue volatility and simplifies fleet and schedule planning.

Infrastructure considerations also favor regional deployment. Existing airports, underutilized regional airfields, and purpose-built facilities can often accommodate RAM operations with fewer spatial, noise, and community constraints than dense urban environments. Access to established aviation services, including ground handling, air traffic procedures, and weather support, reduces operational variability and accelerates integration. Where new infrastructure is required, regional settings typically offer greater flexibility in site selection and layout, enabling designs that better support efficient ground handling, rapid turnaround, and reliable scheduling.

Despite these advantages, RAM also introduces passenger-specific constraints that are largely absent in freight operations. Seats must be filled at fare levels passengers are willing to pay, reintroducing sensitivity to per-departure demand concentration. Schedule reliability becomes critical, as delays or cancellations impose direct economic penalties and erode passenger trust. Cabin comfort, noise exposure, and perceived safety influence adoption in ways that do not apply to cargo and require operational conditions that maintain acceptable ride quality, noise levels, and passenger confidence.

Competition further shapes the RAM economic landscape. Regional AAM services compete with conventional turboprop aircraft, intercity rail, long-duration automobile travel, and, in some cases, legacy airline networks. Although AAM platforms may offer advantages in point-to-point routing or reduced door-to-door travel time for specific city pairs, these benefits must offset higher capital costs and the absence of economies of scale available to established modes of transportation. For these reasons, economic viability depends on disciplined frequency planning that concentrates demand and on alignment between delivered time savings and passengers' willingness to pay.

Taken together, RAM represents the least constrained passenger application of AAM but one that remains materially more restrictive than freight logistics. Its economic viability rests on the same utilization-driven principles identified in Section 8.1, moderated by passenger-specific requirements that tighten acceptable operating margins. When deployed on routes with meaningful time savings, predictable demand patterns, and reliable infrastructure, RAM can achieve sustained economic performance. Outside of these conditions, RAM's advantages diminish rapidly, demonstrating the importance of disciplined market selection.

8.2.2 Urban Air Mobility

UAM is the most economically constrained passenger application of AAM. Within the passenger spectrum, UAM operates over the shortest mission lengths and within the most infrastructure-intensive environments, and it is subject to the highest degree of public scrutiny (Figure 8.7). Community acceptance, perceived safety, and noise exposure directly influence allowable operating hours, infrastructure approval, and achievable flight frequency. These characteristics compress the range of economically viable operating conditions, independent of vehicle performance capability.

From an economic standpoint, short mission lengths fundamentally alter the cost structure. Fixed per-flight costs associated with passenger processing, dispatch, ground handling, and energy replenishment must be recovered over limited flight time and distance. Accordingly, per-mission cost is highly sensitive to turnaround time, vertiport pad availability, and dispatch reliability, with even small ground delays propagating directly into lost utilization. Once minimum vehicle performance thresholds are met, cadence and its effect on utilization dominate economic performance.

Demand structure further constrains UAM operations. Although urban areas generate substantial aggregate travel demand, that demand is highly fragmented across origins, destinations, and time. Passenger willingness to substitute aerial service for established ground transportation varies widely with price, convenience, and perceived reliability. Therefore, sustaining economic viability requires consistent aggregation of a sufficient number of passengers on specific routes and departure windows, a condition that can be difficult to maintain in discretionary, short-duration travel markets. In contrast to regional services, demand cannot be readily concentrated through schedule discipline alone.

Infrastructure integration presents an additional and structurally limiting constraint. UAM operations must be embedded within dense urban environments characterized by limited available space, competing land uses, strict noise limits, and heightened community sensitivity. Vertiport siting, access, and throughput directly govern achievable flight rates. Opportunities for infrastructure expansion remain constrained by zoning and public acceptance. Unlike regional settings, urban infrastructure constraints cannot be mitigated through relocation or spatial flexibility without eroding the demand base itself.

Figure 8.7 Conceptual depiction of an UAM air taxi operating over Manhattan, illustrating the dense infrastructure, constrained airspace, and proximity to population centers that govern UAM economic viability. Courtesy of Vertical Aerospace.

Moreover, competition from ground transportation presents substantial challenges to the economic feasibility of urban passenger AAM. Urban passenger AAM services must compete directly with already mature, low-cost, high-capacity surface modes, including private automobiles, ride-hailing services, and mass transit. These systems benefit from extensive infrastructure investment, pricing familiarity, and network effects arising from ubiquitous surface connectivity. Within this context, UAM is most plausibly positioned as a complement to existing public transportation, providing time-critical or connectivity-enhancing service on specific segments where surface networks are indirect, capacity-constrained, or poorly synchronized. Even in this complementary role, time savings and service quality must be substantial, communicated clearly, and delivered reliably to justify higher operating costs.

8.3 AAM for Public Good

Public-good applications of AAM occupy a distinct position within the AAM ecosystem because their economic objectives differ from those of commercial passenger and freight services. Public-good missions are wide-ranging and can include medical transport, emergency response, disaster relief, and public safety operations. The value in these missions is not realized through revenue generation, seat occupancy, or price optimization. Instead, value is realized through response time, geographic coverage, service availability, and system resilience. In this way, these applications provide a complementary perspective to the market-driven analyses presented in Sections 8.1 and 8.2 and illustrate how the same underlying cost drivers can support economic viability under alternative optimization criteria.

In public-service contexts, transportation functions as an enabling input within a broader operational system instead of an end product. Economic justification stems from the ability of aerial assets to reduce response time, expand access to time-critical services, and sustain operations when surface networks are impaired. Consequently, comparisons based solely on cost per mile or cost per mission are insufficient. Evaluation, instead, centers on response speed, service coverage, and mission outcomes across the system as a whole.

Medical transport provides a clear illustration of this logic. In emergency medical services, the economic value of rapid aerial response is closely coupled to patient outcomes. Faster access to advanced treatment can reduce mortality, shorten hospital stays, and lower long-term care expenditures, even when aerial transport entails higher up-front operating costs. In this context, the economic contribution of AAM is expressed through outcome improvement and cost avoidance and not transport cost efficiency alone. Electric and hybrid-electric AAM platforms have the potential to further improve this calculus by reducing energy cost, noise exposure, and the maintenance burden relative to conventional helicopters, thereby increasing fleet availability and mission readiness without compromising response capability.

Disaster response and humanitarian operations are two excellent examples of the role AAM can play as a network-level resiliency asset (Figure 8.8). Natural disasters frequently disrupt surface transportation through flooding, debris, or infrastructure failure, imposing severe access constraints precisely when demand for rapid response is the greatest. Aerial mobility provides an independent pathway that bypasses many of these disruptions, enabling rapid delivery of personnel, medical support, and critical supplies under degraded conditions. In these scenarios, economic value is expressed through reduced recovery time, restored access to isolated populations, and preservation of essential services. By decoupling from damaged surface networks, aerial mobility reduces the duration and severity of system disruption and supports coordinated response actions that would

Figure 8.8 Conceptual depiction of AAM vehicles supporting disaster response and humanitarian operations by bypassing damaged surface transportation networks and enabling rapid access to affected areas. Courtesy of NASA / Public domain.

otherwise be delayed or infeasible. Although these benefits are difficult to precisely monetize, they align directly with public-sector decision frameworks that emphasize continuity, coverage, and risk mitigation over unit-cost minimization.

Public safety and law enforcement missions similarly prioritize persistence, coverage, and readiness over throughput. Aerial platforms capable of rapid deployment, efficient loiter, and a reduced acoustic footprint support surveillance, search and rescue, and situational awareness functions that are poorly served by surface assets alone. In these applications, electric AAM systems may offer advantages through lower operating cost and simplified maintenance, enabling more frequent redeployment or longer windows of availability. Utilization remains economically important, but instead of measuring via revenue-generating cycles, it is measured in terms of mission readiness and availability.

Despite these distinct objective functions, public-good AAM applications remain subject to many of the same physical and infrastructural constraints that govern commercial operations. Range and payload limitations bound feasible mission profiles, and access to reliable charging or refueling infrastructure governs redeployment speed and readiness. Weather sensitivity and airspace coordination impose additional operational challenges that must be managed through system design and planning. The key distinction from commercial passenger and freight operations is how these constraints are evaluated. Public-sector operators typically assess performance across a portfolio of missions and time horizons, emphasizing aggregate coverage and resilience over the economics of individual missions.

As such, utilization is a central consideration across public-good use cases in AAM economic assessments. Unlike commercial use cases, high utilization does not imply continuous operation but translates to a high probability that assets are available, ready, and capable when called upon. Reduced maintenance downtime, predictable energy replenishment, and streamlined ground operations directly increase this form of utilization by improving fleet readiness. Accordingly, economic viability hinges on whether such readiness can be sustained at a lower lifecycle cost than incumbent aerial assets. In most cases, the relevant benchmark is a helicopter fleet, which is, in general, typically more maintenance-intensive and carries higher readiness and lifecycle costs.

Public-good deployments can catalyze the broader AAM ecosystem by providing justification for infrastructure development and generating operational learning. Medical, emergency, and humanitarian missions can validate broader investment in vertiports, energy systems, and operational frameworks that later enable expanded commercial services. In this sense, public-service applications provide an additional pathway for economic justification outside of revenue-maximizing markets.

8.4 Emerging and Hybrid Business Models

Emerging AAM business models converge on economic logics that prioritize utilization, infrastructure availability, and realizable demand through operating conditions characterized by predictable dispatch cadence, reliable infrastructure support, and stable demand patterns. These conditions imply that economic sustainability depends on deliberately constraining operating scope to limit demand uncertainty and support the utilization levels required for viability. This section examines four recurring business-model logics through which operators implement these constraints in practice (Figure 8.9).

Consider, first, the *selective deployment* model. This model reflects a recurring observation throughout the chapter that AAM delivers its greatest economic leverage when applied selectively within a broader journey, particularly where surface or conventional aviation modes impose disproportionate time, reliability, or access penalties. In freight networks, this often corresponds to the first or last segment of a route, where terrain, congestion, or water crossings degrade surface performance. In passenger and public-service applications, similar logic applies to connections

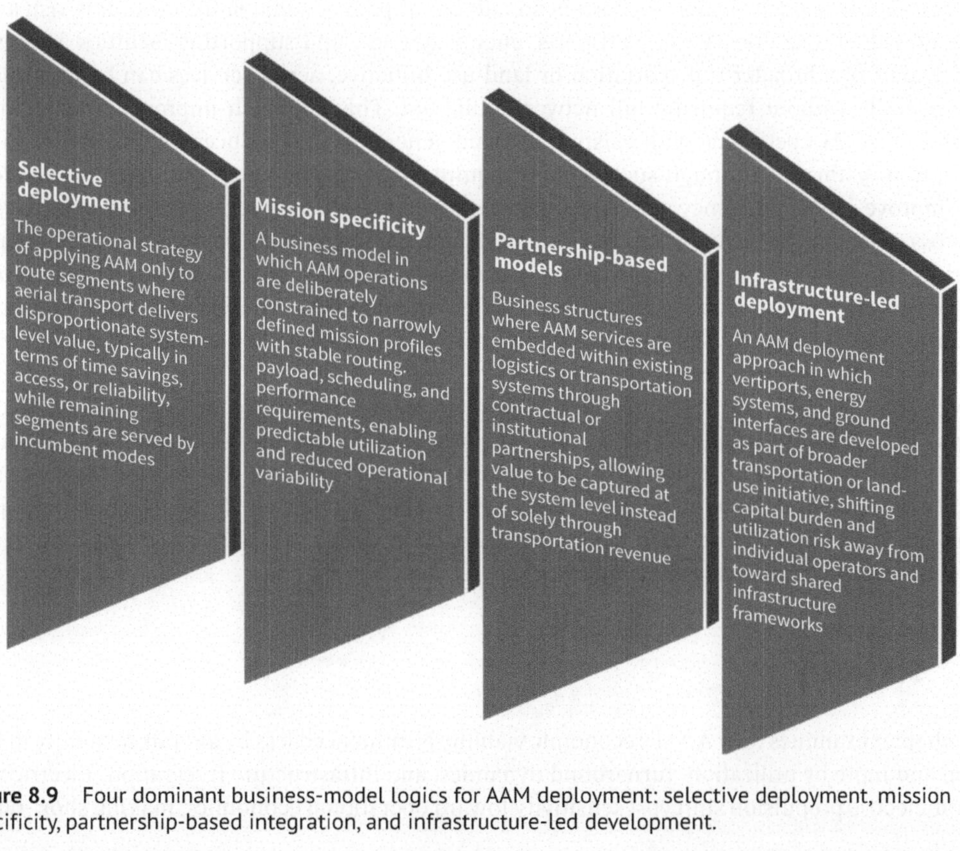

Figure 8.9 Four dominant business-model logics for AAM deployment: selective deployment, mission specificity, partnership-based integration, and infrastructure-led development.

that bypass bottlenecks or extend access to locations poorly served by ground infrastructure. Selective concentration of aerial operations within these segments reduces required vehicle range, simplifies energy and infrastructure requirements, and concentrates utilization where sustained economic performance is most achievable.

A second class of viable models emphasizes *mission specificity*. Under this business model logic, operators define narrow mission envelopes characterized by stable routing, predictable scheduling, and consistent payload characteristics, in contrast to broad general-purpose mobility offerings. Examples include fixed medical transfer corridors, recurring time-critical cargo routes, or event-driven passenger demand tied to specific locations and time windows. Mission specificity reduces operational variability, which, in turn, simplifies planning and enables tighter coupling between fleet sizing, infrastructure placement, and staffing. This coupling improves utilization predictability, reduces fixed-cost exposure, and lowers risk associated with deployment forecasting and capacity commitment.

Partnership-based models further reinforce economic viability by anchoring AAM operations to established demand and institutional frameworks. Integration with logistics providers, healthcare systems, emergency services, or transportation authorities allows aerial services to leverage existing networks in lieu of creating demand independently. In these arrangements, the economic value of AAM is often realized through system-level performance improvements, such as lower sensitivity to schedule disruption, expanded service coverage, or improved reliability, instead of direct transportation revenue alone. Contractual or institutional support reflects the distribution of value across the broader system, reducing reliance on fare or per-mission pricing to recover capital costs.

Infrastructure-led deployment represents an additional pathway that shifts economic risk away from individual operators. When vertiports, energy systems, and supporting facilities are developed as part of a broader transportation or land-use initiative, aerial services can be introduced incrementally without requiring full network build-out. This approach improves utilization by co-locating AAM operations with existing demand generators and reduces duplication of infrastructure investment. Although such models require coordination across multiple stakeholders, they improve capital efficiency and align AAM deployment with long-term planning objectives.

Across all emerging business models, a consistent economic pattern emerges: viable strategies prioritize constrained operating envelopes, predictable demand, and incremental scaling over expansive network ambitions. These choices reflect the dominant role of utilization and infrastructure availability in AAM economics and reinforce that disciplined deployment matters more than maximizing technical capability.

Over time, operational experience gained through these constrained models informs vehicle design, infrastructure standards, and regulatory processes, reducing uncertainty and lowering cost. As these conditions mature, AAM services will continue to expand. The underlying economic logic, however, will remain unchanged. Sustained viability depends on aligning technology, operations, and market structure such that utilization and system-level value consistently outweigh capital intensity and operational complexity.

8.5 Summary

This chapter examines how AAM's economic viability is influenced less by aircraft capability in isolation and more by utilization, turnaround dynamics, and infrastructure integration. Electric and hybrid-electric propulsion shift cost sensitivity toward operational economics, making short-range operations particularly dependent on cadence, ground coordination, and reliable access to supporting infrastructure. Within this framework, AAM creates value where time savings translate directly into higher asset utilization and system-level performance gains, frequently positioning it in closer economic competition with surface transportation than with conventional aviation.

As passenger requirements accumulate, the set of viable operating conditions narrows, with regional passenger services occupying an intermediate regime and UAM constrained to tightly defined contexts. Public-service applications further demonstrate that AAM value cannot be assessed solely through commercial metrics, as response speed, coverage, and operational resilience provide alternative justifications for deployment. Across freight, passenger, and public-good use cases, the most robust AAM business models emphasize selective deployment, mission specificity, institutional integration, and incremental scaling, reinforcing that AAM succeeds by inserting aerial capability where it most effectively reshapes access, timing, and reliability within larger transportation systems.

9

The Future of Advanced Air Mobility

Advanced Air Mobility (AAM) occupies a position within the evolution of transportation systems that is neither notional nor settled. The aircraft, infrastructure concepts, operational frameworks, and economic models examined throughout this textbook demonstrate that AAM is technically plausible and operationally coherent at a limited scale. At the same time, none of these elements, individually or collectively, guarantee economically viable expansion, persistence, or societal embedding as scale increases. The future of AAM will be determined less by whether individual vehicles can perform as intended, given the demonstrated capability across multiple manufacturers, and more by whether AAM systems continue to function effectively as operational scale grows and as coupling across transportation, energy, infrastructure, and governance extends well beyond aviation itself. Concomitantly, public and community acceptance remain a binding condition on this evolution.

Looking beyond the next decade or two does not require imagining radically new vehicle concepts or technological breakthroughs. The dominant factors that will shape long-term AAM outcomes are already observable. These factors emerge in battery and energy infrastructure systems that must accommodate new highly variable, demand-driven electrical loads associated with charging cycles and fleet operations. They also emerge in urban environments where land availability, zoning, and competing uses constrain infrastructure placement; in multimodal interfaces that must coordinate air and surface transportation flows; and in governance structures that must regulate in ways that remain responsive as the operational ecosystem evolves. Public acceptance represents a parallel conditioning factor and is influenced less by technical performance than by cumulative environmental impact, safety, accessibility, and human experience. The long-term trajectory of AAM depends on how these factors interact as scale increases, operational complexity deepens, and coupling across domains becomes more pronounced.

9.1 Conditional Futures and System-level Constraints

Forecasting has limited value in complex sociotechnical systems because system behavior emerges from tightly coupled interactions rather than linear cause-and-effect relationships, and AAM exhibits this coupling to an unusually high degree. Aircraft performance, automation capability, infrastructure availability, energy supply, public trust, and institutional governance evolve on different timescales and under different incentives. No single roadmap can synchronize these dynamics, and no credible analysis can assume that progress in one domain will compel progress in the others. Thus, *the future of AAM is best described by the conditions under which the system remains viable* instead of as a sequence of milestones.

This distinction matters because aviation history is often misread as a story of linear advancement. In reality, commercial aviation stabilized through the emergence of a specific set of operating conditions within which aircraft capability, infrastructure investment, regulatory authority, and public acceptance aligned. This alignment remains contingent on sustained economic demand, acceptable safety performance, manageable community impacts, and institutional capacity to enforce and adapt operational rules. AAM seeks much deeper integration into populated environments, routine interaction with nonusers exposed to the impacts of AAM operations, and an operational density closer to that of surface transportation. These ambitions expand the number and coupling of conditions that must be simultaneously satisfied.

Conditional futures are governed by limiting factors that bound feasible operation. In AAM, many of these limiting factors are already well established. Energy density constrains mission flexibility and utilization. Acoustic emissions and visual presence associated with low-altitude operations and frequent takeoffs and landings constrain where and when operations can occur. Human cognitive limits constrain supervision and intervention regardless of automation level. Urban land use constrains infrastructure placement. Institutional risk tolerance constrains operational authority by limiting how much decision-making can be delegated. None of these constraints disappear with maturity. Instead, each of these constraints becomes more binding as scale increases because small inefficiencies and rare events accumulate into emergent system-level behavior.

Predictive narratives fail in this context because they implicitly treat constraints as manageable secondary effects that diminish with time and scale. A conditional framework assumes the opposite. As AAM scales, coupling increases. Vehicles draw energy from shared grids. Vertiports compete with housing, commerce, and public space. Automated decision systems interact with human oversight structures that must remain accountable. Failures propagate across domains opposed to remaining localized. In these circumstances, a system's viability depends on its ability to absorb stress without reorganizing itself into a configuration that undermines the original value proposition. As scale increases, shared resources and interfaces cause constraints to interact and amplify, showcasing why predictive narratives fail when they assume constraints diminish with growth.

Predictive narratives frame progress as a sequence of milestones.

A *conditional future* describes viability in terms of the conditions that must remain satisfied for operations to persist and scale:

- Progress is not assumed to follow a linear path
- Progress in one domain does not compel progress in other domains
- Constraints are not assumed to diminish with time but, instead, interact and multiply
- System coherence is contingent on sustained alignment between necessary conditions instead of accumulated capability

This conditional framing also offers insight into why technological optimism alone is insufficient. Improved batteries, quieter rotors, and more capable autonomy expand feasible regions within the AAM design space, but they do not eliminate boundaries. For example, greater automation can reduce pilot workload, but it simultaneously increases demands on verification, validation, and oversight. Higher utilization improves economics, but it intensifies peak energy demand and infrastructure congestion. Distributed vertiports shorten the surface travel time required to reach air operations, but they increase exposure to community impacts. So, each gain introduces a compensating pressure elsewhere in the system. This alters the balance of constraints without eliminating them and reinforces that AAM's future is determined by the conditions under which gains are achieved, not by the growth of capability in isolation.

A conditional future is, consequently, governed by the system's ability to manage its trade spaces as external conditions evolve. Long-term AAM viability depends on whether operating, infrastructure, energy, and governance trade spaces remain navigable as urbanization patterns, climate variability, energy market volatility, and demographic change shift system operating conditions. These forces do not act independently, and their combined effects cannot be fully foreseen. The resulting pressures, nevertheless, tend to manifest in recurring forms, such as limits on vertiport siting, restrictions on operating hours, caps on simultaneous departures, constrained access to electrical capacity, and tighter authorization for automation.

In these circumstances, AAM systems fail in predictable ways. Operations become fragile when schedules rely on minimal energy margin, when throughput assumes uninterrupted grid availability, when cumulative noise or visual exposure outpaces community tolerance, or when operational authority depends on uniform regulatory interpretation across local jurisdictions. However, systems persist when trade spaces can be rebalanced as constraints change. Persistence is supported by trade space management, such as possessing multiple admissible operating configurations, the ability to throttle capacity without collapse, and governance mechanisms that allow oversight authorities to revise constraints incrementally without halting service.

Treating the future as conditional also avoids a common conceptual error in discussions of autonomy and scale. It is often assumed in complex sociotechnical systems that once technical feasibility is demonstrated, societal acceptance and institutional adaptation will follow. Historical experience in aviation and other safety-critical systems suggests otherwise as operational acceptance has consistently lagged initial technical feasibility. Extended-range Twin-engine Operational Performance Standards (ETOPS) serve as an aviation example. Here, twin-engine aircraft possessed the technical capability to conduct extended overwater flights well before operating authority expanded. However, acceptance only emerged after sustained evidence of predictable engine performance, failure management, and recovery under abnormal conditions. This example illustrates a broader principle: acceptance tends to emerge only after a system demonstrates predictable behavior under abnormal and off-nominal conditions, not simply when it performs well under nominal ones. As acceptance matures, oversight authorities gain greater discretion to expand operating permissions and to transition oversight from novelty-driven scrutiny toward outcome-based supervision. These acceptance-mediated limits impose dynamic constraints on the pace and form of AAM evolution, even as computational or mechanical capability advances.

> Many discussions of AAM are *often framed as predictive narratives* that define progress as a sequence of milestones. Future assessments of AAM are *better off adopting a conditional framing* that considers the conditions required for AAM to persist.

Thus, the future of AAM will be determined by the ability of system architectures now being established to tolerate the uncertainties that inevitably accompany scale. The discussion now turns to the most consequential of these uncertainties. Each acts as a constraint that shapes AAM's evolution regardless of a specific vehicle, business model, or regulatory regime. Collectively, they define the conditions under which AAM can persist as a meaningful component of future transportation systems or remain a technological experiment.

9.1.1 Scaling Without Centralization

Scaling in conventional aviation has historically been accompanied by increased centralization. Airspace control, infrastructure investment, scheduling authority, and safety oversight consolidated as traffic volumes grew because rising traffic density exceeded the ability of decentralized actors to maintain shared situational awareness of traffic, weather, and system state. As density increased, safe operation required coordinated separation, sequencing, and priority management across many aircraft simultaneously, exceeding the cognitive limits of individuals and making responsibility for safety outcomes ambiguous without centralized coordination. Centralization reduced ambiguity, simplified accountability, concentrated scarce human expertise, and improved system safety, predictability, and coordination. This model proved effective for conventional aviation because operations remained sparse relative to surface transportation, infrastructure nodes were limited in number, and access to airspace was tightly regulated.

AAM challenges this historical operational model. AAM pursues scale via frequency, geographic dispersion, and proximity to population centers, instead of relying on concentrated flows between a small number of hubs. This shift fundamentally alters system geometry. AAM operations occur closer to the ground, at higher densities, and across a wider range of operating contexts. Under these conditions, centralized control encounters practical limits associated with decision latency, information throughput, and conflict resolution at the temporal and spatial scales required for dense operations. As scale increases, delays in sensing, communication, and coordination increasingly constrain real-time responsiveness and separation assurance. As these constraints intensify, maintaining system performance increasingly depends on distributing time-critical decisions closer to the point of execution.

However, decentralized operations still require explicit structure. Scalable AAM systems require coordination mechanisms that distribute decision authority while preserving coherence, safety, and accountability at the system level. Local actors must retain autonomy to respond to rapidly changing conditions, yet their actions must remain predictable and interpretable within a shared operational framework. Accordingly, scaling without centralization requires clearly constrained decision-making latitude that is implemented through common operating rules and predefined operating envelopes.

The difficulty in distributing decision-making stems from the fact that decentralization relocates complexity without eliminating it. When decision-making authority is distributed to local actors, consistency must be enforced through shared rules, standardized interfaces, and verification processes that unambiguously associate decisions and outcomes with responsible entities. Absent such structure, local errors do not remain local; they propagate across shared interfaces and require system-level coordination to correct. As operational density increases, small deviations rapidly propagate into systemic operational inefficiencies instead of remaining isolated coordination issues. Thus, the system's ability to scale depends on whether distributed local decisions, and their effects, can remain bounded within operating envelopes that preserve global stability.

Infrastructure further complicates the balance between local decision-making and system-level coordination. As discussed in Chapter 4 on land use and zoning, vertiports, energy connections, and access corridors are inherently local. Even so, their performance and impacts extend across the broader system. Local optimization, pursued in isolation, risks producing congestion, inequity, or community resistance at the system level. Conversely, excessive centralization of infrastructure planning can freeze adaptation and suppress innovation. Consequently, scalable AAM depends on governance frameworks that enforce system-wide constraints on cumulative impact while permitting limited local variation in execution.

Scaling without centralization does not imply the disappearance of central oversight. As operational scale and system coupling increase, effective oversight depends less on continuous directive control and more on the specification of enforceable operating constraints, system-wide performance monitoring, and structured intervention when locally executed decisions violate defined constraints. Central authorities retain responsibility for defining separation minima, safety thresholds, and system-wide infrastructure and energy limits that bound aggregate demand, throughput, and interaction across shared airspace, infrastructure, and energy systems. Local execution occurs within these bounds. Within this structure, system stability ultimately depends on the consistency and clarity of the shared rules more than on real-time directives. This mode of oversight limits the propagation of local deviations, preserves the traceability of responsibility, and enables targeted intervention when behavior jeopardizes system-level performance or safety.

9.1.2 Trust, Transparency, and Human Oversight

Trust acts as a governing constraint on AAM systems by limiting the operational authority, automation levels, and operating density that oversight bodies and communities are willing to permit under uncertainty. In this context, trust determines operating envelopes, acceptable degrees of autonomy, and the pace at which operations may expand. Trust does not accumulate proportionally with flight hours or exposure. Instead, it solidifies only after system behavior remains predictable under stress.

As AAM operations integrate into cities and interact routinely with nonparticipants, trust becomes a primary regulator of both expansion and delegated authority. Performance under nominal conditions is insufficient. Operational latitude is granted only when off-nominal behavior can be understood, traced, and corrected without ambiguity. When this condition is not met, operational limits tighten regardless of technical capability. Sustaining trust under these conditions requires system properties that make behavior observable, accountable, and correctable as complexity increases.

Transparency is a necessary condition for the transition from a tightly constrained operation to expanded delegated authority. Transparency demands that automated behavior be explainable at levels sufficient for supervision, accountability, and public legitimacy. This requirement precludes the use of opaque or "black-box" decision systems whose internal logic cannot be interrogated, validated, or reconstructed under off-nominal conditions. As automation increases, transparency shifts from individual actions to system logic, decision boundaries, and failure handling. Systems that cannot elucidate these elements under off-nominal conditions will fail to sustain trust even when overall system performance appears favorable.

The role of humans evolves directly from this transparency constraint structure. Automation redistributes responsibility toward supervision, enforcement of defined operating limits, and exception management. This redistribution concentrates human involvement into fewer, greater consequence interventions with higher cognitive demand. For this reason, effective human participation depends on procedures, interfaces, and authority structures that support rapid interpretation and decisive action when uncertainty appears.

Institutional trust follows the same logic. Certification, oversight, and incident or accident response all depend on the ability to trace authority and assign responsibility across layered automation. When decision authority becomes opaque, oversight bodies respond by constraining operational scope to preserve accountability. Consequently, increases in technical capability alone do not translate into greater autonomy or scale. Governance capacity, too, must evolve alongside automation so that responsibility remains enforceable as system complexity increases.

Public trust adds a final binding constraint. Building on earlier discussion of acceptance, public acceptance of routine AAM operations depends less on the absolute absence of incidents than on confidence that operators act with restraint, respond effectively when disruptions occur, and demonstrably learn from failure. Systems perceived as opaque, unresponsive, or dismissive of local impact lose legitimacy even when nominal performance appears strong. Acceptance is, therefore, provisional and continuously shaped by observed operational behavior, not secured through early technical success or favorable safety statistics alone.

Thus, the long-term viability of AAM depends on whether trust can be preserved as operational decision-making authority becomes increasingly automated and as operational frequency and interactions grow. Trust is conditioned by the consequences that off-nominal events impose on passengers, bystanders, and surrounding communities, including the severity and reversibility of those outcomes. Systems that embed transparency, maintain clear authority, and support effective human intervention retain greater operational latitude. Trust, transparency, and human involvement define structural limits on AAM that determine how quickly and far the system can scale.

9.1.3 Energy, Infrastructure, and System Coupling

Energy constitutes one of the most binding system-level constraints on AAM's long-term evolution. Again, unlike conventional aviation, where mature fuel distribution networks largely decouple routine operations from local infrastructure, AAM operations remain tightly coupled to external energy systems at the locations and times dictated by high-frequency use. The exact nature of this coupling varies by propulsion architecture. Battery-electric aircraft depend directly on regional electrical capacity and power delivery. Hybrid-electric configurations redistribute coupling across electrical supply and liquid fuel logistics. Hydrogen-based systems introduce additional dependencies through production, storage, and handling infrastructure. Across all these architectures, vehicle performance, utilization, and operational reliability remain contingent on energy availability and system-level management practices.

At limited scale, this coupling is often masked. Early operations can rely on excess grid capacity, off-peak charging, or on-site energy storage that buffers short-term demand from upstream supply constraints. As operations expand, aggregate electrical load grows nonlinearly because high-frequency departures, synchronized arrivals, and weather-driven schedule compression concentrate energy demand in both time and space. Adverse conditions further intensify this effect by forcing recovery operations into narrower operating windows, increasing concurrent energy demand independent of average utilization. In this operating condition, constraints in distribution capacity, power quality, and resilience emerge within local and regional electrical systems.

The tight coupling between AAM energy demand and operational scaling challenges traditional infrastructure planning assumptions. Historically, aviation infrastructure has been designed around discrete nodes with long development horizons and relatively stable demand profiles. AAM infrastructure must accommodate temporal variability, spatial dispersion, and frequent reconfiguration of operations in response to demand, weather, and energy availability. Energy systems, by contrast, are planned around more predictable and gradual change. Any misalignment between these planning logics produces an incompatibility that incremental upgrades do not fully address.

Technological advances in onboard energy storage do not resolve this coupling challenge. Higher onboard energy storage density expands mission envelopes and reduces sensitivity to charging frequency, enabling higher utilization. As utilization increases, aggregate energy demand intensifies. To sustain these higher operating tempos, charging power must increase, shifting constraints from total energy quantity toward power delivery capacity, thermal management limits, and grid stability. Once again, we see that a technical improvement relocates pressure within the coupled system instead of eliminating it.

Long-term AAM viability depends on treating energy and supporting infrastructure as internal system constraints instead of external inputs. Treating energy availability as an external input assumes that sufficient electrical or fuel energy will be available on demand, independent of operational decisions. The inclination to treat energy availability as an external input reflects a legacy aviation abstraction where fuel supply is assumed to be continuously available once distribution and contractual arrangements are set in place. At limited scale, this abstraction holds for AAM as well. Early operations can be undertaken with available capacity without materially influencing upstream energy systems. However, as scale increases, this assumption fails. Charging demand is determined by aspects such as fleet scheduling, turnaround expectations, and charging or refueling concurrency, each following from operational choices internal to the AAM system. Systems that do not explicitly incorporate energy constraints into scheduling, infrastructure design, and operating rules lack mechanisms to adapt when energy becomes limiting and will exhibit increasing fragility as scale grows. This applies regardless of whether energy transfer occurs through charging, battery exchange, or other refueling architectures, since each approach modifies turnaround timing and increases the simultaneity of energy demand across multiple aircraft, reshaping infrastructure requirements without removing the fundamental constraint. Systems that integrate energy availability into operational control, capacity management, and planning processes retain greater adaptive capacity.

9.1.4 Urban Integration and the Disappearance of the Airport

The integration of AAM into urban environments is a fundamental shift in how aviation infrastructure is situated. The so-called disappearance of the airport does not imply the removal of any aviation facilities; it describes the loss of the airport as a singular, spatially bounded entity that separates aviation activity from everyday urban systems. This shift does not eliminate airports from the AAM system; it redistributes aviation access across a wider set of urban interfaces while reducing reliance on a single dominant facility type. As AAM scales, aviation infrastructure becomes progressively distributed across the urban fabric, shifting aviation activity from a function isolated within a distinct enclave to a routine service embedded within everyday urban systems.

Conventional airports are discrete zones that concentrate noise, risk, access, and security requirements into finite locations. This configuration simplifies governance and confines exposure, but it simultaneously systematizes separation between aviation activity and everyday urban life. This places access points far from typical trip origins and destinations. AAM inverts this logic: AAM's operational value depends on proximity, frequency of service, and continuous coupling to surface transportation networks. These value propositions, while imposing constraints, favor infrastructure with smaller spatial footprints that can be integrated into existing urban forms, yielding a heterogeneous set of aviation access points distributed across transportation corridors, multimodal nodes, and mixed-use developments.

This spatial integration shifts the binding constraints away from primarily aeronautical considerations toward urban ones. In doing so, competition for land intensifies as aviation infrastructure occupies spaces traditionally allocated to housing, commerce, and public amenities. Meanwhile, visual presence, acoustic exposure, and perceived risk become persistent elements of daily life. Because residents continuously interface with AAM infrastructure as part of routine movement through the city, acceptance hinges on cumulative exposure and direct interactions over time.

The recasting of aviation services also revises public expectations. Airports are environments where restricted access, heightened security, and specialized operating rules are socially accepted. However, when aviation activity migrates into ordinary urban settings, those assumptions no longer apply. Communities expect transparency, responsiveness, and restraint consistent with other urban services. These expectations constrain operating practices and influence infrastructure and operational design in ways that technical optimization alone cannot resolve.

Urban integration does not imply uniform solutions. Cities differ in density, institutional governance capacity, culture, and infrastructure maturity. Infrastructure and operating concepts that function effectively in one context may prove untenable in another. This makes long-term viability dependent upon adaptability across varying locales instead of uniform solutions. Systems that rely on rigid infrastructure models or generalized operating assumptions experience degraded system performance as local conditions diverge.

Finally, as aviation infrastructure disperses, governance complexity increases. Authority fragments across municipal planning departments, transportation agencies, utilities, and aviation oversight bodies as a consequence of infrastructure distribution and overlapping jurisdictional responsibility. Decisions about siting, access, and operating hours intersect with zoning, equity, and environmental priorities that vary across jurisdictions. The absence of a single, bounded airport complicates accountability by diffusing responsibility across institutions. Maintaining system coherence under these conditions increasingly depends on coordination mechanisms and shared standards that align decisions across actors with distinct mandates.

The disappearance of the airport is best understood as a structural reconfiguration, not a final endpoint. As AAM becomes embedded within cities, the airport no longer functions as the dominant interface between aviation and everyday urban life, even as existing airports continue to support regional connectivity, intermodal exchange, and high-throughput operations. Aviation increasingly competes with other urban functions for space, attention, and legitimacy across a broader set of locations. The future of AAM depends on whether this integration can occur without overwhelming urban systems or eroding public acceptance. Where integration is managed through deliberate trade-space management, AAM can persist as a routine component of urban mobility. Where integration proceeds without sensitivity to local constraints, resistance accumulates and expansion slows.

9.1.5 Governing a System That Must Keep Changing

Governance is the final binding condition on AAM's long-term viability. As discussed in Chapters 4 and 5, vehicles approach design closure through certification and production standardization; infrastructure solidifies through long-lived physical assets and siting decisions; and automation architectures stabilize through verified authority allocation and assurance requirements. These technical elements trend toward relatively stable configurations once deployed at scale. But, unlike vehicles, infrastructure, or automation architectures, governance does not converge in the same manner. Rather, governance operates as an ongoing control process whose effectiveness depends on the continuous application, revision, and enforcement of constraints as system behavior evolves. In AAM, governance remains coupled to urban environments, energy systems, automation capability, and public tolerance that continue to change as scale increases.

Conventional aviation governance developed under slower-moving conditions, and major reforms generally followed major accidents or series of incidents. Aircraft configurations evolved gradually, infrastructure life cycles were long, and traffic growth followed relatively predictable trajectories. Under these conditions, certification functioned as a stabilizing mechanism because aircraft configurations, control logic, and operating assumptions remained largely fixed after entry into service. Oversight could rely on sustained compliance with certified configurations over extended periods, and operational learning rarely altered system behavior during routine operations. AAM departs from this regime. Automation authority, infrastructure interfaces, and operating contexts continue to evolve after deployment as scale increases. These changes emerge through operational growth, system coupling, and shifts in delegated authority that occur outside the certification process itself. As a result, governance frameworks that treat certification as the primary mechanism for maintaining system stability lose effectiveness, even though certification remains intact and necessary.

Accordingly, effective AAM governance depends on adaptability without the loss of authority. Oversight bodies must maintain enforceable safety and accountability standards that still permit operational practices to evolve. This balance is structurally constrained. Governance that is excessively rigid hinders learning and adaptation. Governance that relaxes constraints without enforceable mechanisms exposes the system to failure modes that undermine legitimacy and public trust. Long-term system viability depends on governance mechanisms that define clear operating boundaries, monitor aggregate system behavior, and adjust constraints in response to observed performance without interrupting service.

Accountability anchors this process. As previously established, as automation deepens and decision-making authority becomes distributed across vehicles, operators, and supervisory systems, responsibility and liability must remain traceable across layered automation. When authority, responsibility, and liability cannot be clearly assigned and made traceable, oversight authorities narrow operational scope to preserve enforceability and institutional control, independent of technical capability. This shift reflects a governance response to ambiguity, not a limitation of technology. Operational latitude expands only when accountability remains intact, enabling intervention, correction, and learning to remain feasible as delegated authority increases.

Public legitimacy further conditions governance effectiveness. In urban environments, governance decisions are continuously evaluated by affected communities. Outcomes that are technically justified yet difficult for communities to understand, question, or engage with weaken confidence and narrow the range of governance actions that can be sustained. Governance that demonstrates measured use of authority, timely response to disruption, and credible learning following adverse events is more likely to preserve institutional latitude under continued public scrutiny. Based on this, public confidence serves as a boundary on how governance authority can be exercised, revised, and delegated.

Governing a system that must keep changing requires accepting that no final equilibrium exists. Thus, AAM governance must remain effective under continuous changes in scale, operating schemes, and system coupling. Governance frameworks grounded in static assumptions or the need for perfect foresight increasingly lose effectiveness as complexity increases. Frameworks that preserve accountability, incorporate feedback, and recalibrate operating boundaries as conditions evolve retain operational relevance. Collectively, these governance dynamics define the conditions under which AAM can persist as a meaningful component of future transportation systems.

9.2 Summary

AAM's future will *not* be defined by singular breakthroughs, definitive timelines, or the maturation of technology alone. It will ultimately be determined by how effectively a tightly coupled sociotechnical system manages such factors as constraint, variability, and scale over time. Across vehicles, infrastructure, energy, governance, and public acceptance, AAM remains viable only when trade spaces remain navigable and authority, accountability, and legitimacy remain intact. All this means AAM succeeds through coherent operation within constraints. AAM's long-term contribution to mobility depends on embedding aerial capability into broader transportation and urban systems without simultaneously destabilizing the conditions that allow it to exist in the first place.

Index

Note: Page numbers in *italics* and **bold** refers to figures and tables respectively.

a

AAM aircraft types 14, *14*
- eCTOL/eSTOL aircraft 19–20, *20*
- eVTOL aircraft 14–19, *15–18*

AAM development
- cultural and behavioral shifts 8–9
- economic conditions 8
- energy infrastructure 7
- environmental concern 7
- hub-and-spoke system 7, *7*
- materials and manufacturing 6–7
- social and policy motivations 8
- strategic and geopolitical considerations 8
- sustainability 7

AAM integration
- automation 66–68, *67*
- ground-based human role 68–73
- onboard sensing modalities **76–77**
- predictable behavior 65–66
- remote supervision 68–73
- strategic flow management 73–75, *74*

AAM Interagency Working Group (IWG) 130
accessibility 107, 120
accountability 230–232, 234, 235
advanced air mobility (AAM)
- chargers *vs.* ground-vehicle chargers 112
- conditional futures. *see* conditional futures and system-level constraints
- defining characteristics 5–6, 209
- development drivers 6–9
- economic performance 209
- emerging and hybrid business models 223–224, *223*
- environmental operating context of. *see* environment
- ethical considerations 148
- evolution 1–3
- freight and logistics 211–217, *211–212, 214, 216*
- implications for 209
- International Regulatory Framework 146–147
- operational architectures *4*
- passenger transport 217–221, *218, 220*
- for public good applications 221–223, *222*
- stakeholders 9–11

Advanced Air Mobility Coordination and Leadership Act 130
aerial mobility 221–222
aeroacoustics
- atmospheric attenuation 191–192, *191*
- definition 188
- measuring noise 193–197, **193**, *194–197*
- noise contour maps 197–198, *198*
- noise reduction 204–207, *205–207*
- origin of sound 189–191, *189–190*
- qualitative nature of sound and noise 192–193
- sources of noise 198–204, *200, 202, 204*

Aeronautical Information Manual (AIM) 132

aero-propulsive coupling 40
air carrier certificate 140
airframe noise 203–204
airline transport pilot (ATP) 140
airside zone 95
airspace integration 152
airspace management 142
air traffic control (ATC) 63, 130
air traffic management (ATM) 6, 61
  communication 63
  controlled airspace 62–63, *62*
  development of 61
  legacy ATC limitations 65
  low-altitude operations 65
  performance-based navigation (PBN) 63–64
  separation minima 63
  surveillance 63
  trajectory conformance 64–65
  uncontrolled airspace 62–63, *62*
airworthiness certification 133
airworthiness directive (AD) 138
airworthiness standards 134
Albatross (airship) 1, *2*
amperes 48
anode 46–47, 58
anomaly detection 68, 162
Applications of Research to Operations at the Mesoscale (AROME) 181
arcing 54
arrival procedures 158
artificial intelligence (AI) 13, 182
atmospheric boundary layer (ABL)
  characteristics 169–173
  defining feature of 168
  temperature inversion 169
  typical diurnal evolution of *169*
authentication 162
automated decision systems 228
Automated Surface Observing System (ASOS) 178
Automated Weather Observing System (AWOS) 178
automatic dependent surveillance-broadcast (ADS-B) 63, 142
automation 22, 66–68, *67*, 156, 231

b

battery 46–50, *46*, *49*
  fuel cells 57–59
  safety 53–55
  state indication and display 55–57
  state parameters **56**
  swapping 117–118
battery electric vehicles (BEVs) 184
battery management system (BMS) 53
  hardware architecture 55
  software architecture 55
Bell Nexus tilt-duct vectored-thrust aircraft 17, *18*
Beta Alia 250 lift + cruise aircraft. 14, *16*
Beta Technologies CX300 *155*
blade self-noise 201, 202
blade–wake interaction (BWI) noise 201
blown-lift 20
boundary layer ingestion (BLI) 40

c

carbon-fiber-reinforced polymers (CFRPs) 35
carbon fibers 35
categorical exclusions 144
cell groups 51
cells 51–55, *51*. *see also* fuel cells
center of gravity (CG) 31–32
charge capacity (Q) 48
charge-control algorithms 55
charge-depleting 43
charger cabinets 114, 120
charge-sustaining 43
charging infrastructure 109–110, *109*
  alternative energy-delivery architectures 117–118
  battery swapping 117–118
  charger design 115
  energy scheduling 117
  fast charging 116
  grid interaction 118–119
  international considerations 119–120
  interoperability 115–116
  managed charging 116
  site-level integration 118–119
charging pedestal 114, 117

circulation
  passenger 107, 121
  vertical 121, *122*, 125
Civil Aviation Administration of China (CAAC) 147
civil aviation authorities 147
clearance, definition 54
collective pitch 28
communication and information exchange 85, *85*
community acceptance 164
community noise equivalent level 196–197, *197*
composite materials 33–35, *34–35*
Composite Materials Handbook (CMH-17) 35
conditional futures and system-level constraints
  defined 228
  disappearance of airport 234
  energy, infrastructure, and system coupling 232–233
  governance 234–235
  scaling without centralization 230–231
  trust, transparency, and human oversight 231–232
  urban integration 233–234
conformance monitoring 154, 157
connectivity 107
containment 54–55, 158–159
continued operational safety (COS) 138
controlled airspace 62–63, *62*
conventional airports 233
conventional aviation 91, 230
  governance 235
conventional rotorcraft 100
conventional route *64*
Convention on International Civil Aviation 1944 (Chicago Convention) 146
cooperative sensing 86–87
cooperative surveillance systems *63*
Coriolis effect 170, 171
corridor design 77–80, *78–79*, *81*
creepage 54
curb access 93, *94*
cybersecurity 161–164
cycle fade 53
cyclic pitch 28, 29

d

day–night average sound level 196
design drivers
  aircraft dimensions and noise 22
  capacity 21
  cruise speed 21
  range 21
designees 137
digital and data-network providers 10
direct operating cost (DOC) 22
disk loading 25–28, *26–27*
  vs. hover efficiency *26*
distributed electric propulsion (DEP) 6, 20, 22, 38–40, *38*
distributed separation assurance 87–88, *87*
doppler radars 178
downwash and outwash (DWOW) 98, *98–99*
downwash caution area (DCA) 95
downwash speed 26
drones 4
ducted-fan eVTOL 17–18, *18*

e

economic conditions 8
economic viability 10, 212–214, 219–222
effective perceived noise level 197
electrical charging fundamentals 110
  AAM *vs.* ground-vehicle chargers 112
  charge transfer and time 111
  communication requirement 112–113
  constant-current charging *112*
  constant-voltage phase *112*
  control 112–113
  current 110
  power delivery 110
  resistive heating 110–111
  safety 112–113
  thermal constraints 110–111
  voltage 110
electric conventional takeoff and landing (eCTOL) 13
electric potential 47
electric propulsion 13
electric short takeoff and landing (eSTOL) 13

electric vertical takeoff and landing (eVTOL) aircraft 3, 13, 100, 130, 168
design architectures *15*
ducted-fan 17–18, *18*
examples of *15*
hybrid 18–19
lift + cruise architectures 14–15
multicopter architectures 14
tilt-duct 17, *18*
tilt-rotor 16, *17*
tilt-wing 16, *17*
vectored-thrust 16
electrification of propulsion 6
electromagnetic interference (EMI) 40
Elroy Air Chaparral C1 eVTOL cargo aircraft *211*
emergency access 100
encryption 162
energy
density 49
scheduling 117–118
and utility providers 9
energy storage and management systems 22, 45, 156
battery 46–50, *46*, *49*
battery safety 53–55
cost 53
cycle life 53
form factor 53
safety 53
voltage matching 52
weight 53
environment
aeroacoustic considerations for 188–207, *189–191*, **193**, *194–198*, *200*, *202*, *204–207*
sustainability 183–188
weather. *see* weather environment
Environmental, Social, and Governance (ESG) framework 186–187
equivalent continuous noise level 196, *196*
eVTOL battery modules and packs 50–52, *50*
eVTOL design 21
composite materials 33–35, *34–35*
control strategies 28–30, *28–30*
disk loading 25–28, *26–27*
drivers 21–22
enabling technologies 22
first-principles approach 23–24
fundamentals 22–23
rotor solidity 25–28, *25*, *27*
wingborne flight 30–33, *31*
eVTOL Integration Pilot Program (eIPP) 131
executive order (EO) 131
Extended-range Twin-engine Operational Performance Standards (ETOPS) 229

f

FAA Reauthorization Act 130
fair-weather cumulus 170
fast charging 116
fault-detection and diagnostic 55
fault-mode management 162
Federal Aviation Administration (FAA) regulations
AAM infrastructure 143–144, *144*
accident and incident investigation 145–146
airspace management 142
airworthiness certification 133
certification of AAM Vehicles 132–133
certification of operators 140–142
light sport aircraft 139
maintenance 142–143
noise standards 137–138
production certification 138–139
safety of approved designs 138
state, local, and tribal governments, role of 144–145
TC 133–136, *135*, *136*
training and certification of pilots 140
final approach and takeoff area (FATO) *83*, 96, *97*
financial and insurance institutions 10
fire protection 100–101
first-principles approach 23–24
fleet-management platforms 162
flight management systems (FMSs) 66
flow stability 154
freight and logistics
aerial logistics 212
characteristics 211
conventional aerial freight

economics 213, *214*
demand aggregation 211
electric and hybrid-electric AAM platforms 211
hybrid logistics networks and selective aerial integration 215
network structure and routing economics 214–215, *214*
requirements 211
surface-based freight 212–213, *212*
timing and reliability enabler 215–217
front-retreating/rear-advancing (FRRA) 30
fuel cells 57–58
hydrogen storage considerations 58–59

g

glass fibers 34
graceful degradation 29
ground-handling coordination 125
ground-vehicle chargers *vs.* AAM chargers 112

h

helicopters 23
High-Resolution Rapid Refresh (HRRR) 181
high stiffness-to-weight ratio 35
high strength-to-weight ratio 35
hosting capacity 119
hover efficiency 21–22, 24
vs. disk loading *26*
and downwash speed *27*
hub-and-spoke system 7, *7*
hybrid electric vehicles (HEVs) 184
hybrid eVTOLs 18–19
hybrid propulsion systems 41–45, *41–42*
hydrogen storage considerations 58–59

i

inertial measurement units (IMU) 86
inertial sublayer 177
inflow turbulence noise 201–203
infrastructure 91, 143–144, *144*
charging and refueling 109–120, *112–114*, *116*, *118*
developers 9
multimodal transport hubs 120–127, *122*
urban integration and land-use planning 101–109, *102–103*, *108*
vertiports 92–101, *92–100*
infrastructure-led deployment 224
institutional trust 231–232
instrument flight rules (IFR) 62
Integrated Weather Observation System (IWOS) 180
interface zone 97
International Civil Aviation Organization (ICAO) 146
International Regulatory Framework 146–147
interoperability 115–116
Issue Paper process 135

j

joule 47

k

Kalman filtering 86, *86*

l

landside zone 97
land-use planning. *see* urban integration and land-use planning
life-cycle analysis (LCA) 185–186
lift + cruise vehicles 14–15
light detection and ranging (LiDAR) 75, 179
light sport aircraft 139
Lilium Jet ducted-fan vectored-thrust aircraft 18, *18*
lithium-ion (Li-ion) batteries 45
logistics 157. *see also* freight and logistics

m

machine learning (ML) 13, 182
managed charging 116
management systems 22
mechanical turbulence 171
medical response mission 157–159, *158*
medical transport 221
microclimates 65, 174, 177, 178
mission specificity 224
modules 51, *51*

multicopters 14, 31
multifunction control and display unit (MCDU) *67*
multimodal transport hubs 120
  continuous passenger circulation 121–122
  external interface 123
  operational coordination 124–125
  operational separation 122–123
  physical and spatial integration 121
  safety zoning 122–123
  urban and regional integration 125–127
  urban interface 123

n

national airspace system (NAS) 130
National Environmental Policy Act of 1969 (NEPA) 144
National Fire Protection Association (NFPA) 99–101
National Strategy and Comprehensive Plan 131
natural disasters 221
navigation performance 85–86, *86*
Neighborhood Weather Observation System (NWOS) 180
network integration 126
network segmentation 162
nocturnal low-level jet (LLJ) 173
noise
  A-weighted dB 194, *194*
  community noise equivalent level 196–197, *197*
  day–night average sound level 196
  designing to reduce 204–207, *205–207*
  effective perceived noise level 197
  equivalent continuous noise level 196, *196*
  maximum sound level 194–195
  noise contour maps 197–198, *198*
  perceived noise level 197
  qualitative nature of sound 192–193
  single-event noise exposure level 195, *195*
  sound exposure level 195
  sound pressure level 193–194
  sources of 198–204, *200*, *202*
  standards 137–138
numerical weather prediction (NWP) models 180–183, *180*

o

operational considerations
  community and public acceptance 163–164
  medical response 157–159, *158*
  regional cargo 155–157, *155*
  security and cybersecurity 159–163, *160*
  urban passenger mobility 152–155, *153–154*
operational coordination 124
  airside interface 125
  ground-handling 125
  passenger-flow management 124–125
  staging 124–125
  temporal alignment 124
organization designation authorization (ODA) 137
original equipment manufacturers (OEMs) 9

p

packs 51–54, *51*
parallel-hybrid sub-architecture 42
Part 380, 141–142
partitioning 54
partnership-based models 224
passenger circulation 121
passenger-flow management 124–125
peak shaving 43
perceived noise level 197
percentage of charge 56
performance-based navigation (PBN) 63–64
performance-based standards 135
pilots, in early AAM deployments 70–73
pitching moments 32–33
planetary boundary layer (PBL) 168
polyacrylonitrile (PAN) 35
polymer matrix 34
power (P) 24
  density 49
  propulsion technologies 36–38
  quality 119
Powered-Lift Part 194 SFAR 140
predictable behavior 65–66
predictive narratives 228, 229
prepregs 35
primary batteries 46
production certification 138–139
propagation 54

propeller–wing integration 20
propulsion-controlled aircraft (PCA) 40
propulsion technologies
 distributed electric propulsion 38–40, *38*
 hybrid propulsion systems 41–45, *41–42*
 power, torque and RPM 36–38
public acceptance 164
public-good missions 221–223, *222*
public legitimacy 235
public service missions 151
public trust 232

r

radar 63, 86, 178
radiative cooling 170
radiosondes 178
Ragone plot 49, *49*, 58
reduction–oxidation (redox) process 46, 47
reference aircraft 95
regional air mobility (RAM) 3–4, 132, 151, 218–219
regional cargo mission 155–157, *155*
regulations 129. *see also* Federal Aviation Administration (FAA) regulations
remote sensing systems 178
remote supervision 68–73
required navigation performance (RNP) 64, *64*
revolutions per minute (RPM) 36–38, *37*, 40
RNAV routes *64*
rotor
 phasing 200
 solidity 25–28, *25*, *27*
roughness sublayer 176–177

s

screening protocols 159–161, *160*
secondary batteries 46
security by design 162
selective deployment model 223
sensor-fusion architecture 86, *86*
separation management 63
series-hybrid sub-architecture 43
service providers 9, 74
shaft torque 36
short takeoff and landing (STOL) aircraft 20
signal
 intermittent 192
 occlusion and multipath effects 75, *76*
simplified vehicle operations (SVO) 22, 69–70
single-event noise exposure level 195, *195*
siting process 103, *103*
 area refinement 104
 multicriteria evaluation 104–105
 site evaluation 104
 strategic screening 104
small uncrewed aircraft systems (sUAS) 4, 13
solid-state battery 46
sound
 exposure level 195
 pressure level 193–194
spatial zones 95
S–P convention 52
special class process 135
special conditions 135
Special Federal Aviation Regulation (SFAR) 140
special flight permit 133
staging 124–125
stakeholders 9–11
Standards and Recommended Practices (SARPs) 146
state-estimation algorithms 55
state of energy (SOE) 56
state of health (SOH) 55
state of power (SOP) 56
strategic flow management 73–75, *74*
stratocumulus clouds 170
structured routing 77–80
substantial damage 146
supercooled liquid water (SLW) 168
supervised autonomy 68–69, *69*
surveillance 86–87
sustainability
 operational 183–187, *184*, *186*
 system-level 187–188
system-level scalability 152
system of systems 5

t

temperature inversion 169
temporal alignment 124
thermal buildup 111

thermal-management algorithms 55
thermal runaway 53
  primary initiators and cycle *54*
thermal turbulence 172
thermoplastics 34
thermosets 34
thrust 23–24, 29
thunderstorms 170
tilt-duct eVTOLs 17, *18*
tilt-rotor aircraft 16, *17*
tilt-wing aircraft 16, *17*
tonal noise
  broadband noise 201–204, *202*, *204*
  BVI noise 199–201, *200*
  loading noise 199
  thickness noise 199
torque 32, 36–38
touchdown and lift-off (TLOF) 81, *83*, 96, *97*, 154
trajectory conformance 64–65, 85–86
trajectory integrity 161
transformer 113, 119
transparency 231
trust 231–232
turbulence 171
turning effect 32
type certificate (TC) 133–136, *135*, *136*

u

uncertainty modeling 88
uncontrolled airspace 62–63, *62*
urban
  canopy layer 175–176, *175*
  metabolism 174
  passenger mobility 152–155, *153–154*
urban air mobility (UAM) 3, 132, 151, 220–221, *220*
urban airspace 75
  corridor 77–80, *78–79*, *81*
  physical and operational characteristics 75–77, **76–77**
  structured routing 77–80
  vertiport-linked airspace geometry 81–83
urban and regional integration 125, 234
  catchment structure 126
  community outcomes 127
  development patterns 126–127
  equity considerations 127
  land use 126–127
  long-term adaptation 127
  network role 126
  station-area form 126–127
urban boundary layer (UBL)
  characteristic temperature difference 174
  inertial sublayer 177
  roughness sublayer 176–177
  stability and turbulence characteristics 175
  structure and properties 174
  thermodynamic environment 175
urban integration and land-use planning 101–102
  accessibility 107
  aeronautical compatibility 103
  community compatibility 103
  community considerations 105
  compatibility 102–103, *102*
  connectivity 107
  environmental considerations 105
  and evaluation process 108–109, *108*
  functional compatibility 103
  physical compatibility 103
  regulation 105–106
  siting 103–105, *103*
  zoning 105–106
US Department of Transportation (DOT) 130
US Federal Aviation Administration (FAA) 130
US National Transportation Safety Board (NTSB) 145–146
US Regulatory Framework
  FAA regulation. *see* Federal Aviation Administration (FAA) regulations
  federal AAM legislation 130–132
  federal AAM regulations 132

v

vacuum-assisted resin transfer molding (VARTM) 35
vectored-thrust aircraft 16
vehicle-to-ground (V2G) paths 85, *85*
vehicle-to-vehicle (V2V) links 85, *85*
Vertical Aerospace VX4, 18–19, *19*
vertical circulation 121, *122*, 125

vertical takeoff and landing (VTOL) 134
vertiport 5, 92–93, *92–93*, 187–188
  core functions 93–94, *94*
  electrical architecture 113–115, *114*
  energy domain 93–94, *95*
  functional and spatial architecture 95–97
  information domain 94, *95*
  structural and environmental design 98–101, *98–100*
  transportation domain 93, *95*
vertiport-linked airspace geometry 81–83
Vertiport Weather Observation System (VWOS) 179
visibility 170
visual flight rules (VFR) 62
voltage 47
volumetric energy density 53

W

watt (W) 47
watt-hours (Wh) 48, 53
weather environment
  ABL 168–173, *169–170*, *172*
  NWP 180–183, *180*
  spatiotemporal weather observations 177–180, *178*
  UBL 173–177, *175–176*
winds 170–171
wingborne flight 30–33, *31*
work (W) 47, 48

Wiley The manufacturer's authorized representative according to the EU General Product Safety Regulation is Wiley-VCH GmbH, Boschstr. 12, 69469 Weinheim, Germany, e-mail: Product_Safety@wiley.com.

Printed and bound by CPI Group (UK) Ltd, Croydon, CR0 4YY
30/07/2026
02181221-0002